Environmental Security in Africa

Environmental Security in Africa

Conflicts, Politics, and Development

Edited by Elisha Jasper Dung,
Leonard Sitji Bombom, and
Augustine Avwunudiogba

LEXINGTON BOOKS
Lanham • Boulder • New York • London

Published by Lexington Books
An imprint of The Rowman & Littlefield Publishing Group, Inc.
4501 Forbes Boulevard, Suite 200, Lanham, Maryland 20706
www.rowman.com

86-90 Paul Street, London EC2A 4NE

British Library Cataloguing in Publication Information Available

Library of Congress Cataloging-in-Publication Data Available

978-1-66693-635-3 (cloth)
978-1-66693-636-0 (electronic)

This book is dedicated to the casualties of environmental conflicts in Africa.

Contents

Illustrations

FIGURES

TABLES

Preface

The initial idea for this book came from a discussion on the so-called farmers–herders conflict on the Jos Plateau, which was gathering much momentum at the time. There were too many mischaracterizations of the issues, and we could focus more on the underlying factors that commentators glossed over.

Dr. Leonard Bombom's experience teaching Conflict Analysis to graduate students at the Centre for Conflict Management and Peace Studies at the University of Jos, Nigeria, bolsters this point. Going through examples of conflicts in different countries over Africa, it dawned on us that, for the most part, environmental resources are either the root causes or play essential roles in sustaining or prolonging the conflicts. From the farmers–herders conflicts in Nigeria, the unending war in the Congo; the long-suffering civil war in Angola; the civil war in South Sudan; apartheid in Southern African states; banditry and communal violence in Nigeria, Kenya, and Sudan; the Rwandan genocide; the international tension between Kenya and Uganda over Lake Victoria and between Uganda and Congo over mineral resources along the Congo borders; the water conflict between Ethiopia and Egypt; and tension between fishing communities of Nigeria, Cameroon, and Chad at the Lake Chad Basin and Nigeria–Cameroon over Bakassi, to name a few, all had elements of environmental resources access involved.

Issues of environmental security, we realized, are often not treated with the degree and gravity they deserve probably because we are unconscious of the impact the environment has on our day-to-day living and we have become accustomed or complacent to our "environment" or ignorant of them or are just taking them for granted since our focus magnetizes the seemingly more critical social and political issues that are prevalent in the print and social media.

So, with these thoughts in mind, we mooted the idea of a book and thought, why not look at environmental security in Africa? It came from the realization that many of our political, social, and development problems are tied to environmental security. In fact, some are merely environmental issues disguised in different garbs. Environmental preservation and conservation issues, droughts, and floods have exacerbated political and social tensions or even birthed them. These require reexamination, recasting a light on them to forge a new awareness of environmental concerns. That is the hope this book represents—refocusing our minds and energy on the basics of our human existence, sustenance, and development—the environment.

There is a plethora of issues around the environment in Africa. First, the physical environment is the backbone of livelihoods in Africa and, by implication, the mainstay of the African economy. The primary economy, especially agriculture, mining, and forestry, employs over 60 percent of the African workforce. Climate and environmental changes have tasked the continent, rendering her vulnerable to their impacts. For example, studies have indicated that Africa is warming up faster than the rest of the world, with potentially dire consequences on food security, habitation, and human security. Droughts and floods have become more frequent and devastating. It is also a fact that Africa's contribution to these changes is not significant. Nevertheless, the continent bears the brunt of the impacts. With high vulnerability and low adaptive capacity and resilience for various reasons, the exposure of Africa to many environmental problems has become alarming and attracted global attention.

Second, globalization and modernism have besieged the African social, cultural, and political environments for better or worse. They have opened up the economies to global forces and drastically changed the landscapes of engagement with the physical environment. The quest for development has added pressure on environmental resources and unleashed an avalanche of problems in almost all spheres of life: environmental degradation, loss of biodiversity, deforestation, desertification, droughts, floods, conflicts of varying degrees and types, political tension, and instability within and between countries. At the center of these and other existential threats is the health and security of the African environment.

Third, efforts to combat these issues yield different results than the desired and expected favorable fruits. With probably the weakest economy and political institutions in the world, Africa has struggled to find lasting solutions to its many environmental problems and issues, which, rather than shrink, have persisted and expanded in both scope and complexity. Environmental resources have become scarcer, and the pressure on human society and the environment has increased.

There is a growing consensus among scholars that Africa's development is inextricably tied to its environmental health and security. Environmental security concerns, therefore, have taken center stage and assumed greater significance in the development debate on Africa. The need to preserve and conserve the physical environment, re-examine the social, cultural, and political paradigms for development, and re-engineer tools for environmental problem-solving has become more urgent. Therefore, there is an impetus to cast new light on Africa's environmental security issues. The need to shine new light on this problem requires a multidisciplinary approach because of the multidimensional nature of environmental issues.

This book thoroughly assesses environmental security issues in Africa, especially regarding environmental conflicts, politics, and national development. Despite the recent global attention to the problem of environmental security and conflict in Africa, attempts at finding an enduring solution to the phenomenon have yet to be discovered. Part of the reason for this situation is the dynamic and multifaceted nature of conflicts constantly evolving within the context of global climate change, globalization, and the recent anti-globalization and nationalism movements worldwide. Consequently, the book brings together a group of scholars with varying backgrounds and experiences from the African continent, including geographers, environmental scientists, historians, sociologists, political scientists, anthropologists, legal practitioners, philosophers, and other social scientists, to provide a wider lens through which these issues are investigated, interrogated, and discussed.

Elisha Jasper Dung
Department of Advancement Studies
Alabama State University, Montgomery, Alabama.

Leonard Sitji Bombom
Department of Geography & Planning
University of Jos, Jos, Nigeria.

Augustine Avwunudiogba
Department of Anthropology, Geography and Environmental Resources
California State University, Stanislaus, California.

Acknowledgments

It is almost only possible to complete an edited book of this nature with the help and wise counsel of colleagues and friends. Therefore, the successful completion of this project was only possible with the efforts, contributions, and encouragement of several individuals. This is more so as we engaged in this project after the COVID-19 pandemic and global inflation, posing severe challenges to everyone involved. First, we would like to extend our appreciation and gratitude to the contributors to this volume who delivered their completed manuscripts despite the challenges posed by the COVID-19 pandemic and inflation. We would also like to thank our acquisition editor, Mrs. Sydney Williams, and the assistant acquisition editor Ms. Emilia Rivera for their patience and commitment to completing this project. In addition, we express our sincere appreciation to the anonymous reviewers whose constructive comments improved the quality of the book. We also wish to thank our colleagues at the Department of Advancement Studies, Alabama State University (ASU), the Department of Geography and Planning, University of Jos, Nigeria, and the Department of Anthropology, Geography and Environmental Resources, California State University, Stanislaus (CSUS), for their stimulating academic discussions over the years and at various points during the completion of this project. Thanks to Dr. James Tuedio, Dean, College of the Arts, Humanities and Social Sciences (CAHSS), CSUS, for financial support to Dr. Augustine Avwunudiogba through the CAHSS Research, Scholarship, and Creative Activity (RSCA) and Dean's Travel Awards that enabled him to participate in this project. We also acknowledge the contribution of Brian Ochieng Okoth in the co-authored chapter "Environmental Security in Kenya: A Case of Geothermal Projects in Samburu and Narok," who died in a car accident in Nairobi, Kenya, in 2023. Finally, this book would not have been possible without our families' and friends' support and encouragement.

Hence, thanks to Mrs. Comfort Gabriel Bombom, Ms. Lela Elmore, Dr. Paul Erhunmwunsee, Dr. Evelyn Hodge, Dean of University College, and Dr. Parichart Thornton, chair of the Department of Advancement Studies, ASU. We appreciate their understanding and accommodation of our absence from family and social occasions to meet our obligations in editing this book.

Abbreviations

ACLED	Armed Conflict Location and Event Data
ADB	African Development Bank
AfCFTA	African Continental Free Trade Agreement
AGOA	African Growth and Opportunity Act
APS	Amsterdam Port Services
ASALs	Arid and semiarid lands
ASEAN	Association of Southeast Asian Nations
AU	African Union
bcm	Billion cubic meters
CBO	Community-Based Organization
CFA	Communauté financière d'Afrique (Eng. African Financial Community)
COD	Chemical oxygen demand
COMESA	Common Market for Eastern and Southern Africa
CPA	Comprehensive Peace Agreement
CSR	Corporate social responsibility
DALYs	Death and disability-adjusted life years
DFID	Department for International Development
DOPRI	Director of Public Relations and Information
DRC	Democratic Republic of the Congo
EAC	East Africa Community
EC	European Commission
ECOWAS	Economic Community of West African States
EEC	Eurasian Economic Union
EIB	European Investment Bank
EPA	Economic partnership agreements
EPRDF	Ethiopian People's Revolutionary Democratic Front

ESN	Eastern Security Network
EU	European Union
EU–UN	European Union and the United Nations
FAO	Food and Agriculture Organization
FCT	Federal Capital Territory
FDI	Foreign direct investment
FIAS	Foreign Investment Advisory Service
FTA	Free Trade Agreement
GATT	General Agreement on Trade and Tariffs
GBD	Global Burden of Disease
GDC	Geothermal Development Company
GDP	Gross Domestic Product
GERD	Grand Ethiopian Renaissance Dam
GNI	Gross National Income
GSP	Generalised System of Preferences
GWh	Gigawatt hours
IBRD	International Bank for Reconstruction and Development
ICS ID	International Center for Settlement of Investment Disputes
IDA	International Development Association
iDMC	Internal Displacement Monitoring Centre
IDPs	Internally Displaced Persons
IFC	International Finance Corporation
IMDGC	Maritime Dangerous Goods Code
IOCs	International oil companies
IPCC	Intergovernmental Panel on Climate Change
IRIN	Integrated Regional Information Networks
IPOB	Indigenous Peoples of Biafra
JICA	Japan International Cooperation Agency
JSTF	Joint Security Task Force
JTF	Joint Task Force
KFW	Kreditanstalt für Wiederaufbau (Eng. Reconstruction Credit Institute)
LAPSSET	Lamu Port–South Sudan–Ethiopia Transport Corridor
LASLA	Large-scale land acquisitions
LDCs	Least developed countries
LGA	Local government area
MAF	Ministry of Agriculture and Forestry
MAG	Medium Airlift Group
MARPOL	International Convention for the Prevention of Pollution from Ships
MDCs	Medium developed countries
MEAs	Multilateral Environmental Agreements

MEND	Movement for the Emancipation of the Niger Delta
MIGA	Multilateral Investment Guarantee Agency
MNCs	Multinational cooperations
MOCs	Multinational Oil Companies
MOUs	Memoranda of Understanding
MPCI	Patriotic Movement of Côte d'Ivoire
MW	Megawatt
NAAEC	North American Agreement on Environmental Cooperation
NAF	Nigerian Air Force
NAFTA	North American Free Trade Agreement
NASAC	Network of African Science Academies
NBC	Niger River Basin Commission
NBI	Nile Basin Initiative
NBS	National Bureau of Statistics
NCD	Nigerian Civil Defense
NCN	North Central Nigeria
NDA	Niger Delta Avengers
NDDC	Niger Delta Development Commission
NDV	Niger Delta Vigilante
NGOs	Nongovernmental organizations
NLNG	Nigerian Liquefied Natural Gas
NNPC	Nigerian National Petroleum Corporation
NP	Nigerian Police
NPE	National Policy on Environment
NPS	National Park Service
NSRP	Nigeria Stability and Reconciliation Programme
OCHA	Office for the Coordination of Humanitarian Affairs
OMO	Oil Mineral Ordinance
OMVS	Organisation pour la mise en valeur du fleuve Sénégal (Eng. Senegal River Basin Development Authority)
OPA	Ouagadougou Peace Agreement
OSNA	Oyo State Security Network Agency
OTS	Operation Thunder Strike
PACAC	Presidential Advisory Committee against Corruption
PAP	Priority Actions Programme
PDCI	Parti Démocratique de Côte d'Ivoire (Eng. Democratic Party of Côte d'Ivoire)
PMI	Purchasing Managers' Index (a trading LTD company)
PPP	Public–private partnership
RAC	Regional Activity Centre
RAP	Resettlement Action Plan
RDR	Rassemblement des républicains (Eng. Rally of Republicans)

REC	Regional Economic Community
RISDP	Regional Indicative Strategic Development Plan
RSN	Regional security networks
RTA	Regional Trade Agreement
SADC	Southern African Development Community
SAN	Senior Advocate of Nigeria
SDGs	Sustainable Development Goals
SGR	Standard gauge railway
SLF	Sustainable Livelihoods Framework
Sp.	Specie (Singular)
SPLM/A	Sudan People's Liberation Movement/Army
Spp.	Species (Plural)
SSA	Sub-Saharan Africa
STS	Operations: ship-to-ship operations
TPLF	Tigray People's Liberation Front
UN	United Nations
UNDP	United Nations Development Programme
UNEP	United Nations Environment Programme
UNESCO	United Nations Educational, Scientific and Cultural Organization
UNICEF	United Nations Children's Fund
UNO	United Nations Organization
UNOCI	United Nations Operation in Côte d'Ivoire
USAID	U.S. Agency for International Development
USD	U.S. Dollars
WBG	The World Bank Group
WHO	World Health Organization
WRT	With regard to (BV Holdings, a Dutch oil chemical company)

Introduction

Elisha Jasper Dung, Leonard Sitji Bombom, and Augustine Avwunudiogba

This book investigates the nature, scope, and dimension of environmental security in Africa from a multidisciplinary perspective. This investigation examines the history, theories, spatial patterns, sociocultural and socioeconomic consequences, and legal ramifications of environmental concerns and conflicts, politics, and development, which are important correlates of environmental security in Africa. National and regional considerations are discussed, inferences are derived, summaries are made, and wide-ranging solutions are offered.

There is no straightforward definition of environmental security. However, there is general agreement that the concepts of environmental security arose from the synergetic combination of two important terms—environment and security. In this context, the environment is broadly defined as the milieu of natural and human-constructed spaces that form human production's organizational fabric. Security, on the other hand, can be generally seen as the condition that safeguards the environment to ensure optimum human development. Thus, environmental security, crucial to human development, emanates from different disciplines and schools of thought. As Barnett (2009) puts it, "(T) he way a person or a group understands each of these concepts informs their understanding of the combined concept of environmental security."

In recent years, environmentally induced insecurity or related problems have taken more severe dimensions in scope and intensity, especially on the African continent. Climate change, environmental degradation, loss of biodiversity, overuse or overexploitation of nonrenewable environmental resources, scarcity, lack of or inhibited access to resources, deforestation, and many other factors have directly or indirectly caused, aided, and abetted or prolonged conflicts in Africa. As a result, the political space has become congested with environmental security issues, which invariably bear tremendous influence on national development.

Environmental security is broad and includes concerns about both natural and human impacts on the environment and its consequences on social and political order at local, national, or regional levels. The environment is that component of the Earth that supports and provides necessary life-sustaining resources for human development. Therefore, the health of the human enterprise is closely tied to the health of the environment. Similarly, human security is equally interrelated with the security of the environment. This human–environmental relationship is, therefore, important to human development at all levels. The consensus among researchers and scientists is that the human agency is the most important driver of environmental change and is, therefore, at the center of environmental security concerns. However, environmental change could also be a natural process. The significance of the human element is buttressed by the fact that at the heart of environmental security is politics—environmental policy, governance, and institutions. Issues of vulnerability, resilience, and adaptation are also important considerations in environmental security. Together, they all form part of the human response to environmental change and security that determines, to a large extent, the occurrence, nature, scale, intensity, and duration of environmental conflicts.

Violent conflicts arising from environmental concerns are numerous on the African continent and range from farmers–herders conflicts, and intercommunal conflicts, to international conflicts and tensions, often based on issues of access to scarce environmental resources. Politics of environmental security, therefore, determines the extent to which these conflicts may be contained to create the space for local, national, and regional development.

There is no overriding theory for this edited volume. However, due to the multidisciplinary nature of environmental security and conflict, the book is informed by contemporary social, economic, and political theories in the social, behavioral, and humanities disciplines. Several authors have adopted the neo-Malthusian theory to investigate various environmental security and conflict issues in Africa. For example, Elisha Dung employs neo-Malthusian and political–ecological approaches to address the linkage of the environment and society in examining Africa's environmental security geography. This nature–society relationship is critical because environmental conflicts are fundamentally social, economic, political, and cultural problems.

Similarly, Faith Iwejingi Sajini uses the Malthusian theory to investigate the linkage between population growth and resource exploitation in Nigeria and its implications for environmental security. In the same vein, Bombom, Amupitan, and Edino discusses farmers–herders clashes, drawing on the strands of the models of climate change, the concept of environmental resources scarcity, neo-Malthusian theory, push–pull migration theory, and aspects of political ecology. The conceptual framework these authors have presented includes climate and environment; population dynamics; and historical, cultural, political, and socioeconomic factors, using the Middle Belt region of Nigeria as a case study. Furthermore, Onah, Ogwuche, and Dahiru utilize the basic tenets of the neo-Malthusian approach as a framework for understanding the drivers of environmental conflict. These drivers are competition over scarce environmental resources (internal), poor governance of environmental resources, and environmental resources dynamics and pressure (external). Additionally, Fagbeba Bakoyema uses the theory of realism to discuss the environmental causes of war and the conditions of peace between and among states, while Zainab Nuhu and Tasi'u Yalwa Rilwanu demonstrate how the basic needs theory explains how access to and the provision of land, which is fundamental for food production, grazing of animals, and space for human habitation for different actors in the conflict, represent their needs as threats, triggers, and tipping points of conflicts. The employment of these theories, concepts, and approaches helps to appreciate that the intersectional action of various state and non-state actors has been implicated in the global problem of environmental insecurity, leading to conflicts over resources and political instability that impede economic and social development.

In this edited volume, the authors adopted two basic data analysis methods—theoretical and empirical—for the respective chapters. The theoretical approaches involved secondary data mainly collected from literature and analyzed in three ways. Several chapters employed triangulation methods to advance their positions on African environmental security issues. Drawing from a wealth of data and information from several sources, they paint scenarios of environmental concerns on conflicts, politics, and development. The chapters on the relationships between environment and conflict and the chapter on the impacts of Ethiopian dams on the environment are examples of this.

The second set of articles in this category adopted a rigorous thematic analysis of essential topics, highlighting specific areas of interest for environmental security. Each theme is carefully selected and thoroughly discussed, with ample evidence from empirical studies. These include the chapters on the geography of environmental security in Africa; globalization, climate change, and environmental security; geothermal projects in Kenya; toxic waste dump in Cote D'Ivoire; forests as theaters of security operations and mangrove forests management in Africa; and the review of literature on environmental conflicts, politics, and development in Africa. The third set of chapters undertook a content analysis. Chapters on population growth and resource exploitation: implications for environmental security in Nigeria, coverage, implementation, and enforcement of sustainability provisions in regional trade agreements in Africa, and dimensions of environmental conflicts in the Niger Delta of Nigeria fall into this category.

Some chapters employed an empirical approach involving primary data collection and analysis. The chapter (17) on land grabbing in South Sudan extensively conducts field studies over five years. The author conducts in-depth interviews and thematically analyzes data with supporting evidence from the literature. The chapter (13) on the socioeconomic impacts of land reclamation in the Niger Delta region of Nigeria also collected and analyzed primary data. Using a combination of satellite imagery and field studies to collect data on land reclamation, the authors use descriptive statistics and thematic analyses to analyze the data. Interviews were also conducted for the chapter (14) on environmental security and geothermal projects in Kenya. The interviews conducted in chapters 13, 14, and 17 are anonymized.

The book comprises seventeen chapters—with many spreading across multiple academic disciplines, including international relations, security studies, history, geography, environmental law, and environmental resources management. Considering this, in chapter 1, "The Geography of Environmental Security in Africa," Elisha Dung contends that environmental security is the arena that integrates promoters of security and advocates of environmental change. Elisha argues that environmental conflict does not occur in a vacuum and has a geographic component: location, spatial distribution, scale, and temporal dimensions. The author examines the nature–society relationship in environmental conflicts because they are fundamentally social, economic, political, and cultural in nature. Elisha suggests that a multilevel approach comprising global, regional, subregional, and national policymakers can potentially enhance environmental security in Africa. He reveals that partnerships, and collaborations of nongovernmental organizations and corporations, individual governments of countries, regional blocs, and international agencies, are now seemingly realizing the need to take action to avert looming environmental disasters. These environmental disasters would then lead to environmental insecurity generated by conflicts. The author examines the spread of environmental security concerns in Africa and the types and scopes of responses and measures under consideration to address or prevent them. In addition, he analyzes the spatial and temporal patterns of environmental conflicts and determines their hotspots in Africa. Augustine Avwunudiogba and Ufuoma Akpovwovwo, in chapter 2, "Globalization, Climate Change, and Environmental Security in Africa," assert that environmental security is a major concern in the African continent because of its link to social, economic, and political stability. They argue that in recent decades, a combination of human and natural induced processes has exerted pressure on Africa's ecosystems resulting in the degradation of ecosystem services and environmental instability. Environmental instability often leads to environmental insecurity and its attendant negative consequences on community cohesion and function. Environmental insecurity has resulted in the economic and social dislocation of the population in many countries of Africa, leading to mass migration and the persistent problems of refugees and internally displaced persons. Available data suggest that this trend will likely worsen in the coming decades in Africa. There is an urgent need to interrogate the drivers of present and future conditions of environmental security in Africa as a basis for formulating adaptive and mitigative policy measures at the national and transnational levels. This task is imperative because economic development and improvement of the human condition in Africa are predicated on a secure environment. The authors explore the role of two major forces—globalization and climate change as drivers of ecosystem changes linked to environmental security in Africa. In the first part of the chapter, they examine the role of globalization in promoting the overexploitation of Africa's natural resources, natural resources access conflict, and ecosystem degradation and how this increases the vulnerability of communities to the negative externalities associated with primary resource extraction. The second part of the chapter examines the role of climate-induced environmental change, the occurrence of high-magnitude events that has become the norm. Climate-driven events such as floods, prolonged drought, desertification, and wildfire are now frequent in many African countries. African countries must address the environmental security implications of globalization and climate change to create an enabling environment for progress.

Environmental resources provide ecosystem services that are categorized into provisioning, cultural, regulatory, and supporting services. Thus, in chapter 3, "Access to Environmental Resources: A Framework for Understanding the Drivers of Environmental Conflicts," Sabastian Onah, Jonathan Ogwuche, and Mohammed Dahiru maintain that natural resources are stocks from which people derive vital ecosystem services, insisting that the range of ecosystem services shows how critically important these processes are to the functioning of the Earth's systems and human survival, livelihoods, and lifestyles and vice versa. The authors identify three main drivers responsible for environmental damage and scarcity. They are competition over scarce environmental resources (internal), poor governance of environmental resources, and environmental resources dynamics and pressure (external). They argue that these internal and external stress factors, combined with weak institutions and governance, create the nature of access to environmental resources. Also, environmental scarcity stems from three primary sources—supply-induced (environmental degradation), demand-induced (population growth), and structural (unequal distribution). These three sources create two common patterns of access to environmental resources: unequal availability and status (quantity and quality) of environmental resources. These drivers and sources of environmental scarcity create nature and patterns of access to environmental resources, exacerbating environmental conflict.

Fagdéba Bakoyéma, in chapter 4, "Environmental Conflicts, Politics, and Development in Africa: A Review," believes that the high incidences of environmentally triggered conflicts are not new. Bakoyema opines that history and contemporary events have demonstrated that environmental stress often results in conflict, frequently along ethnic schisms. This reality, he argues, is exacerbated by population pressure, resource shortages, environmental change, and developmental challenges. Evidence suggests that this trend will persist because environmental change will continue to stress marginal environments, especially in places with weak governance, clarifying the relationship between regional stability and environmental factors. The author highlights the linkages between people, governance, and environmental stressors and argues that the environment, resources, security, and Africa are inextricably interwoven in human interactions. Bakoyema further believes that security, conflicts, politics, and development have meanings only within an environment and a geographical location. Scholars of conflict studies, international law, and environmental security have explored direct and indirect links between armed conflicts and the environment. In addition to conflicts, deforestation for survival reasons has been viewed as negatively affecting the environment. Bakoyema focuses on conflicts stemming from environmental changes, politics, and socioeconomic development in Africa and the policies guiding local and global efforts to address environmental issues on the continent.

Environmental conflicts in many parts of the universe have become a critical issue that challenges local, regional, national, and global coexistence. Thus, in chapter 5, "Climate and Environmental Changes: A Nexus of Environmental Conflicts in Sub-Saharan Africa," Ahmad Abdullahi and Abdulkadir Maharazu establish a relationship between the twin phenomena of environment and security and how they trigger environmentally induced conflicts in many parts of sub-Saharan Africa (SSA). The authors assert that climate and environmental changes are already affecting SSA as in other places. The changes include higher temperatures, droughts, floods, wildfires, and soil degradation. These were reported to have brought about distress in migration, poor health conditions, domestic crises, and water and food insecurity. The authors recommend including knowledge of climate and environmental changes in the region's education policy to help reduce conflicts associated with the twin phenomena. In the same vein, Zainab Nuhu and Tasi'u Yalwa Rilwanu, identify the threats, triggers, and tipping points of conflicts in North Central Nigeria (NCN) in chapter 6, "Threats, Triggers, and Tipping Points of Environmental Conflicts: a Case Study of North Central Nigeria, 2002–2022," where they point out that NCN is home to diverse multiethnic groups and an area rich in environmental resources with prevalent conflict events. They claim that more than half of all reported conflict events were caused by disputes over access to land or resources and can metamorphose into ethnic, religious, or ethnoreligious conflicts. While they reveal that most conflicts are violent and armed in nature, cattle rustling and destruction of farmlands are the most prevalent. According to the authors, an increase in population, high demand for environmental resources, dwindling resources, and competition are at the forefront of triggers and serve as tipping points for conflict escalations.

Domagoj Krpan points out in chapter 7, "The Impact of Ethiopian Dams on the Environment in East Africa," that Ethiopia is home to many rivers, which are crucial for the survival of the communities from Egypt to Kenya. He stresses that the northeastern part of the country is the water source of the Blue Nile or Abay River, which supplies about 80 percent of the water in the Nile during the rainy season. In Khartoum, Sudan, the Blue Nile and the White Nile combine into the Nile. The largest Ethiopian river outside the Nile Basin is the Omo River. It empties into Lake Turkana on the border with Kenya. The other rivers flow from Ethiopia into South Sudan, Somalia, and Djibouti (Lake Abbe). In the last couple of decades, Ethiopia has been building dams to use its water supplies as a source of hydroelectric energy. The dams are raising concerns not only among ecologists but also among the governments in the region. The most (in)famous case is the construction of the Grand Ethiopian Renaissance Dam (GERD). The GERD is built on the Blue Nile and can influence the amount of water flowing downstream. This dam can be very dangerous for Sudan and Egypt and their population of over 100 million. The situation is so complex that, in 2013, Egyptian President Mohammad Morsi suggested destroying the dam. A similar problem is with the Gilgel Gibe III Dam on the Omo River, which can endanger the population around the river and Lake Turkana. So, the dams are affecting the whole region. Domagoj Krpan analyzes how the construction of dams impacts the environment in Ethiopia. In addition, it explores the transboundary impacts of the Ethiopian dams on other countries such as Kenya, Egypt, and Sudan. Domagoj Krpan also analyzes a similar situation with the River Senegal and what Ethiopia, Egypt, and Sudan can learn from the multilateral organization, the Senegal River Basin Development Authority. On the other hand, in chapter 8, "Revisiting the 2006 Toxic Waste Dump Incident in Cote d'Ivoire," Balla Keita revisits an unfortunate incident that has left a deep scar in Côte d'Ivoire. According to Balla, in September 2002, political differences based on ethnic, regional, tribal, and religious origins led to a civil war that lasted for a decade (ten years). Four years after the start of the civil war, the Probo Koala, a ship registered in Panama and chartered by Trafigura, a Singapore-based oil and shipping company, off-loaded toxic waste to an Ivoirian waste handling company. The toxic waste was dumped in at least eighteen sites in Abidjan's vibrant and populated city, creating a health crisis that has impacted the country today. The author assesses the environmental impact of this toxic waste dump and its implication for environmental security and health.

In chapter 9, "Coverage, Implementation, and Enforcement of Sustainability Provisions in Regional Trade Agreements in Africa," Moses Ogutu draws the linkage between free trade agreements (FTAs) in Africa and environmental security and the impacts that it has on positive developmental consequences for the citizens of its member countries. He argues that the successful realization of the developmental outcomes of FTAs largely depends on measures to ensure environmental protection, equitable and sustainable growth, and respect for human rights. In this respect, the author reviews coverage of sustainability and inclusivity provisions in the design framework of RTAs and implementation and enforcement mechanisms. Additionally, the author analyzes the coverage and implementation of sustainability provisions in key African regional trade policy frameworks—the EAC, COMESA, SADC, and AfCFTA agreements. Finally, the author concludes by providing guidelines on how trade agreements can promote and implement environmental security. Moses argues for the inclusion of a comprehensive and consolidated Protocol on Trade and Sustainable Development in the design framework of the AfCFTA in line with global trends. Such inclusion will ensure that member states commit to promoting inclusivity and sustainable development, in addition to free trade. Resource-based conflicts have historically engaged academic interest, with environmental conflicts given less attention, especially in the Niger Delta region of Nigeria. In this regard, Emmanuel Obuah and Reginald Keke, in chapter 10, "The Dimensions of Environmental Conflicts in the Niger Delta of Nigeria, 1956–2015," examine the dimensions of environmental conflicts in Nigeria's oil-rich and gas-surplus Niger Delta region. The authors argue that how people lived before the discovery and exploitation of oil and gas has changed significantly. They allege that sixty-five years of oil exploration have resulted in environmental conflicts of immense proportions. In Ogoni land, as of 2006, they claim that environmental conflicts have reduced crop yields by 10 percent, farmers' incomes by 5 percent, and reduced microbial load by 20–61 percent. The consequences are hunger, poverty, and a decrease in the life expectancy of the people. The authors assess the causes of resource-based conflicts, the various dimensions, and their impact on people and their environment. The authors advocate using institutionalized interactive conflict management mechanisms, capacity-building, transparency, environmental justice, and people-centered projects to ensure proper resource governance, sustainable resource exploration, and exploitation procedures based on sound principles of global resource governance and best practices, which will end resource-based conflicts. In chapter

11, "Herdsmen–Farmers Conflict in Nigeria's Middle Belt Region," Leonard Sitji Bombom, Racheal Yemisi Amupitan, and Marcus Ogwu Edino argue that the conflict between herdsmen and farmers in many parts of SSA has reached an unprecedented dimension in both scope and intensity. The authors present a conceptual framework for understanding the farmers–herders conflict in Nigeria's Middle Belt region, including climate and environmental, population dynamics, historical, cultural, political, and socioeconomic factors. The authors categorize the factors into core causes (climate and environmental changes), allied causes (population dynamics), and proxy causes (historical, cultural, political, and socioeconomic) and argue that the intense competition for environmental resources by herders and farmers, which the allied and proxy causes have exacerbated, is the leading cause of the conflict. Leonard, Racheal, and Marcus opine that the historical, cultural, political, and socioeconomic relationships (proxy causes) between the two general groups, however, have overshadowed the core and allied causes. These proxy causes, they argue, have become the face and manifestation of the conflicts. The authors make several policy options, including herders' sedentarization, grazing reserve creation, and proactive environmental education to devise adaptation strategies for herders and farmers.

In chapter 12, "Forests: The Environmental Theater for Internal Security Operations in Nigeria," Oluchukwu Onianwa investigates and details how forests in Nigeria have become the epicenter of internal security operations against non-state actors who have made these forests a base or newly found environment for the planning and execution of numerous operations such as kidnapping and terrorists' attacks against citizens and the state. The author identifies different locations of these forests occupied by armed groups and how the security agencies have continuously engaged them in armed confrontations. Onianwa interrogates the environmental impact of armed groups versus military battles in the Nigerian forests. He claims that forests cover more than one-third of Nigeria's land area. The author argues that forest-induced battle has become a new dimension of internal conflict in Nigeria with substantial implications for the entire West African subregion. Despite the heavy bombardment and ground military offensives unleashed against them, the armed groups have remained undeterred in their confrontation against the security forces in the forests. While fleeing from the heavy military onslaught, they have formed the habit of relocating from one forest to another, both in Northern and Southern Nigeria, where they continue their nefarious activities.

The need to mitigate the land shortfall in the Niger Delta region to accommodate infrastructural, urbanization, and industrial development is essential to the economy of the region and Nigeria as a whole. This argument is made in chapter 13, "Socioeconomic Impact of Land/Swamp Reclamation:

Implications on Environmental Security in the Niger Delta Region of Nigeria," by Michael Didia and Temple Abali, who chronicle the region's techniques and socioeconomic impact of land reclamation. The region is a low-lying, poorly drained coastal plain traversed by a maze of rivers. In order to meet the ever-increasing land use needs of the teeming population, several techniques, including hydraulic sand-filling, the use of chikoko mud, the use of concrete breakwater, and the dumping of solid wastes, are variously used to reclaim land in the region. The development of infrastructure and higher wages, access to professional or unqualified employment, the creation of recreational centers, employment opportunities, the expansion of land capacity, improved citizens' living standards, the creation of new land, the primary foundation for human development and survival of any active community are benefits derived from this venture. A vital step for directing investors and governmental entities toward effective resource allocation decisions should include the entire socioeconomic benefit of reclamation initiatives. Also, a functional institutional framework should be put in place to regulate reclamation activities across the region to enhance environmental security in the Niger Delta Region.

In chapter 14, "Environmental Security in Africa: A Case of Geothermal Projects in Samburu and Narok in Kenya," Clifford Okwany, Evelyne Owino, and Brian Okoth examine how "frontier entrepreneurs" impact the socio-ecological environment of the Maasai and the Samburu pastoral groups, North and South Rift, Kenya. The authors compare the effects of development initiatives of Kenya's Geothermal Development Company (GDC), the sacrificed zones for geothermal energy—the proposed geothermal projects in Suswa, Narok, and Baragoi, Samburu, in Kenya to the completed five units of the Olkaria geothermal energy in Naivasha that have transformed the socio-ecology of the Maasai pastoral community, creating environmental risks and insecurity such as noise, water, and air pollution in Naivasha, Nakuru County. They contend that the threats have affected the wildlife in Hell's Gate National Park bordering the Olkaria geothermal units, which has led to human displacements by the Kenya Electricity Generating Company (KenGen) and forecast that the sacrificed zones in Baragoi and Suswa risk the same. The authors demonstrate that conflicts, organized violence, and insecurities in these spaces are due to the historical neglect and exploitation of the colonial and postindependence governments.

The uniqueness of mangroves is unparalleled in terms of its adaptability and capability. Ufuoma Akpovwovwo made this assertion in chapter 15, "Mangrove Forest Management in Africa," claiming that mangroves serve as a nursery, shield, and filter for biodiversity along the coastal areas, with implications for the hinterland. Akpovwovwo argues that mangroves' functions are indicative of their ability to ensure a holistic sustainability of life across terrestrial and aquatic environments. A significant chunk of the global mangrove forest is in Africa, with Nigeria ranking first in land coverage in Africa and third in the world. Globally, forest resources have been negatively impacted due to population explosion and urbanization. This negative impact has, in turn, gradually contributed to reducing its functionality of attaining environmental sustainability. Consequently, the implications for human survival are dire. Similar pressure has been mounted on the mangrove forest. The author asserts that until recently, mangrove forests were regarded as environments of little worth. The rugged terrain of the mangrove forest environment further contributes to this narrative. The author examines the mangrove forests' status across Africa while appraising the management strategies adopted over time.

In chapter 16, "Population Growth and Resource Exploitation: Implications for Environmental Security in Nigeria," Faith Sajini examines the nexus between population growth and resource exploitation and how this dual phenomenon threatens environmental security in Nigeria. According to Sajini, the country's population has grown from 11 million in 1901 to 140 million in 2006 and currently stands at 216 million. She argues that this phenomenal increase places a heavy demand on environmental resources, leading to their overexploitation to satisfy the basic needs of the teeming population with severe consequences on the environment. Sajini examines the impact of population growth in Nigeria on the environment's resources and highlights the implications of population growth and resource exploitation for environmental security. The chapter anchors on one prominent theory examining the relationship between population growth and resources—the Malthusian theory of population growth. The author cites factors such as improvement in medical health care, better sanitation, and improvement in transportation which enhanced food circulation, as the initial factors of population growth; however, sociocultural and religious factors, among others, perpetuate this growth in Nigeria via a high birth rate. Also, increased population leads to an increase in resource exploitation, leaving the environment degraded and hazardous for human existence. Sajini recommends stringent population regulatory measures and sustainable environmental resource management measures in Nigeria.

South Sudan, the youngest country in the world, has a rural population of about nine million people—80 percent of its total population, who depend on access and use of the land for their livelihood strategies in a country shattered by decades of war. Sara de Simone makes this claim in chapter 17, "'Land Grabs' and the 'Local Community': The Search for Human Security between Participation and Patronage in South Sudan." Sara argues that while physical security could never be taken for granted by the population of the region—where violent clashes continue to claim dozens of victims each month even during the "post-conflict" phase, human security has been even harder to achieve. Sara insists that South Sudan has occupied the lowest ranks on the Human Development Index (HDI) since its appearance on the international scene. At the same time, its environment has borne the brunt of heavy fighting, mass displacement, war-related exploitation of natural resources, and general rainfall trends modifications linked to climate change. In this scenario, in search of means of diversifying the economy, the newly established Government of South Sudan started encouraging foreign investments in the agroforestry sector in 2006–7, emphasizing the abundance of land, its underutilization, and the relatively low population density of the region. The author believes that opening South Sudan's land to foreign investments resulted in what has been described as a scramble for South Sudan's land. The author shows how this scramble for South Sudan land exacerbated the vulnerability of local communities, their exposure to environmental injustice, and the worsening of their human security by alienating the most vital resources for their livelihoods.

Finally, Elisha Dung, Leonard Bombom, and Augustine Avwunudiogba conclude the book, "Environmental Security in Africa: The Way Forward," by summarizing the significant findings and recommendations of the various contributors to this edited volume. The editors shine a light on the severity of environmental security in Africa and conclude that the future of Africa appears to tether on the brink of an abyss of uncertainty unless they deal with the burgeoning population increases and dwindling resources in the continent through efficient management. They offer suggestions on how to ensure a secure environment for a secure Africa.

Part I

CONCEPTUAL AND THEORETICAL CONSIDERATIONS

Chapter 1

The Geography of Environmental Security in Africa

Elisha Jasper Dung

Environmental security is the arena that integrates promoters of security and advocates of environmental change. Across Africa, environmental security concerns have taken center stage as the continent reels under tremendous environmental pressures. Effects of climate change, deforestation, desertification, erosion, environmental degradation, floods, plastic waste pollution, fossil fuel pollution, depletion of resources, scarcity, and armed conflicts, among many other environmental security issues, have intensified or registered their presence by becoming too evident and too frequent. Environmental conflict does not occur in a vacuum and has a geographic component: location, spatial distribution, scale, and temporal dimensions. The chapter relies on the Malthusian and neo-Malthusian theory to address the sources of conflicts. These include demand-induced sources that stem from population growth, supply-induced causes of environmental conflicts, drivers of environmental degradation and change, and structural inequalities affecting the distribution of resources.

The political–ecological approach is also employed to address the linkage of nature and society. This nature–society relationship is critical because environmental conflicts are fundamentally social, economic, political, and cultural. Indeed, a multilevel approach comprising global, regional, subregional, and national policymakers has the potential to enhance environmental security in Africa. Partnerships, and collaborations of nongovernmental organizations (NGOs) and corporations, individual governments of countries, regional blocs, and international agencies, are now seemingly coming to realize the need to take action to avert looming environmental disasters. These environmental disasters would then lead to environmental insecurity generated by conflicts. This book chapter examines the spread of environmental security concerns in Africa and the types and scopes of responses and measures under consideration to address or prevent them. The chapter analyzes the spatial and temporal patterns of environmental conflicts in Africa and determines the hotspots of environmental conflicts in Africa. The chapter concludes with some recommendations and suggestions to improve environmental security in Africa.

THEORETICAL FRAMEWORK

In this chapter, we employ the Malthusian/neo-Malthusian theory and political–ecological approach framework to explore the spatiotemporal dimensions of environmental security in Africa. Combined, the Malthusian and neo-Malthusian theory and the political ecology approach provide a broad framework for understanding the geography of environmental security issues in Africa. This strategy is necessary because environmental security has both natural and human drivers and manifests itself in complex multidimensional and multiscale scenarios. Despite its shortcomings, the Malthusian and neo-Malthusian theory provides a useful framework for the analysis of environmental security in Africa.

From the Malthusian perspective, population pressure on the existing environmental resource base will, in the long run, lead to scarcity as more people compete to use the finite resources within a geographic area.[1] Dwindling resources and the ever-increasing population will, therefore, eventually lead to conflict and the ultimate collapse of society.[2] Although the Malthusian approach has been criticized for its inability to recognize the possibility of technological advancement to moderate the impact of population on environmental resources, available archaeological evidence suggests the collapse of advanced civilizations on the account of resource overuse.[3] In modern societies, therefore, it is plausible to expect increasing conflicts resulting from the diminishing resource base. This analysis is relevant in the African context where demographic trends suggest that African countries by and large are still in the late part of stage two and the early part of stage three of the demographic transition, where a decline in death rate due to improved food supply, health care, sanitation, urbanization, and education leads to population growth.[4] The implication is that African countries will continue to witness high population increase in the foreseeable future. Indeed, figure 1.1 shows (a graph that shows population trends of Africa relative to other continents) the average population trend for Africa including projected total population growth from 2022 to the year 2041[5] confirming the continuous rapid population growth in the region, surpassing those in other regions, including Asia, North America, Europe, Latin America, and the Caribbean.

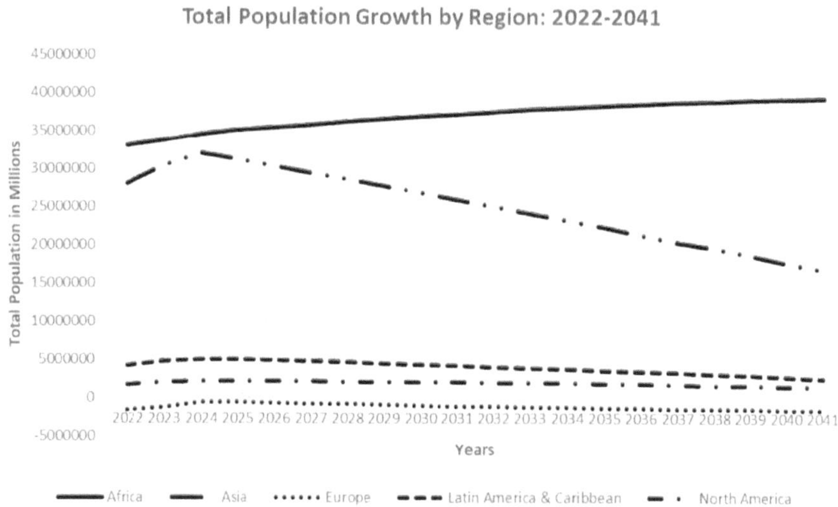

Figure 1.1 Total Population Growth by Region: 2022–2041. *Source*: Created by the author.

At the same time, many African countries' primary resource base is dwindling, creating frequent internal conflicts among different ethnic groups and international conflicts over access to transnational boundary resources such as international rivers, lakes, and maritime water bodies.[6] Thus, the analysis of environmental security in Africa must take into account rapid population growth and its various ramifications. For example, rapid population growth at the local level can lead to conflicts between ethnic groups over access to farmlands, grazing lands, and other shared resources. However, population pressure alone is not enough to explain the complex issue of environmental security in Africa. In this regard, the political–ecological approach is a useful framework for addressing the issue of environmental insecurity in the continent. Political ecology argues that environmental resource use is necessarily situated in political policy wherein different actors and agents play crucial roles at different temporal and spatial scales in terms of who has access and control of environmental resources.[7]

The divergence of interests among the various agents and actors will, therefore, result in conflicts, which will amplify environmental insecurity. We utilize the political ecology of mineral exploitation in Africa as a classic example to elucidate the critical role that political ecology can play in illuminating the conditions of environmental security in the continent. In this case, the natural minerals are located in local communities who have little or no say in their exploitation and utilization. For example, the actors in this scenario may include multinational or transnational corporations, national governments, local governments, and disenfranchised communities on whose land the mineral resources are located.

Every so often in Africa, the national governments habitually work in cahoots with transnational corporations to extract the mineral resources to the detriment and interest of the local population. In most cases, the monetary value from such exploitations go to the multinational corporations and the national governments based on the sharing agreements between them while the local populations are left to deal with the negative externalities such as pollution, degraded soils and agricultural lands, and the dislocation of the community's social fabric associated with the mining operations. It is, therefore, not surprising that regions and locales blessed with mineral resources in Africa are often the hotspots of social and environmental conflicts leading to environmental insecurity. To understand the problem of environmental security in Africa, therefore, it is imperative to analyze the role played by international and national actors/agents and communities in the negotiations and disputes over resource landscapes in Africa. This will help to illuminate the geography of environmental security as discussed in the remaining section of this chapter.

ENVIRONMENTAL SECURITY PRESSURES AND CONCERNS IN AFRICA

Several interrelated factors pose environmental pressures and concern in Africa. Although these factors may not necessarily lead to incidents of environmental conflicts, their presence and occurrence increases the vulnerability of localities where they occur to the incidence of environmental conflicts and insecurity. In other words, these pressures acting alone or in conjunction with other social and economic challenges can easily transform a vulnerable community into an environmental conflict zone. At the forefront of these pressures are processes that lead to the degradation of the environmental resource base on which community livelihood depends. This is particularly predominant in agrarian communities or communities where the primary sector is often the main or only source of income generation for the sustenance of the livelihood of such communities, as is common in many rural communities in Africa.

Climate Change

Climate change has the potential to exert pressure on various aspects of a community's natural resources, making them become more vulnerable to environmental instability and potential conflicts.[8] For example, climate change could have potential negative effects on the water resources of communities that depend on natural water bodies and lack water supply from municipal water authorities. This may lead to conflict at the individual and community levels over access to dwindling water supply.

Deforestation

Anthropogenic and naturally induced deforestation can potentially reduce forest resources and the associated services that they provide to local communities. In Africa, rural communities often rely on the forest for medicinal plants, food, firewood, and building material. As the extent of forest stock decreases, either due to deforestation because of legal and illegal lumbering operations or conversion of forests to tree crop plantation that is geared toward the export economy, local communities that depend on these forest resources become marginalized and disenfranchised and inhibited in their quest to use these resources. This has a lasting effect on the social structure of the communities as century-long livelihood practices are no longer sustainable given the restricted access to the ever-dwindling forest resources. Sometimes, national governments may use the influence of power to designate large tracts of land as national forest reserves without consideration for local communities that have been living in these areas since the beginning of time. The combined effects of these dwindling forest resources and restrictions often lead to community dislocation and communal clashes between ethnic groups, creating situations that often escalate into environmental conflict issues.

Desertification

Also related to deforestation is the process of desertification, which is a well-documented problem in the Sahelian zone and other regions of Africa.[9] Desertification in the Sahelian zone has affected agricultural practices, especially pastoralism in those countries in the Sahelian zone, leading to the southward migration of herders with more frequent encounters with sedentary cultivators in the wetter portion of the region.[10] This changing transhuman activities have led to an increase in a number of conflicts between herders and sedentary cultivators in the Sahelian countries such as parts of northern Senegal, southern Mauritania, central Mali, northern Burkina Faso, the farthest south of Algeria, Niger, the outermost north of Nigeria, Cameroon, Central African Republic, central Chad, central and southern Sudan, the extreme north of South Sudan, Eritrea, and Ethiopia.[11] Therefore, desertification is a major contributor and consideration to regional environmental security in Africa.

Erosion

Soil erosion like desertification can lead to the reduction of soil health, which may ultimately have effects on soil fertility and agricultural productivity. Unlike desertification that is mainly the problem of the Sahelian zone, soil erosion can occur even in forested areas where poor land use management practices involving agriculture, deforestation, and road construction can accelerate the problem of soil erosion. In some regions, such as the eastern part of Nigeria, massive irreversible gully erosion has led to the loss of communal land and the dislocation of settlements, creating additional pressure on the already precarious livelihoods of the people such that the potential environmental conflicts become amplified. In the Sahelian zone, wind erosion associated with desertification leads to the removal of tons of topsoil, exposing infertile subsoil and hindering efforts of programs aimed at checking the southward advance of the Sahara Desert. Soil erosion increases the cost of environmental remediation in magnitudes beyond the resources of local communities. It is, therefore, not uncommon for villages and settlements to disappear as the inhabitants are forced to relocate to more favorable locations, which heightens the potential for conflict with other ethnic groups and tribes in those new locations.

Pollution

Environmental pollution, especially those associated with resource extraction, has generated community conflicts in many regions of Africa, especially in the Niger Delta region of Nigeria.[12] Resources extraction such as minerals by international, multinational, and transnational corporations in Africa often result in situations where the local communities are left to bear the brunt of the negative externalities including the byproducts of the mining operations that often degrade and pollute the local environments. This is because most African countries often lack serious environmental protection laws, and in cases where such laws exist, they are rarely enforced. In short, national environmental regulatory laws and policies are often designed to benefit these multinational corporations as an inducement for them to invest their technology and capital to maximize the exploration and exploitation of mineral resources without serious consideration to the environmental impacts.

Because local communities are often politically powerless, they are left to bear the burden of the environmental effects of pollution that affect not only their livelihood, but also their health and economic stability among other environmental challenges. In response to the neglect from the national governments and multinational corporations, and in the absence of alternative sources of livelihood, communities in mineral-producing areas, for example, often resort to protests and armed conflicts with elements of the national governments and representatives of the multinational corporations. These community agitations are seen in the Niger Delta region of Nigeria, for example. Mining operations in Africa are, therefore, a major source of environmental conflict and insecurity.

Depletion of Resources and Armed Conflicts

Although environmental insecurity may result in armed conflicts, armed conflicts, which result from other causes, may lead to environmental degradation and eventual environmental security issues. Activities associated with armed conflicts, including the use of weapons of war, the deliberate destruction of settlements, farmlands, community water sources, and other community environmental resources, may lead to environmental pressure and degradation of affected areas with long-term negative impacts that will prevent the utilization of those resources even with the cessation of the armed conflicts. For example, landmines in Northern Africa, especially Libya from the Libyan–Egyptian war of 1977 and in 2011 because of the Libyan civil war, are testaments to this assertion.

Also, landmines in Western Sahara instituted by Morocco, where thousands of people and numerous livestock (goats & Camels) have been killed or injured, are another example.[13] In addition, landmines exist in Angola because of the twenty-seven-year-old civil war that ended only in 2002, yet landmines that are scattered in fields, villages, towns, and entire communities are still amassing casualties of war, including women and children who were not born when the war began or even politically ended.[14]

THE GEOGRAPHY OF ENVIRONMENTAL SECURITY IN AFRICA

As the second largest continent in the world with diverse geologic, climatic, and biogeographical zones, environmental security issues in Africa will vary according to this spatial variability in the biophysical environment. Added to this fact is the diverse nature of the ethnic and tribal makeup of the countries whose boundaries were the arbitral creation of colonial legacy. Furthermore, Africa is blessed with a variety of specific geographical mineral resources, some of which are unique and crucial to specific global industries. For example, Africa is the major source of strategic minerals such as Coltan (columbite–tantalite), Chromium, Platinum, Manganese, and valuable minerals such as gold, diamond, Cobalt, Uranium, Lithium, and crude oil. This variety of natural and human factors and processes create spatial variability of environmental pressure and environmental security concerns. This section shows some examples of the geographic incidence and potential occurrence of environmental security related conflicts on the African continent.

The Sahel region in Africa is a belt that is about 1,000 kilometers (620 mi) from the continent's northern tip in the southerly direction and spans 5,400 kilometers (3,360 mi) from the Atlantic Ocean in the west to the Red Sea in the east. The Sahel is a transitional ecoregion of semiarid grasslands, savannas, steppes, and thorn shrublands lying between the wooded Sudan savanna to the south and the Sahara Desert to the north.[15] The Sahel region is one of the regions with ongoing and potential conflicts resulting from the impact of global climate change.[16] As a transitional zone between the Sahara Desert and the more humid forest region of the western and central African region, the Sahelian zone has always been a marginal ecoregion that sustains nomadic pastoral ethnic groups that cut across the region. This lifestyle was sustainable so long as the region did not experience prolonged drought. In more recent years, the incidence of prolonged drought has become a permanent feature of the Sahelian zone, which has been further exacerbated by the impact of global climate change. The consequence is that the Sahelian zone has been migrating further south as desertification intensifies (see figure 1.2).

The loss of vegetation and water resources that support pastoral livestock has resulted in not only conflicts for the limited available grasslands, but also means that pastoralists are increasingly migrating further south in search of pasture and water for their livestock, a situation that has generated increased conflicts with the sedentary agriculturalists in the southern transitional zone. Therefore, the southern boundary of the Sahelian zone has become a zone of environmental conflict. It is now commonplace in the news of conflict between livestock herders and sedentary farmers resulting in the sacking of villages and the destruction of lives and property in Sahelian countries such as Mali, Burkina Faso, Niger, Nigeria, Sudan, and Ethiopia. In the Sahelian zone of Nigeria, for example, there is frequent conflict between Fulani herdsmen migrating from the Sahelian zone and the more sedentary farmers of the North Central region, including the Tiv, Berom, Ngas, Nupe, Igala, and the Idomas. In extreme years, the herdsmen have even migrated further south to the southwestern and southeastern forest belts of Nigeria, with conflicts with traditional forest farmers who complain that livestock, especially cattle herded by the pastoralists from the Sahelian zone, frequently destroy their crops causing them major financial losses.[17] The severity of this conflict has been heightened because the pastoralists are often armed with military-grade rifles and weapons and are not hesitant in using them to forcefully gain access into communal lands for their cattle to graze upon. Given the projections about the impact of climate change, it is most likely that the Sahelian zone of Africa will continue to be a flashpoint of environmental conflict and

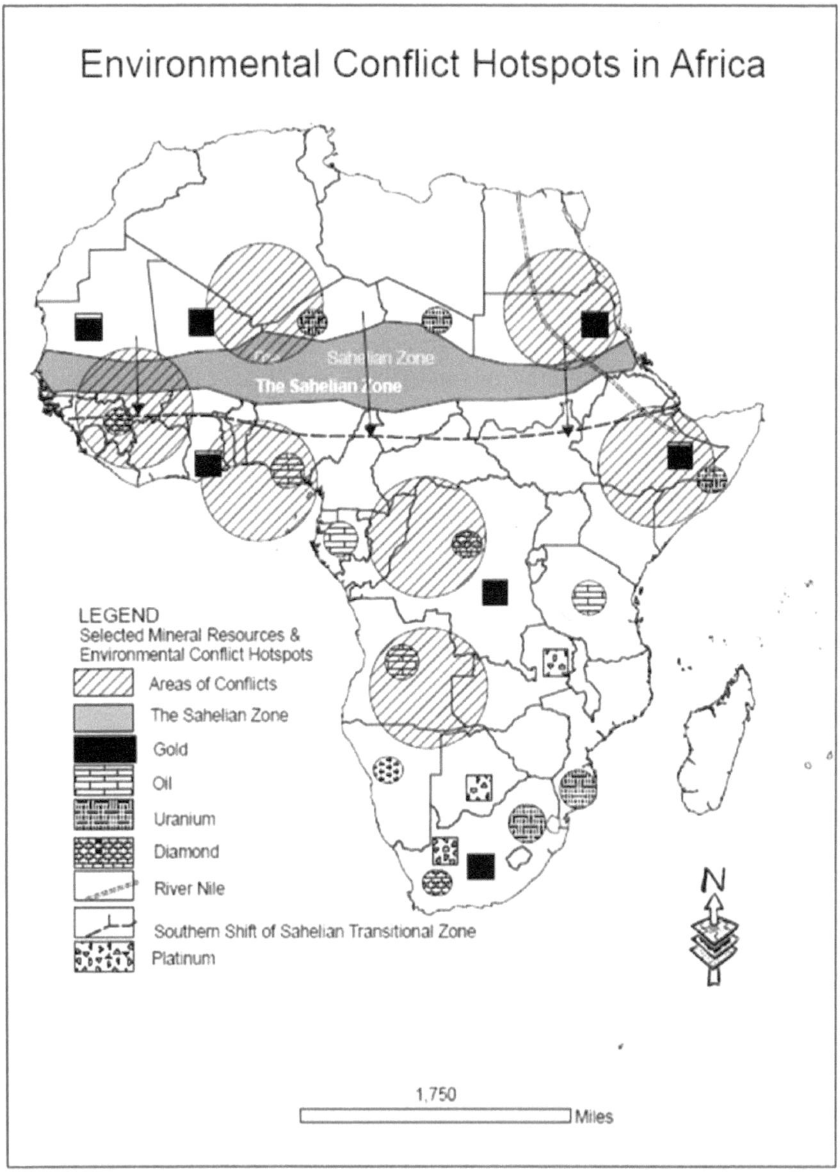

Figure 1.2 Environmental Conflict Hotspots in Africa. *Source*: Created by the author.

environmental security concerns for the foreseeable future. The environmental conflicts along the Sahelian zone will not only result from pastoralist migration activities but also the general human migration further south as the region becomes more adverse and hostile to permanent settlement. This will further pose potential conflict between the migrants and their host communities in the more humid south.

Figure 1.2 shows that the majority of the environmental conflict hotspots in Africa are situated in regions with strategic mineral deposits. These regions are often the theater of operation by multinational/transnational corporations, national governments, and local communities with different and sometimes conflicting interests. The conflicting interests and actions of these agents are often the precursor to many of the conflicts experienced in Africa. In the Niger Delta Basin, for example, the exploration and exploitation of crude oil since the 1960s has generated not only large sums of money for the Nigerian state but has also resulted in the major environmental conflicts between host communities, the corporations operating in the region, and the Nigerian state.

These conflicts result from the imbalance of the power structure and relationships between the three actors/agents, whereby the host communities are left to bear the negative consequences of the crude oil extracting process, including oil pollution of riverine resources, damage of communal farmlands, and air and water pollution by effluence associated with oil production. It is, therefore, not surprising that the environmental insecurity created by the exploitative tendency of the crude oil industry has led to major environmental conflicts with major national security implications for the Nigerian state.[18]

Similarly, other regions of Africa (see figure 1.2) have examples of these environmental conflicts resulting from mineral exploitation. In the Congo-Zaire region, the ongoing civil war is rooted and fueled by the need to control the source of the strategic mineral deposits such as coltan, gold, diamond, and crude oil. In addition to the enormous human loss, much damage has been done to the ecology of the river basin of the region, including the destruction of the habitat for the central African primates and the increase in poaching activities, all of which has further exacerbated the environmental pressure in the region creating environmental security issues.[19] Similar conflicts are ongoing in the disputed crude oil regions of Sudan and Southern Sudan, a situation that is very delicate and could easily degenerate into a major and prolonged environmental conflict situation. All these examples underscore the need for a regional approach to environmental security issues in Africa as argued in the previous section.

Another potential hotspot area is the Nile River basin. Here, the water resources of the Nile basin are a critical resource to eleven countries, but the Blue Nile is essential to Egypt, Sudan, and Ethiopia. Most of the landmass of these countries lie in the Sahelian zone of the Sahara Desert region of Africa. The recent trend in climate change has meant that these countries are receiving less rainfall and rely heavily on the water resources of the river Nile. The ongoing construction of a major water dam, the Grand Ethiopian Renaissance Dam (GERD), in the headwaters of the Blue Nile has raised concern among the downstream counties of Sudan and Egypt, with the major concern being that the dam construction will change the hydrological regime of the Nile River and hence the available water resources. This situation, which has been a source of diplomatic standoff among the affected countries, can translate into an environmental conflict if a conscientious regional approach is not adopted to resolve the issue.

SOCIOECONOMIC DIMENSIONS OF ENVIRONMENTAL SECURITY IN AFRICA

Environmental security considerations should be at the forefront of government policy in Africa because of its several ramifications for national security. Indeed, it has been widely accepted that national security is ultimately linked to the environmental security of nations. In countries that experience environmental instability, even on a temporal scale due to natural disasters such as earthquakes, volcanoes, flooding, and prolonged droughts, often grapple with national security issues. This is because environmental insecurity has social, cultural, economic, and political implications.

Social/Cultural

From a social and cultural consideration, an insecure environment would lead to the breakdown of the social and cultural fabric of communities, which then leads to a myriad of social problems, including the destabilization of community relationships and feelings of communal belonging, as well as affecting communal way of life, and maintenance of security and order. The destabilization and breakdown of community order will create a situation where there is an increase in conflict, arson, burglary, theft, and a host of other illegal activities. The ethnic tensions and periodic conflicts that led to the Rwandan civil war between the Tutsis and Hutus are examples. The crises between the two ethnic groups began when the Belgium colonizers created these two artificial ethnic groups that began to compete for resources and political domination between 1959 and 1961 after Rwanda gained independence. This search for domination led to discriminatory behavior and mistrust between the two ethnic groups, leading to fear of one another. The crisis reached its climax on April 6, 1994, with the assassination of then-President Juvenal Habyarimana, a Hutu. Suddenly, genocidal killings of the Tutsis exploded. Within 100 days, the Hutu Military and Militia had massacred between 500,000 and 800,000 Tutsis and some moderately minded Hutus. These figures represent about 75 percent of the entire Tutsi population and approximately 20 percent of the total Rwandan population.[20]

Economic

Environmental security also has economic ramifications. Generally, conditions of insecurity do not promote economic progression. This is because environmental insecurity creates uncertainty and discourages investments in long-term enterprises. This is crucial for industries and sectors in African countries that require foreign investments. Besides its impact on dampening direct foreign investments, domestic initiatives are also impacted. The overall consequence is a decline in productivity, diminishing employment opportunities for the citizens of the affected countries, which then amplifies existing social problems. At the national level, the reduction in economic activities also leads to a reduction in gross domestic product (GDP) and gross national income (GNI), both of which translate to limited revenue available to execute governments' social and development programs.

Political

From a political perspective, environmental insecurity can create national instability along ethnic or geographic lines, posing challenges to effective governance. In this regard, conflicts resulting from environmental insecurity may degenerate into prolonged civil war, and in some cases into international armed conflicts between countries. Environmental insecurity is not only limited to issues of national security but also ultimately leads to international security concerns on the African continent.

Thus, environmental security has political implications at the community, national, international, and multinational levels. For example, the Boko Haram terrorist group, which began in Nigeria, became a political hot potato for neighboring countries such as Cameroon and the Chad Republic. Abundant evidence shows that terrorist activities have resulted in social and moral panic among Nigeria, Cameroon, and Chad populations with calamitous consequences, particularly on educational development, food security, economy, national cohesion, and human rights.[21] This development shows that terrorism seriously threatens human security on many fronts. The policy implication is that social capital, participatory politics, and trust between the military/police and community partnership remain suspects.[22] The kidnapping of the Chibok girls and the disruption of and prevention of people from voting in Nigeria and the other affected areas are just a few examples. These kidnapped girls only returned home seven years after their abduction in 2014.[23]

Some pundits claim that Boko Haram exists because they have sponsors, political godfathers, and top military officers in their ranks. This assertion fuels a belief that Boko Haram is a conspiracy involving the government and top brass of the military. Participants believe the war is deliberately prolonged because it is a money-making machinery for the political and military elites. This development has implications for policy in the war on terrorism as there are claims that the disbursement of weapons to the military personnel fighting Boko Haram varied by unit and the agency of commanders.[24]

APPROACHES TO ENVIRONMENTAL SECURITY ISSUES IN AFRICA

The foregone discussion indicates that environmental issues in Africa occur at multiple spatial and temporal scales. Given the crucial role that a secure environment plays in the economic development of African countries, it is imperative to adopt a multifaceted approach in the formation of policies, programs, and implementation of plans aimed at reducing the pressure and triggers of environmental security in Africa.

Global

Certain elements of environmental security in Africa necessitate a global approach. The effects of climate change on environmental security cannot be effectively addressed within the continent alone because climate change, which is one of the drivers of environmental security, is a globally driven process. In this respect, African countries need to participate in global efforts by international organizations to fight climate change, bearing in mind the peculiar needs of African countries in economic development. Other pressures such as pollution of maritime waters and pollution associated with mining by international multinational corporations would also require a global approach. For example, multinational corporations operating in Africa should be made to adopt the strict environmental standards observed in their home countries rather than lobbying to weaken the existing environmental laws on the African continent. On their own part, African countries can lobby and advocate for these higher environmental standards using the instruments of the United Nations (UN) organizations, such as United Nations Development Programme (UNDP), United Nations Educational, Scientific and Cultural Organization (UNESCO), and so forth. Also, African countries should actively engage in bilateral negotiations with the host countries of multinational corporations to ensure that they apply the same standards as their home countries at every level of their activities in Africa. To ensure that the corporations hold their own end of the bargain, the high standard requirements should be clearly stated in the Memoranda of Understanding (MoUs) as well as the contractual obligations to compel them to address any negative externalities arising from their operations.

Regional

As already discussed, the potential environmental security hotspots in Africa are also regional in nature. This is the case where two or more countries share common resources, such as the major river basins of Africa, including the Nile, Niger, Congo, and Zambezi. Similarly, existing civil conflicts are regional in nature such as the ongoing conflicts in the Democratic Republic of the Congo (DRC) region, the Horn of Africa, and the Northwest region of Africa. Thus, a regional approach is also needed to tackle conflicts that would translate into environmental security concerns. It is important to acknowledge here that in some cases there are existing regional bodies designed to facilitate the cooperative use of the transnational resources while minimizing conflicts. These include the Niger River Basin Commission (NBC) established in 1963 that is made up of all the nine countries that share part of the River Niger Basin, including Benin, Burkina Faso, Cameroon, Cote d'Ivoire, Guinea, Mali, Niger, Nigeria, and Chad.[25] Such bodies should be strengthened and take an active role in the joint management of these environmental resources. Similar organizations could be established within the East African nature or game reserve corridor to help reduce potential environmental conflicts between conservation and indigenous communities that have relied on these resources for their livelihood. Other regional initiatives may also be necessary to jointly tackle environmental degradation problems such as desertification or desert encroachment. Given the nature of the porous boundaries between African countries, the need for a regional approach in resolving potential situations that could create environmental insecurity cannot be overemphasized. For example, Boko Haram, a designated terrorist group, operates in a forest zone that lies as wide as Northeastern Nigeria, Northern Cameroon, and Niger Republic and is a good candidate for a regional approach to reduce the insecurity associated with their activities (There is a multinational Joint Military Task Force in place to tackle Boko Haram . . . it is only a matter of commitment to the task).

National

Most environmental issues in Africa are often framed within individual national governments. Thus, the level of environmental security within the continent reflects the aggregate environmental security policies of the different countries. Unfortunately, most African countries' governments have not explicitly made the link between environmental security and national security in contrast to what is obtained in countries of Western Europe and North America. Although African countries may have laws designed to safeguard the environment as a general matter, most African countries lack explicit environmental security policies. There is an urgent need to sensitize policymakers to recognize the importance of environmental security in national security. Strong environmental policy in the individual countries in Africa will encourage regional, continental, and global cooperation in this regard.

Local

Every environmental security issue has a local component. Thus, the incorporation of the local community into any framework designed to reduce environmental insecurity cannot be overstated. This is because at the very crux of the matter, those who bear the brunt and feel the immediate impact of the consequences of environmental insecurity are the local communities. Moreover, the local communities may lack political clout and representation at the national government level, yet they are the first to notice any negative effects of environmental insecurity. Getting the local communities involved will help to assuage their fears, reduce potential conflicts, and help funnel resources to those that need it most. Unfortunately, most African countries operate "top heavy" political systems, where most of the power is concentrated at the federal level. The result is that the local communities are often neglected or considered as afterthoughts in the political discourses and the formulation of national environmental security policies. A case in point is the Ogoni community of the oil-producing area of the Niger Delta of Nigeria whose agitation for government action to reduce the polluting effect of the crude oil industry was met with brutal putdown by the agents of the federal government of Nigeria, a situation that has led to perpetual environmental conflicts in that region of Nigeria.

SUMMARY AND CONCLUSION

The African continent has grappled with a myriad of problems that has slowed its developmental process despite the abundance of natural resources and human capital. Except for a few countries, African countries still fall under the low-income countries according to the World Bank classification. African development has been hampered by several reasons including national and international conflicts, poor governance, military incursion into the democratization process and the truncation of democratic institutions, and natural environmental challenges, to mention but a few. The combination of these forces and current environmental security issues, which have impacted national security issues in many countries in Africa, will continue to inhibit the abilities of African countries to fully realize their potentials and transform the continent into a major economic player. Thus, it is important that urgent attention be paid to addressing the issue of environmental insecurity and its various ramifications to ensure national security, a condition required for economic development. While the enactment of national environmental policies is a step in the right direction, there is need for more efforts at a regional approach to the problem of environmental security as indicated by the spatial nature for environmental hotspot regions in the continent.

NOTES

1. Finkbeiner, Elena M., Nathan J. Bennett, Timothy H. Frawley, Julia G. Mason, Dana K. Briscoe, Cassandra M. Brooks, Crystal A. Ng et al. "Reconstructing overfishing: Moving beyond Malthus for effective and equitable solutions." *Fish and Fisheries* 18, no. 6 (2017): 1180–1191.
2. Von Weizsäcker, Ernst U., and Anders Wijkman. *Come on!*. New York, NY, USA: Springer, 2018.
3. Middleton, Guy D. *Understanding collapse: Ancient history and modern myths*. Cambridge: Cambridge University Press, 2017.
4. Price, Marie, Lester Rowntree, Martin Lewis, and William Wyckoff. *Diversity amid globalization: World regions, environment, development*, 7th edition. Hoboken, NJ, Pearson College Division, 2019.
5. United Nations. World Population Prospects, 2022. Department of Economic and Social Affairs, Population Division. World Population Prospects - Population Division - United Nations.
6. Chellaney, Brahma. *Water, peace, and war: Confronting the global water crisis*. Lanham, MD, Rowman & Littlefield, 2013.
7. Ahlborg, Helene, and Andrea J. Nightingale. "Theorizing power in political ecology: The 'where' of power in resource governance projects." *Journal of Political Ecology* 25, no. 1 (2018): 381–401.
8. Gleditsch, Nils Petter. "Whither the weather? Climate change and conflict." *Journal of Peace Research* 49, no. 1 (2012): 3–9.
9. Brandt, Martin, Compton J. Tucker, Ankit Kariryaa, Kjeld Rasmussen, Christin Abel, Jennifer Small, Jerome Chave et al. "An unexpectedly large count of trees in the West African Sahara and Sahel." *Nature* 587, no. 7832 (2020): 78–82.
10. Chukwuemeka, Emma EO, A. Aloysuis, and Maximus I. Eneh. "The logic of open grazing in Nigeria: Interrogating the effect on sustainable development." *International Journal of Family Business and Management* 2, no. 1 (2018): 1–17.
11. Gaona Prieto, Rodrigo, Andrés De Castro, and José-Ignacio Antón. "Evaluation of a training program of African gendarmes' in Spain under the European Union GAR-SI Sahel project." *Policing: An International Journal* 45, no. 2 (2022): 266–281; See also, Banunle, Albert, and Richard Apau. "Environmental and natural

resources consequences of armed conflicts and violent extremism in Sahelian countries in Africa." *on Terrorism,* (2019): 73.

12. Bamidele, Seun, and Nicholas Idris Erameh. "Environmental degradation and sustainable peace dialogue in the Niger delta region of Nigeria." *Resources Policy* 80, (2023): 103274.

13. Ahmida, Ali Abdullatif. *Genocide in Libya: Shar, a hidden colonial history.* London, Routledge, 2020.

14. Alden, Chris. "From soldiers to citizens: The social, economic and political reintegration of UNITA ex-combatants." *Institute for Security Studies Monographs* 2007, no. 130 (2007): 153; See also, Porto, J. Gomes, Imogen Parsons, and Chris Alden. "From soldiers to citizens the social, economic and political reintegration of unita ex combatants." Institute for Security studies (ISS) monograph series • no 130, March 2007.

15. Mahmud, Sani., Ibraheem Alhassan, and J. Ishaku Dantata. "Efficacy study on the influence of manure and Tillage on Cowpea performance in a Sahel Savannah region of Nigeria." *International Journal of Agriculture Environment and Food Sciences* 3, no. 4 (2019): 204–210.

16. Freeman, Laura. "Environmental change, migration, and conflict in Africa: A critical examination of the interconnections." *The Journal of Environment & Development* 26, no. 4 (2017): 351–374.

17. Ajala, Olayinka. "New drivers of conflict in Nigeria: An analysis of the clashes between farmers and pastoralists." *Third World Quarterly* 41, no. 12 (2020): 2048–2066.

18. Ezirim, Gerald Ekenedirichukwu. "Oil crimes, national security, and the Nigerian state, 1999–2015." *Japanese Journal of Political Science* 19, no. 1 (2018): 80–100.

19. Price, Steven. *War and tropical forests: Conservation in areas of armed conflict.* Boca Raton, Florida: CRC Press, 2020.

20. Scull, Nicholas C., Christophe D. Mbonyingabo, and Mayriam Kotb. "Transforming ordinary people into killers: A psychosocial examination of Hutu participation in the Tutsi genocide." *Peace and Conflict: Journal of Peace Psychology* 22, no. 4 (2016): 334.

21. Zenn, Jacob. "Boko Haram and the kidnapping of the Chibok Schoolgirls." *Combating Terrorism Center* 7, no. 5 (2014): 1–28.

22. Nnam, Macpherson Uchenna, Cyril O. Ugwuoke, Vivian Chizoma Njemanze, and Francis Azalahu Akwara. "Boko Haram terrorism and human security in Nigeria: Matters Arising." *Journal of Aggression, Maltreatment & Trauma* 29, no. 10 (2020): 1257–1278.

23. Busari, Stephanie. "Hundreds of Nigerian schoolgirls released days after kidnapping." *CNN.* March 2, 2021. Hundreds of Nigerian schoolgirls released days after kidnapping | CNN.

24. Oriola, Temitope B. "Nigerian soldiers on the war against Boko Haram." *African Affairs* 120, no. 479 (2021): 147–175.

25. Barlow, Aaron. "One hand does not catch a buffalo: 50 years of amazing peace corps stories." Knutsford, England: Traveler's Tales, Volume one: Africa (2011).

REFERENCES

Ahlborg, Helene, and Andrea J. Nightingale. "Theorizing power in political ecology: The 'where' of power in resource governance projects." *Journal of Political Ecology* 25, no. 1 (2018): 381–401.

Ahmida, Ali Abdullatif. *Genocide in Libya: Shar, a hidden colonial history*. Routledge, 2020.

Ajala, Olayinka. "New drivers of conflict in Nigeria: An analysis of the clashes between farmers and pastoralists." *Third World Quarterly* 41, no. 12 (2020): 2048–2066.

Alden, Chris. "From soldiers to citizens: The social, economic and political reintegration of UNITA ex-combatants." *Institute for Security Studies Monographs* 2007, no. 130 (2007): 153.

Bamidele, Seun, and Nicholas Idris Erameh. "Environmental degradation and sustainable peace dialogue in the Niger delta region of Nigeria." *Resources Policy* 80, (2023): 103274.

Banunle, Albert, and Richard Apau. "Environmental and natural resources consequences of armed conflicts and violent extremism in sahelian countries in Africa." *on Terrorism,* (2019): 73.

Barlow, Aaron. "One hand does not catch a buffalo: 50 years of amazing peace corps stories." Knutsford, England: Traveler's Tales, Volume one: Africa (2011).

Brandt, Martin, Compton J. Tucker, Ankit Kariryaa, Kjeld Rasmussen, Christin Abel, Jennifer Small, Jerome Chave et al. "An unexpectedly large count of trees in the West African Sahara and Sahel." *Nature* 587, no. 7832 (2020): 78–82.

Busari, Stephanie. Hundreds of Nigerian Schoolgirls Released days after kidnapping. CNN. March 2, 2021. Hundreds of Nigerian schoolgirls released days after kidnapping | CNN.

Chellaney, Brahma. *Water, peace, and war: Confronting the global water crisis*. Rowman & Littlefield, 2013.

Chukwuemeka, Emma EO, A. Aloysuis, and Maximus I. Eneh. "The logic of open grazing in Nigeria: Interrogating the effect on sustainable development." *International Journal of Family Business and Management* 2, no. 1 (2018): 1–17.

Ezirim, Gerald Ekenedirichukwu. "Oil crimes, national security, and the Nigerian state, 1999–2015." *Japanese Journal of Political Science* 19, no. 1 (2018): 80–100.

Finkbeiner, Elena M., Nathan J. Bennett, Timothy H. Frawley, Julia G. Mason, Dana K. Briscoe, Cassandra M. Brooks, Crystal A. Ng et al. "Reconstructing overfishing: Moving beyond Malthus for effective and equitable solutions." *Fish and Fisheries* 18, no. 6 (2017): 1180–1191.

Freeman, Laura. "Environmental change, migration, and conflict in Africa: A critical examination of the interconnections." *The Journal of Environment & Development* 26, no. 4 (2017): 351–374.

Gaona Prieto, Rodrigo, Andrés De Castro, and José-Ignacio Antón. "Evaluation of a training program of African gendarmes' in Spain under the European Union GARSI Sahel project." *Policing: An International Journal* 45, no. 2 (2022): 266–281.

Gleditsch, Nils Petter. "Whither the weather? Climate change and conflict." *Journal of Peace Research* 49, no. 1 (2012): 3–9.

Mahmud, Sani., Ibraheem Alhassan, and J. Ishaku Dantata. "Efficacy study on the influence of manure and Tillage on Cowpea performance in a Sahel Savannah region of Nigeria." *International Journal of Agriculture Environment and Food Sciences* 3, no. 4 (2019): 204–210.

Middleton, Guy D. *Understanding collapse: Ancient history and modern myths.* Cambridge University Press, 2017.

Nnam, Macpherson Uchenna, Cyril O. Ugwuoke, Vivian Chizoma Njemanze, and Francis Azalahu Akwara. "Boko Haram terrorism and human security in Nigeria: Matters Arising." *Journal of Aggression, Maltreatment & Trauma* 29, no. 10 (2020): 1257–1278.

Oriola, Temitope B. "Nigerian soldiers on the war against Boko Haram." *African Affairs* 120, no. 479 (2021): 147–175.

Porto, J. Gomes, Imogen Parsons, and Chris Alden. "From soldiers to citizens the social, economic and political reintegration of unita ex combatants." Institute for Security studies (ISS) monograph series • no 130, March (2007).

Price, Marie, Lester Rowntree, Martin Lewis, and William Wyckoff. *Diversity amid globalization: World regions, environment, development*, 7th edition. Pearson College Division, 2019.

Price, Steven. *War and tropical forests: Conservation in areas of armed conflict.* CRC Press, 2020.

Scull, Nicholas C., Christophe D. Mbonyingabo, and Mayriam Kotb. "Transforming ordinary people into killers: A psychosocial examination of Hutu participation in the Tutsi genocide." *Peace and Conflict: Journal of Peace Psychology* 22, no. 4 (2016): 334.

United Nations. World Population Prospects, 2022. Department of Economic and Social Affairs, Population Division. World Population Prospects—Population Division—United Nations.

Von Weizsäcker, Ernst U., and Anders Wijkman. *Come on!*. Springer, 2018.

Zenn, Jacob. "Boko Haram and the kidnapping of the Chibok Schoolgirls." *Combating Terrorism Center* 7, no. 5 (2014): 1–28.

Chapter 2

Globalization, Climate Change, and Environmental Security in Africa

Augustine Avwunudiobga and
Ufuoma Efe Akpovwovwo

Environmental security is a major concern in the African continent because of its link to social, economic, and political stability. In recent decades, a combination of human- and nature-induced processes has exerted pressure on Africa's ecosystems, resulting in degradation of ecosystem services and environmental instability. Environmental instability often leads to environmental insecurity and its attendant negative consequences on community cohesion and function. Environmental insecurity has resulted in economic and social dislocation of the population in many countries of Africa, leading to mass migration and the persistent problems of refugees and internally displaced persons (IDPs). Available data suggest that this is a trend that is likely to worsen in the coming decades in Africa.

There is an urgent need to interrogate the drivers of present and future conditions of environmental security in Africa as a basis for formulating adaptive and mitigative policy measures at the national and transnational levels. This task is imperative because economic development and the improvement of human condition in the continent of Africa is predicated on a secure environment. This chapter explores the role of two major forces—globalization and climate change—as drivers of ecosystem and environmental changes linked to environmental security in Africa. In the first part of the chapter, we examine the role of globalization in promoting the overexploitation of Africa's natural resources, natural resources access conflict, and degradation of the ecosystem and how this increases the vulnerability of powerless communities to the negative externalities associated with primary resource extraction. The second part of the chapter examines the role of climate-induced environmental change and the occurrence of high-magnitude events that have become the norm. Climate-driven events such as flooding, prolonged drought, desertification, and wildfire are now a frequent occurrence in many African countries. This will in turn worsen the environmental security status of many African countries in the future. African countries must address the environmental security implications of globalization and climate change to create an enabling environment for progress on the continent.

THEORETICAL FRAMEWORK

The foundation of human civilization is intimately intertwined with the biophysical environment. The biophysical environment performs two critical roles in this regard. First, it creates the condition and habitat necessary for Homo sapiens to exist. Second, it provides the resources necessary for human survival and development.[1] It is not surprising, therefore, that most ancient civilizations were located where the environmental conditions to fulfill these two critical functions were at optimum.[2] The environment, as used here, broadly refers to the product of the interaction of the four major spheres of the earth, that is, the atmosphere, the lithosphere, the biosphere, and the hydrosphere. The interaction of these spheres creates the conditions and the resources that are part of the complex open earth systems that are fundamental to the emergence of humans as a dominant force on our planet. By their nature, earth systems are open systems that are delicately balanced but are influenced by processes external to their boundaries.[3] Under natural conditions, earth systems are self-regulating, creating predictable long-term stability and environmental security that promote the progressive development of human civilization. In this regard, earth systems are incredibly resilient due in part to their complexity and tendency to self-regulate. Thus, the interaction between the four spheres of the Earth has created recognizable broad biomes whose spatial patterns are clearly defined on continental scales. Nested within these broad biomes are landscapes and ecosystems, which also exhibit characteristics of the broad earth system albeit at a lower spatial scale. The rise of humans, a dominant force within our earth system which began since the Holocene,[4] was amplified by the agricultural and industrial revolutions, which ushered in a period marked by declining death rate and increasing birth rate with the consequent population explosion.[5]

Today, the human species can be found in all nooks and corners of our planet, as the world population continues to rise to the projected 8.5 billion by 2030.[6] The result is that there are only a few areas of our earth that do not have a physical human presence. In short, today's natural environments are nothing short of coupled human–environment systems. There are hardly any pristine natural environments. More importantly, the pressure exerted by human activities on earth systems through the utilization of ecosystem resources and negative externalities associated with production and consumption has had a profound influence on the delicate balance within earth systems from the local to the global scales.[7] The consequence is that in many regions of the world, ecosystems and landscapes have undergone simplification and reduced ability to be self-regulating. It is now generally accepted within the scientific community that most of our earth system is now in the so-called tipping points, which might result in the creation of completely new environmental conditions, the full ramifications of which cannot be adequately predicted even with the most complex computer models.[8]

Nevertheless, archaeological evidence suggests that many ancient civilizations met their demise as a result of environmental degradation and instability often caused by misuse and overuse of the natural resource base.[9] Thus, there is a crucial link between environmental stability and environmental security. Environmental instability creates environmental insecurity, which then increases both perceived and real vulnerability to the impacted population. It is in this context that the twin processes of globalization and global climate change, both driven by human civilization, have had profound and continued change on our earth systems at both local and global scales.[10] Nowhere is this exemplified more so than the continent of Africa. As the population continues an upward trajectory, the demand on the natural resource base will also increase by many folds. At the same time, the linking of the African economy to the global world economy via the process of globalization has meant that more pressure has been brought to bear on Africa's natural resources through global demand. The overall effects of these processes on Africa's environments have become increasingly fragile and delicate, increasing the probability of environmental insecurity. This is because environmental insecurity is likely to expose already existing fractured points such as ethnic, religious, and tribal divisions, which abound on the African continent. In the next section, we explore the role of globalization and climate change on environmental security on the African continent.

DRIVERS OF ENVIRONMENTAL SECURITY IN AFRICA

As discussed in the preceding section, earth systems are incredibly resilient and dynamically balanced, creating environmental stability. Changes brought to bear on this balance either by natural processes, natural processes amplified by humans, or human processes can result in disturbance in the components that make up the system. If such changes occur in high frequency and high magnitude and persist over time, it could lead to a change in the balance of the environmental system and create instability. Whether an environment remains stable or unstable depends on the frequency and magnitude of the operating drivers and the ability of the environmental system to adjust. However, where the frequency and magnitude persist, environmental systems have little or no relaxation time and thus are more likely to result in instability and environmental insecurity. In this regard, two major drivers that have persistently impacted the Earth's environments on a global scale, and African environments in particular, are the forces of globalization and climate change.[11]

Globalization and Environmental Security

Globalization, the force that has opened and integrated the world economy, has had a profound effect on global environments.[12] The forces of globalization removed or reduced economic barriers to the movement of capital, mostly from industrial countries of Europe and North America to the global south to take advantage of the market and investment opportunities in the natural resources industry. The demand for agricultural and natural resource products from the global south, where the demand by the middle class in the minority north has incentivized local communities and governments in the global south to embark on exploitative programs aimed at maximizing returns from their natural resource base with little or no consideration for sustainable use.[13] For example, large-scale mining operations controlled by multinational and transnational corporations and investment in large-scale commercial plantation agriculture that is geared toward export crops have all had impacts on environmental security in the developing countries, especially Africa.

In Nigeria, for example, it is well documented that the multinational corporations operating in the crude oil industry in the Niger Delta have been exploiting the oil with less attention to the negative externalities, that is, environmental stress and degradation caused by their exploitative activities.[14] The result is that the fragile ecosystem of the Niger Delta can no longer support the local population's traditional way of livelihood, including fishing and farming. Thus, this environmental degradation has created environmental insecurity, which has manifested in the current state of constant agitation, civil unrest, and frequent community clashes with the agents of the multinational corporations. In some instances, inappropriate and insensitive government response has led to the abandonment of communities and the migration of the populace to the surrounding urban areas. Similar situations abound all over Africa, sometimes resulting in civil wars and international conflicts; for example, in Eastern and Southern Africa mining regions as well as in the Horn of Africa, such as Sudan, Southern Sudan, and Somalia, where the struggle for the control of natural resources occasioned by the demand from the Global North has exacerbated environmental insecurity.[15]

Another element of globalization which involves access to markets in Africa also means that African small-scale farmers are not able to compete with the cheap agricultural produce from industrialized north flooding the markets. The result is that agricultural production of certain crops in certain communities is negatively impacted, while in other areas, large prime, productive agricultural lands have been bought off by multinational corporations[16] to produce export crops to the detriment of crops meant for the local economy. Local farmers in these regions have been pushed away further into marginal lands, creating an insecure environmental situation for the populace leading the communities to forcefully migrate to urban areas. Overall, globalization has increased the pressure on the extraction of natural resources in the African environment with little or no strict environmental regulations, or lukewarm enforcement of existing environmental protection laws.[17] African environments face the risk of continuous degradation, which will eventually lead to environmental insecurity. In summary, globalization has contributed to ecosystem changes in Africa such as deforestation, land degradation, and water and air pollution, all of which reduce the ability of ecosystems to provide services to the local populations that depend upon them. Hence, environments that were once seen as stable and sustainable are increasingly becoming fragile, intensifying both perceived and real sense of insecurity.

Climate Change and Environmental Security

Climate change is now recognized as a major driver of current and future changes in global ecosystems.[18] The consequences of anthropogenically induced climate change are felt in local ecosystems, regional landscapes, and global biomes. There is a general consensus that climate change will significantly alter the geographic pattern of rainfall, temperature, and humidity, and hence current climatic zones of the world. Ecosystems, landscapes, and biomes are a product of the interaction between abiotic and biotic factors, which give rise to a stable environment based on the interactions. A change in the abiotic environment induced by human-accelerated climate change is already causing disruptions and imbalances in local ecosystems, regional landscapes, and global biomes.[19] One of the consequences of this disruption is the risk to ecosystem services provided by the environment upon which human populations depend.

Indeed, ecosystems and the services they provide are the foundation of human civilization. Threats posed by climate change on ecosystems, therefore, pose both direct and indirect consequences on the livability of the world's global environment and, thus, the long-term sustainability of human civilization. The consequences of climate change are being felt and will be felt more in sensitive and marginal environments. In this regard, the African continent faces severe risks from human-induced climate change because of the number of countries whose land area lies astride marginal environments. A cursory examination of the map of Africa indicates that several countries have the majority of their landmass in what can be classified as arid and semi-arid lands under the Koppen classification system. These countries with communities that are already living under marginal conditions will suffer more from global climate change-induced events such as severe and prolonged drought, which will impact the capacity to produce food and fiber, creating environmental insecurity.

Although there remains uncertainty in terms of the total possible effects of climate change, there is certainty in terms of some of the current impacts which are observable at both global and local scales. In general, global climate change is now leading to more frequent high-magnitude extreme meteorological and hydrological events.[20] These events have further positive feedback effects on global and local ecosystems. In addition, climate change is already causing a spatiotemporal shift in weather patterns, especially rainfall distribution and global and local climatic zones. The overall implications of this are that climate change is a threat to ecosystem stability, which forms the foundation of local communities. This is because climate change in the short and long run will result in environmental change, and environmental change will result in environmental insecurity as local communities struggle to adapt to its numerous negative ramifications. The severity of any environmental hazard on a population depends on the ability of that population to either mitigate or adapt to the hazard.[21] This in turn is related to the range of available resources and the ability of the impacted communities to access those resources.

Here lies the problems of communities in most African countries where resources are scarce and therefore the ability to adapt to environmental hazards associated with climate-driven high magnitude extreme events is low. Most African countries lack the institutional support for normal climate events that occur within their national borders. This is because most African countries are still grappling with the need to provide basic essential social services to their populations and therefore lack the financial wherewithal to provide communities the required resources to either adapt or mitigate the effects of climate change. In addition, current infrastructural facilities in most African countries in both urban and rural areas are woefully inadequate to accommodate extreme weather events, leading to more environmental uncertainty and perceived and real insecurity.[22] In African countries, climate change would impact environmental security in several ways. This includes threats to food production and food security and increased frequency of high-magnitude events such as flooding in some areas and prolonged drought phenomena in others.

Other impacts include changes to natural ecosystems; for example, prolonged drought could encourage the acceleration of the ongoing desertification process along the Sahelian zone of Africa and many regions with arid and semi-arid climates, leading to vegetation loss and land degradation. The loss of vegetation and land degradation would amplify the current impact on pastoral (nomadic) livelihoods and sedentary farmers, which form the bulk of the agricultural productive population. In other words, the environmental security of many local communities, whose livelihood depends on the agrarian economy, cannot be guaranteed under current and future scenarios of projected global impacts of climate change.[23] It can be argued that global climate change is possibly the greatest challenge to environmental security in Africa.

The Intergovernmental Panel on Climate change[24] report has confirmed previous conclusions that anthropogenic climate change is here to stay. Among the various facets of this change is global warming, on which the report concludes irrevocable evidence of a gradual but consistent increase in average global temperature of 1.5 degrees Celsius (34.7 degrees Fahrenheit) since the industrial revolution (1800s). This trend is likely to accelerate in the future if the current rate of emission of greenhouse gases (GHGs), the major culprit of atmospheric warming, continues unabated. Concomitant increases in average atmospheric temperature will have direct and indirect impacts on sea level changes and global sea level rise through its effects on the melting of arctic glaciers.[25] An additional possible effect is the overall increase in surface temperature with its transformative effects on global climate patterns and possible shift in existing global climatic classifications. The IPCC report notes a persistent trend in the frequency of extreme climatic and climate-related events such as the intensification of hurricanes and cyclones, flooding, prolonged and extensive drought, and widespread wildfire.[26] These impacts are local, regional, and global in scale.

THE CASE FOR ENVIRONMENTAL SECURITY IN AFRICA

The foregone discussion indicates that many communities and regions in Africa that currently face environmental challenges will become increasingly vulnerable as the twin processes of globalization and climate change continue to impact local ecosystems and regional biomes. Uncertainty created by environmental instability will reduce the ability of these natural systems to provide the sustenance that the impacted populations have come to depend on with the long-lasting impact on the security of human populations becoming a major issue in many African countries. Thus, the case for urgent attention and policies geared toward environmental security on the continent cannot be overemphasized. This is because an increasingly insecure environment has several unpalatable ramifications on the African continent and beyond. The consequences of environmental insecurity include socioeconomic, sociocultural, political, national, and international dimensions.

There is no doubt that environmental security is intimately related to socioeconomic prosperity and stability of a country. This is particularly profound in Africa, where many of the countries are still in the category of least developed countries (LDCs) and medium developed countries (MDCs) with a substantial percentage of the population still engaged in the primary sector of the economy. An insecure environment resulting in the uncertainty of the economic production base will not only dampen productivity and economic output but will result in significant social dislocation. Already most African countries are grappling with the issue of rural–urban migration under normal environmental conditions. Additional insecurity will no doubt amplify this process through the dislocation of the rural population that depend on the natural resource base for their survival. Thus, within national boundaries, there will be heightened rural–urban migration putting pressure on African cities that often lack the basic social services and amenities to sustain their urban population let alone the influx of rural population into the urban periphery exacerbating the phenomenon of urban slum, which has blighted many African urban metropolises. Social dislocation could also amplify regional and international migration. For example, within the African continent, there is a growing trend of population movement from countries in the Sahelian zone to countries in the wetter regions of Africa. This movement is a product of the impact of prolonged drought, which has devastated the pastoral economy on which many ethnic groups in these countries rely upon.

A variety of multiple social problems that are associated with this forced migration include high levels of urban poverty as African urban centers do not have the industrial capacity to absorb the migrant population from the rural areas.[27] Moreover, city and national governments do not have the resources to provide needed programs to ameliorate the impact of these population increases on African cities. The result is a high level of youth employment and underemployment of a highly young potentially productive population. In the absence of economic opportunities, and lack of clear government policies to address the situation, young unemployed African urbanites have turned to international migration to Europe and North America in search of economic opportunities under extreme and often hazardous conditions.[28] The consequences of environmental insecurity in Africa are not only confined to the continent, but rather are a global issue as more people move away from countries that do not have the capacity to adapt to the consequences of environmental insecurity caused by globalization and climate change to more affluent countries that have the capacity to mitigate and moderate the effects of global climate change.

Environmental insecurity in Africa also has a political dimension. In the first place, the negative impact of transboundary resources such as rivers and waterways resulting from climate change may result in political conflicts over who controls and how such resources are shared. For example, waterworks such as dam construction to divert water from rivers that flow through several countries may serve as flashpoints for national contests in the face of dwindling water resources occasioned by climate change. Similarly, shared resources such as coastal waters with fishing resources may become grounds for local and international contests as fishing stock dwindles. In other words, environmental insecurity in Africa may breed international and transnational conflicts within the continent.[29] Within national boundaries, an increasingly insecure environment is also related to national security because an increasingly restless population will pose a challenge to the governance of many African countries.[30] Social unrest can easily translate into political unrest and amplify the already existing cleavages in many African countries with weak democratic institutions, especially in the so-called "failing" or "failed" states.[31] Thus, conflicts based on ethnic, religious, and tribal affiliations may become the norm as every group struggles for limited resources to adapt to the debilitating consequences of global climate change. This will no doubt put stress on the already fragile institutions of the emerging democratic countries in Africa with the potential of more countries degenerating into failed states in the worst-case scenario.

The breakdown of communities through population dislocation could also have a cultural impact. In many African countries, tradition and cultural institutions play major roles in maintaining the cultural fabric and stability of the society. This is often maintained through sets of belief systems and norms that are rooted in the environmental milieu in which such societies have developed over millennia.[32] The changing environmental milieu caused by climate change and the consequent dislocation of the population may result in the breakdown of these century-long customs and traditions, which have served as a glue that holds these societies together. Overall, globalization and global climate would no doubt contribute to cultural loss and the disintegration of cultural identity and harmony in many traditional African societies.

The impact of environmental insecurity in Africa is not only confined to the continent, but also has elements of international and transnational dimensions beyond the continent. Environmental insecurity has resulted in the increase in the number of IDPs,[33] refugee population, and continued streams of economic migrants to Europe and North America.[34] The influx of migrant populations from Africa and other global climate change-impacted regions of the world to Europe and North America has created social and political tensions in the destination countries in Europe and North America. Recently, anti-immigration rhetoric has dominated public political debates in many western countries of Europe and North America and has been partially attributed to the rise of anti-immigration parties and the revival of nationalistic movements.[35] Nevertheless, it should be mentioned that the influx of migrants from Africa to Europe and North America has also resulted in the phenomenon of brain drain, where educated young university graduates who have received education in most cases under generous government investments move to Europe and North America in search of better opportunities.[36] The impact is double whammy because, on the one hand, African countries lose highly trained and qualified manpower that is required to drive the economy of African countries through innovation. On the other hand, African countries also lose the returns on the enormous financial outlay they invest in higher education.

Toward an Environmental Security Policy in Africa

In the light of undeniable current and projected environmental consequences of global climate change as amplified by globalization, African countries urgently need to put in place an environmental security policy. Such a policy should aim at addressing the aforementioned issues by putting in place programs and policies for local populations to be able to prevent, mitigate, and/or adapt to the undesirable consequences of global climate change. An effective environmental security framework will be the one that is dynamic, technologically feasible, economically viable, and socially acceptable and responsive to the unique need of the African condition. Such a framework must take into account national interest as well as international commitment toward the fight and solution for the reduction of the impact of global climate change.

Elements of Environmental Security Framework

The first element in this framework should address the causative factors associated with the human drivers of global climate change at the local level within the economic capacity of each country. In this regard, African countries should make efforts as much as possible to reduce their contribution of GHG emissions, especially carbon dioxide and methane, two greenhouse gases that have been implicated in the rapid warming of the lower atmosphere and setting up positive feedback loops that perpetuate global climate change.[37] Thus, processes such as tropical deforestation and the use of fossil fuel as a major source of energy should be moderated as much as possible. However, it must be noted that African countries should not be expected to bear an undue burden in this regard because available data clearly show that advanced industrial countries of the west and North America, Japan, South

Korea, China, and emerging Asian economies such as Malaysia, Indonesia, and India, as well as Australia and New Zealand, are collectively responsible directly for about 70 percent of the global GHG emission.[38] Furthermore, these same countries are indirectly or partially responsible for GHG emissions in Africa through the process of globalization, whereby their multinational and transnational corporations engage in natural resource exploitation to satisfy the consumption pattern of the population. The activities of these countries all together put a strain on the natural environment of African countries, resulting in increased emission of GHGs. This is because the African economy must embark on a development agenda in order to uplift the population from poverty. Therefore, a fair, equitable, and just global GHG emissions reduction regime must consider the development needs of African countries. Thus, the success of any preventive element must of necessity depend on global initiatives and efforts at reducing GHG emissions since the impacts are not confined to African national borders.

The second element in the framework encompasses mitigation programs and processes. This involves programs and policies designed to mitigate environmental insecurity resulting from globalization and the impacts of global climate change. Mitigating measures include the adoption of new technological innovations that cut down GHG emissions in the emerging agricultural and industrial sectors of Africa. Nevertheless, such technologies must be economically feasible. To this extent, the international community via the organs of the United Nations Organization (UNO) must play a critical role in ensuring the technological transfer of climate-smart technologies. Climate-smart technologies should be made accessible at little or no cost to African countries to incorporate into their programs and policies.

The third element of the framework for environmental security involves adaptation. It is now becoming very clear that even under the most optimistic scenario, African countries will have to contend with the impact of global climate change in the short, medium, and long term. As already discussed in the previous sections, environmental insecurity resulting from the effects of globalization and climate change is already being felt in many African countries. Some of the effects, such as prolonged droughts and desertification, have become a permanent feature of the African landscape.[39] Therefore, a forward-thinking approach that focuses on adaptation measures is needed as part of a framework of environmental security in Africa. The central question that needs to be addressed in this regard is, how do local communities and populations adapt to the changing environmental conditions that induce environmental insecurity? What resources are needed? And what roles do local communities, national, and international organizations play in providing resources for adaptation to vulnerable communities? The considerations for adaptation policy as part of the framework are imperative because most rural African communities, in general, often lack the social and infrastructural capital to absorb the impacts of extreme weather events, let alone global climate change-driven events. A subcomponent of this focal point of adaptation will be to build community resilience by providing multiple economic opportunities in cases where environmental insecurity may result in the loss of livelihoods. Other measures may include the provision of basic critical infrastructure that will enable impacted communities to adapt; for example, large-scale water projects that provide water for irrigated agriculture and domestic use, rapid investment in crop varieties that can thrive within the new environments resulting from global climate change. This may also include the substitution of crops that were formerly grown under normal environmental conditions with new crops that will thrive in the new climate change-imposed environmental conditions.

SUMMARY AND CONCLUSION

Our 4.6-billion-year-old planet has witnessed several environmental changes over the course of its history, most of which can be ascribed to natural processes. There is now a consensus in the scientific community that we now live in the age of the Anthropocene, an age dominated by global human impact on earth systems that has resulted in profound and continued change.[40] The global footprint of the human impact on the earth's environment is now very visible and the role of anthropogenic processes as a driver of global climate change is indisputable. Globalization coupled with global climate change have combined to exert pressure on the Earth's ecosystems, landscapes, and biomes. In the African continent, this has brought profound changes in the environment of many communities, resulting in environmental insecurity and its attendant consequences.

As we have argued in this chapter, environmental security in Africa is ultimately linked to national security. Conversely, the political stability of African countries is linked to the ability of African environments and ecosystems to continue to provide a productive resource base for local populations. Changes to the environment will accentuate the problem of insecurity and set up a domino effect with socioeconomic, political, and cultural ramifications within national and international boundaries. Thus, African countries must address current and emergent issues of environmental insecurity on the continent by putting a solution framework in place. Such a framework should be anchored around preventive, mitigative, and adaptive policies and programs. It should also include elements of the continent's contribution toward the reduction of global GHGs within the constraints of available resources and socially acceptable burden. The developed industrial countries of Europe,

North America, and Asia must contribute their part by reducing the phenomenon of translocating outdated technologies that do not contribute climate change solutions to Africa.

NOTES

1. Rees, William E. "Ecological economics for humanity's plague phase." *Ecological Economics* 169, (2020): 106519.

2. Kalis, Arie J., Josef Merkt, and Jürgen Wunderlich. "Environmental changes during the Holocene climatic optimum in central Europe-human impact and natural causes." *Quaternary Science Reviews* 22, no. 1 (2003): 33–79.

3. Masselink, Gerd, Michael Hughes, and Jasper Knight. *Introduction to coastal processes and geomorphology*. Routledge, 2014.

4. Syvitski, Jaia, Colin N. Waters, John Day, John D. Milliman, Colin Summerhayes, Will Steffen, Jan Zalasiewicz et al. "Extraordinary human energy consumption and resultant geological impacts beginning around 1950 CE initiated the proposed Anthropocene Epoch." *Communications Earth & Environment* 1, no. 1 (2020): 1–13.

5. Gordon, Cameron. "Global demographic change." In *Many Possible Worlds: An Interdisciplinary History of the World Economy Since 1800*, Gordon Cameron (ed), 443–466. Singapore: Springer Nature Singapore, 2023.

6. United Nations Organization. "Securing a sustainable future". United Nations Sustainable Development Goals (SDG) Summit. United Nations HQ. New York, 18–19 September 2023. UN 2023 SDG Summit - United Nations Sustainable Development.

7. Watson, Robert, Ivar Baste, Anne Larigauderie, Paul Leadley, Unai Pascual, Brigitte Baptiste, Sebsebe Demissew et al. "Summary for policymakers of the global assessment report on biodiversity and ecosystem services of the Intergovernmental Science-Policy Platform on Biodiversity and Ecosystem Services." *IPBES Secretariat: Bonn, Germany* (2019): 22–47.

8. Dobner, Petra, and Jasper Finkeldey. "Natural resources and the tipping points of political power—A research agenda." *Sustainability* 14, no. 22 (2022): 14721.

9. Middleton, Guy D. *Understanding collapse: Ancient history and modern myths*. Cambridge University Press, 2017.

10. Angus, Ian. *Facing the anthropocene: Fossil capitalism and the crisis of the earth system*. NYU Press, 2016.

11. Connolly-Boutin, L., and Smit, B. "Climate change, food security, and livelihoods in sub-Saharan Africa." *Regional Environmental Change* 16, no. 2 (2015): 385–399. https://doi.org/10.1007/s10113-015-0761-x. See also; Hendrix, C. S., and Salehyan, I. "Climate change, rainfall, and social conflict in Africa." *Journal of Peace Research* 49, no. 1 (2012): 35–50. https://doi.org/10.1177/0022343311426165.

12. Brauch, Hans Günter. "Concepts of security threats, challenges, vulnerabilities and risks." In *Coping with global environmental change, disasters and security:threats, challenges, vulnerabilities and risks*, Brauch, Hans Günter, Úrsula

Oswald Spring, Czeslaw Mesjasz, John Grin, Patricia Kameri-Mbote, Béchir Chourou, Pál Dunay, and Jörn Birkmann, (eds), 61–106. Springer, New York, 2011.

13. McKay, Ben M., Ruth Hall, and Juan Liu, eds. *Rural transformations and agro-food systems: The BRICS and agrarian change in the Global South.* Routledge, 2018.

14. Obuah, Emmanuel E., and Reginald Chikere Keke. "Mitigating the negative externalities of oil drilling activities in the Niger Delta, Nigeria." *Open Access Library Journal* 9, no. 5 (2022): 1–21.

15. Kabandula, Abigail, and Timothy M. Shaw. "Rising powers and the horn of Africa: Conflicting regionalisms." *Third World Quarterly* 39, no. 12 (2018): 2315–2333.

16. Carmody, P., and Taylor, D. "Globalization, land grabbing, and the present-day Colonial State in Uganda." *The Journal of Environment & Development* 25, no. 1 (2016): 100–126. https://doi.org/10.1177/1070496515622017.

17. Peša, Iva. "Mining, waste and environmental thought on the Central African Copperbelt, 1950–2000." *Environment and History* 28, no. 2 (2022): 259–284.

18. De Frenne, Pieter, Jonathan Lenoir, Miska Luoto, Brett R. Scheffers, Florian Zellweger, Juha Aalto, Michael B. Ashcroft et al. "Forest microclimates and climate change: Importance, drivers, and future research agenda." *Global Change Biology* 27, no. 11 (2021): 2279–2297.

19. Sage, Rowan F. "Global change biology: A primer." *Global Change Biology* 26, no. 1 (2020): 3–30.

20. Abram, Nerilie J., Benjamin J. Henley, Alex Sen Gupta, Tanya JR Lippmann, Hamish Clarke, Andrew J. Dowdy, Jason J. Sharples et al. "Connections of climate change and variability to large and extreme forest fires in southeast Australia." *Communications Earth & Environment* 2, no. 1 (2021): 8.

21. Agrawal, Arun. "Local institutions and adaptation to climate change." *Social dimensions of climate change: Equity and vulnerability in a warming world* 2, (2010): 173–178.

22. Verhoeven, Harry. "Climate & water in a changing Africa: Uncertainty, adaptation & the social construction of fragile environments." *Dædalus* 150, no. 4 (2021): 260–277.

23. Kurukulasuriya, Pradeep, and Shane Rosenthal. "Climate Change and Agriculture: A Review of Impacts and Adaptations", Climate Change Series 91. Environment Department Papers, The World Bank, Washington, DC 2003.

24. IPCC. "Summary for policymakers." In Climate Change 2023: Synthesis Report. A Report of the Intergovernmental Panel on Climate Change. Contribution of Working Groups I, II and III to the Sixth Assessment Report of the Intergovernmental Panel on Climate Change [Core Writing Team, H. Lee and J. Romero (eds.)]. IPCC, Geneva, Switzerland, 36 pages. (in press).

25. Ibid.

26. Ibid.

27. Arndt, Channing, Rob Davies, and James Thurlow. "Urbanization, structural transformation, and rural-urban linkages in South Africa." *South African Urbanisation Review, Cities Support Programme (CSP) of the National Treasury* (2018).

28. Grant, Ursula. "Urbanization and the employment opportunities of youth in developing countries." *Background paper prepared for EFA Global Monitoring Report* (2012).

29. Ertor, I., Zoe W. Brent, D. Galler, and Thibault Josse. "Situating small-scale fisheries in the global struggle for agroecology and food sovereignty." *Amsterdam: Transnational Institute* (2020).

30. Hendrixson, Anne, and Betsy Hartmann. "Threats and burdens: Challenging scarcity-driven narratives of "overpopulation"." *Geoforum* 101, (2019): 250–259.

31. Clausen, Maria-Louise, and Peter Albrecht. "Interventions since the Cold War: From statebuilding to stabilization." *International Affairs* 97, no. 4 (2021): 1203–1220.

32. Dung, Elisha J. "The role of traditional institutions in the mediation of social issues in Africa, Latin America, and the Caribbean." In *Comparative assessment of social issues in Africa, Latin America, and the Caribbean,* Gill, Brenda, I and George Danns (eds.), Lexington Books: Lanham: Maryland, USA , 2023.

33. Avwunudiogba, Augustine, and Elisha J. Dung. "The spatial distribution of internally displaced persons (IDPs) in Africa." In *The Challenges of Refugees and Internally Displaced Persons in Africa,* Sabella O. Abidde (ed.), Springer Nature: Switzerland, 2021.

34. Dung, Elisha J., and Augustine Avwunudiogba. "The human cost of the refugee crisis in Africa." In *African Migrants and the Refugee Crisis,* Abegunrin, O., and Sabella O. Abidde (eds.), 295–313. Springer Nature: Switzerland, 2021.

35. Mamonova, Natalia, and Jaume Franquesa. "Populism, neoliberalism and agrarian movements in Europe. Understanding rural support for right-wing politics and looking for progressive solutions." *Sociologia Ruralis* 60, no. 4 (2020): 710–731.

36. Dung, Elisha, J. "The geography and the patterns of migration." In *Africans and the Exiled Life*, Abide, Sabella O., and Brenda I. Gill (eds.), 23–50. Lexington Books, Lanham: Maryland, USA, 2018.

37. Shakoor, Awais, Fatima Ashraf, Saba Shakoor, Adnan Mustafa, Abdul Rehman, and Muhammad Mohsin Altaf. "Biogeochemical transformation of greenhouse gas emissions from terrestrial to atmospheric environment and potential feedback to climate forcing." *Environmental Science and Pollution Research* 27, (2020): 38513–38536.

38. Climate Watch. Global Greenhouse Gas Emissions. World Resources Institute, 2022. World | Total including LUCF | Greenhouse Gas (GHG) Emissions | Climate Watch (climatewatchdata.org)

39. Glantz, Michael H., and Nicolai S. Orlovsky. "The phenomenon known as desertification received widespread attention recently, as witnessed by the creation of the United Nations Conference on Desertification (UNGOD) in Nairobi in 1977, mainly as a result of the impact of extended drought in the West African Sahel in the early 1970s. That drought led to the loss of human lives." *Natural Resources And People: Conceptual Issues In Interdisciplinary Research* Glantz, Michael H., and Nicolai S. Orlovsky. "Desertification: A review of the concept." *Desertification Control Bulletin* 9 (1983): 15–22 (2019).

40. Hamilton, Clive. "The Anthropocene as rupture." *The Anthropocene Review* 3, no. 2 (2016): 93–106.

REFERENCES

Abram, Nerilie J., Benjamin J. Henley, Alex Sen Gupta, Tanya JR Lippmann, Hamish Clarke, Andrew J. Dowdy, Jason J. Sharples et al. "Connections of climate change and variability to large and extreme forest fires in southeast Australia." *Communications Earth & Environment* 2, no. 1 (2021): 8.

Agrawal, Arun. "Local institutions and adaptation to climate change." *Social Dimensions of Climate Change: Equity and Vulnerability in a Warming World* 2, (2010): 173–178.

Angus, Ian. *Facing the Anthropocene: Fossil capitalism and the crisis of the earth system*. NYU Press, 2016.

Arndt, Channing, Rob Davies, and James Thurlow. "Urbanization, structural transformation, and rural-urban linkages in South Africa." *South African Urbanisation Review, Cities Support Programme (CSP) of the National Treasury,* (2018).

Avwunudiogba, Augustine, and Elisha J. Dung. "The spatial distribution of internally displaced persons (IDPs) in Africa." In Sabella O. Abidde (ed.), *The challenges of refugees and internally displaced persons in Africa.* Springer Nature, 2021.

Brauch, Hans Günter. "Concepts of security threats, challenges, vulnerabilities and risks." In *Coping with global environmental change, disasters and security: threats, challenges, vulnerabilities and risks*, Brauch, Hans Günter, Úrsula Oswald Spring, Czeslaw Mesjasz, John Grin, Patricia Kameri-Mbote, Béchir Chourou, Pál Dunay, and Jörn Birkmann, (eds), 61–106. Springer, New York, 2011.

Carmody, P., and Taylor, D. "Globalization, Land Grabbing, and the Present-Day Colonial State in Uganda." *The Journal of Environment & Development* 25, no. 1 (2016): 100–126. https://doi.org/10.1177/1070496515622017

Climate Watch. Global Greenhouse Gas Emissions. World Resources Institute, 2022. World | Total including LUCF | Greenhouse Gas (GHG) Emissions | Climate Watch (climatewatchdata.org)

Clausen, Maria-Louise, and Peter Albrecht. "Interventions since the cold war: From statebuilding to stabilization." *International Affairs* 97, no. 4 (2021): 1203–1220.

Connolly-Boutin, L., and Smit, B. "Climate change, food security, and livelihoods in sub-Saharan Africa." *Regional Environmental Change* 16, no. 2 (2015): 385–399. https://doi.org/10.1007/s10113-015-0761-x

De Frenne, Pieter, Jonathan Lenoir, Miska Luoto, Brett R. Scheffers, Florian Zellweger, Juha Aalto, Michael B. Ashcroft et al. "Forest microclimates and climate change: Importance, drivers and future research agenda." *Global Change Biology* 27, no. 11 (2021): 2279–2297.

Dobner, Petra, and Jasper Finkeldey. "Natural resources and the tipping points of political power—A research agenda." *Sustainability* 14, no. 22 (2022): 14721.

Dung, Elisha J. "The role of traditional institutions in the mediation of social issues in Africa, Latin America, and the Caribbean." In Gill, Brenda, I and George Danns (eds.), 249272. *Comparative Assessment of Social Issues in Africa, Latin America, and the Caribbean.* Lanham, MD: Lexington Books, 2023.

Dung, Elisha J., and Augustine Avwunudiogba. "The human cost of the refugee crisis in Africa." In Abegunrin, O, and Sabella O. Abidde (eds.), 295–313. *African Migrants and the Refugee Crisis.* Springer Nature, 2021.

Dung, Elisha, J. "The geography and the patterns of migration." In Abide, Sabella O. and Brenda I. Gill (eds.), *Africans and the Exiled Life*. Lexington Books, Lanham, 2018.

Ertor, I., Z.W. Brent, D. Galler, T. Josse. (2020). Situating Small-Scale Fisheries in the Global Struggle for Agroecology and Food Sovereignty. Amsterdam: Transnational Institute. https://scholar.google.com/scholar?cluster=236853875543731749&hl=en&as_sdt=2005&sciodt=0,5

Glantz, Michael H., and Nicolai S. Orlovsky. "The phenomenon known as desertification received widespread attention recently, as witnessed by the creation of the United Nations Conference on Desertification (UNGOD) in Nairobi in 1977, mainly as a result of the impact of extended drought in the West African Sahel in the early 1970s. That drought led to the loss of human lives." *Natural Resources And People: Conceptual Issues In Interdisciplinary Research*. Glantz, Michael H., and Nicolai S.Orlovsky. "Desertification: A review of the concept." *Desertification Control Bulletin* 9 (1983): 15–22. (2019)

Gordon,Cameron. "Global Demographic Change." In *Many Possible Worlds: An Interdisciplinary History of the World Economy Since 1800*, Gordon Cameron (ed), 443–466. Singapore: Springer Nature Singapore, 2023.

Grant, Ursula. "Urbanization and the employment opportunities of youth in developing countries." *Background paper prepared for EFA Global Monitoring Report* (2012).

Hamilton, Clive. "The Anthropocene as rupture." *The Anthropocene Review* 3, no. 2 (2016): 93–106.

Hendrix, C.S., and Salehyan, I. "Climate change, rainfall, and social conflict in Africa." *Journal of Peace Research* 49, no. 1 (2012): 35–50. https://doi.org/10.1177/0022343311426165

Hendrixson, Anne, and Betsy Hartmann. "Threats and burdens: Challenging scarcity-driven narratives of "overpopulation." *Geoforum* 101, (2019): 250–259.

IPCC. "Summary for policymakers." In *Climate change 2023: Synthesis report. A report of the Intergovernmental panel on climate change*. Contribution of Working Groups I, II and III to the Sixth Assessment Report of the Intergovernmental Panel on Climate Change [Core Writing Team, H. Lee and J. Romero (eds.)]. IPCC, Geneva, Switzerland, 36 pages. (in press).

Kabandula, Abigail, and Timothy M. Shaw. "Rising powers and the horn of Africa: Conflicting regionalisms." *Third World Quarterly* 39, no. 12 (2018): 2315–2333.

Kalis, Arie J., Josef Merkt, and Jürgen Wunderlich. "Environmental changes during the Holocene climatic optimum in central Europe-human impact and natural causes." *Quaternary Science Reviews* 22, no. 1 (2003): 33–79.

Kurukulasuriya, Pradeep, and Shane Rosenthal. "Climate Change and Agriculture: A Review of Impacts and Adaptations," Climate Change Series 91. Environment Department Papers, The World Bank, Washington, DC 2003.

Mamonova, Natalia, and Jaume Franquesa. "Populism, neoliberalism and agrarian movements in Europe. Understanding rural support for right-wing politics and looking for progressive solutions." *Sociologia Ruralis* 60, no. 4 (2020): 710–731.

Masselink, Gerd, Michael Hughes, and Jasper Knight. *Introduction to coastal processes and geomorphology*. Routledge, 2014.

McKay, Ben M., Ruth Hall, and Juan Liu, eds. *Rural transformations and agro-food systems: The BRICS and agrarian change in the Global South*. Routledge, 2018.

Middleton, Guy D. *Understanding collapse: Ancient history and modern myths*. Cambridge University Press, 2017.

Obuah, Emmanuel E., and Reginald Chikere Keke. "Mitigating the negative externalities of oil drilling activities in the Niger Delta, Nigeria." *Open Access Library Journal* 9, no. 5 (2022): 1–21.

Peša, Iva. "Mining, waste and environmental thought on the Central African Copperbelt, 1950–2000." *Environment and History* 28, no. 2 (2022): 259–284.

Rees, William E. "Ecological economics for humanity's plague phase." *Ecological Economics* 169 (2020): 106519.

Sage, Rowan F. "Global change biology: A primer." *Global Change Biology* 26, no. 1 (2020): 3–30.

Shakoor, Awais, Fatima Ashraf, Saba Shakoor, Adnan Mustafa, Abdul Rehman, and Muhammad Mohsin Altaf. "Biogeochemical transformation of greenhouse gas emissions from terrestrial to atmospheric environment and potential feedback to climate forcing." *Environmental Science and Pollution Research* 27 (2020): 38513–38536.

Syvitski, Jaia, Colin N. Waters, John Day, John D. Milliman, Colin Summerhayes, Will Steffen, Jan Zalasiewicz et al. "Extraordinary human energy consumption and resultant geological impacts beginning around 1950 CE initiated the proposed Anthropocene Epoch." *Communications Earth & Environment* 1, no. 1 (2020): 1–13.

United Nations Organization. "Securing a sustainable future." United Nations Sustainable Development Goals (SDG) Summit. United Nations HQ. New York, 18–September 19, 2023. UN 2023 SDG Summit - United Nations Sustainable Development.

Verhoeven, Harry. "Climate & water in a changing Africa: Uncertainty, adaptation & the social construction of fragile environments." *Dædalus* 150, no. 4 (2021): 260–277.

Watson, Robert, Ivar Baste, Anne Larigauderie, Paul Leadley, Unai Pascual, Brigitte Baptiste, Sebsebe Demissew et al. "Summary for policymakers of the global assessment report on biodiversity and ecosystem services of the Intergovernmental Science-Policy Platform on Biodiversity and Ecosystem Services." *IPBES Secretariat: Bonn, Germany* (2019): 22–47.

Chapter 3

Access to Environmental Resources

A Framework for Understanding the Drivers of Environmental Conflict

Sebastian Okechukwu Onah, Jonathan Ali Ogwuche, and Muhammad Kabiru Dahiru

The environment plays a significant role in contributing to development and human well-being. The term environment generally refers to a natural resource base that provides environmental services.[1] Environmental resources support human beings to produce goods and services and so contribute significantly to development in different ways. Environmental resources can be categorized into natural and cultural resources.[2] While natural resources comprise of land, soil, water, forest, plant and animal diversity, energy sources, and ecosystem services,[3] the cultural resources constitute the man-made features of social and physical infrastructure.[4]

The world has a considerable abundance of natural resource endowment, yet its distribution is uneven across countries. While some countries are richly endowed with a wide range of natural resources, others barely have them. It is widely recognized that environmental resources contribute significantly to development in different ways: as economic activity and source of growth; as means of livelihood, by providing jobs for people; and as provider of environmental services that can have both good and bad outcomes.[5]

The complex and dynamic interlinkages between environmental resources, management, and development have long been recognized by national and international research and development organizations. However, in spite of the abundance of research, publications, and development projects and policies in Africa, a lot still needs to be understood about the nexus between access to environmental resources and environmental conflict.

ENVIRONMENTAL RESOURCES AND ECOSYSTEM SERVICES

The meaning of environmental resources has continued to change over the years. The nature of the change has been characterized by divergences in the disciplinary conceptualization of the meaning of resources.[6] According to Ochola et al.,[7] several schools of thought, especially in ecology, sociology, and environmental science, define environmental resources from disciplinary perspectives. For instance, while sociology views environmental resources from a humanistic sense, economic perspectives consider the commercial value of environmental resources. In their view, in line with the economic theory, private ownership of environmental resources tends to produce efficient consumption because the profit motive imposes realistic pricing. This materialistic and exploitative dimension of resources is the basis of conflicts associated with the use of environmental resources.

The emergence of ecosystem concepts[8] and considerations of environmental resources as complex sources of interrelated elements existing in a state of ecological balance gave credence to the ecosystem theory.[9] An environmental resource,[10] from the perspective of ecosystem theory, is a material occurring in nature, which has environmental or potential value in a natural system that supports living organisms. Such material resources consist of biotic (living) organisms (such as plants, microbes, and animals) and abiotic (nonliving) or physical and chemical components of the environment (such as atmosphere, hydrosphere, and lithosphere), and their nonliving components, which are critical to life processes. Environmental resources are regarded as stocks from which people derive vital ecosystem services. Ecosystem services are therefore considered as the benefits that humans obtain from ecosystems. The concept of ecosystem services involves the direct and indirect benefits that humans derive from natural and managed ecosystems and include the following: (i) provision of services, for example, food, freshwater, wood, fiber, and fuel; (ii) cultural services, for example, aesthetics, spiritual, educational, and recreational; (iii) regulation of services, for example, climate regulation, flood regulation, disease regulation, and water purification; and (iv) support services of nutrient cycling, soil formation, and primary production.[11] The support services are the foundation for the other categories of ecosystem services that are directly used by society.

In addition, the goods harvested by people are influenced by landscape processes, which include regulatory services and, in turn, influence people's connection to the land and sea (cultural services). The resources and opportunities for livelihoods are provided to humans through ecosystem functions, hence they are called ecosystem services. However, ecosystems are linked to social systems since humans depend on the ecosystem functions to fulfilll their needs and aspirations. Maintaining healthy ecosystems to ensure the availability of environmental resources and other ecosystem services is critically essential for sustainable livelihoods.

ECOSYSTEM SERVICES AND LIVELIHOODS

The nexus between environmental change and the well-being of populations who depend on environmental resources has given rise to a new paradigm called environmental entitlements.[12] This paradigm emphasizes the capacity of environmental resources accessible to the people to produce streams of products and environmental services essential for livelihoods and well-being. These environmental entitlements, according to Ochola et al.,[13] refer to the alternative sets of benefits derived from environmental goods and services over which people have legitimate effective access and command and which are instrumental in achieving well-being. These benefits may include direct uses, in the form of commodities, the market value of such resources or the right to them, and the benefits derived from environmental services.

Livelihoods, according to Bishnu and Ulrike,[14] comprise of the capabilities, assets (both material and social resources), and activities required as well as the means through which people make a living. A conceptual framework for understanding the ecosystem functions and human well-being is the Sustainable Livelihoods Framework (SLF). It was first used by the Department for International Development (DFID) in 1998. This approach is an analytical tool for understanding people's livelihood strategies as shaped by interactions between livelihood assets, vulnerability, transforming structures (institution and processes), and livelihood strategies and outcomes. The SLF is increasingly important in environmental resource management discourses. It shows how, in different contexts, sustainable livelihoods are achieved through access to a range of environmental or livelihood resources, which are combined in the pursuit of different livelihood strategies.[15] In the SLF, resources are referred to as "assets" or "capitals" and are often categorized between five different asset types owned or accessed by family members: human capital (skills, education, and health), physical capital (produced investment goods), financial capital (money, savings, and loan access), natural capital (land, water, trees, etc.), and social capital (networks and associations). For further emphasis, while the natural capital includes the environmental resources of land, forest, water, pastures, wildlife, biodiversity, and so forth, the physical capital consists of the economic infrastructure such as roads, transport, water, energy, communication, and electricity supply and social infrastructure such as schools, hospitals, and so forth. These assets/capitals are central but need to be analyzed in terms of the vulnerability context and transforming structures. The interaction of these three components shapes livelihood strategies to achieve the livelihood outcome. The sustainable livelihood framework, therefore, attempts to capture not just what people do in order to make a living, but the resources that provide them with the capability to build a satisfactory living, the risk factors that they must consider in managing their resources, and the institutional and policy context that either helps or hinders them in their pursuit of a viable or improving living.

Risk to livelihoods occurs when there are vulnerabilities, threats, and a lack of coping mechanisms and exists in three dimensions. According to Chambers,[16] these dimensions are: access to environmental resources; ecological setting; and availability of common property resources. While access to environmental resources is the main factor affecting the risk exposure of individuals or households, ecological setting determines the type of resources available in a certain locality. The SLF considers access to assets and transforming structures and processes as a starting point and sees people as operating within a context of vulnerability, and within this context, they have access to certain assets and resources. According to DFID,[17] the livelihoods of people will improve if they secure access to and better management of environmental resources. It, therefore, means that once capabilities and resources or assets are disturbed, people face livelihood insecurity, and hence the need for environmental resource governance.

ENVIRONMENTAL RESOURCE GOVERNANCE

Livelihood security cannot be achieved without addressing the problems of poverty, structural inequality, political and social exclusion, discrimination, and bad governance. In this context, political economy and political ecology are used as conceptual frameworks for analyzing the livelihoods of people who depend on environmental resources.[18] Rights and ownership over environmental resources are essentially a phenomenon of the political economy, especially when they involve the adverse interests of different individuals, groups, and the state. While the political economy takes into account the power structure of society, the political ecology accommodates ecological considerations, referring to the relationship between nature and society. These two concepts are necessary in analyzing resource policies and, in particular, access to and control over environmental resources for livelihoods. The political economy encompasses the power structure in society, which determines the institutions and policies that frame the allocations of access to and control over environmental resources. Policies and institutions for environmental resource management need to ensure that existing institutions do not infringe on the rights of people over environmental resources. It, therefore, means that theoretically the political economy begins only when access to environmental resources and the means of production are affected.

Political ecology focuses on the political causes of environmental change and the struggle for access to and control over environmental resources.[19] However, from the perspective of poverty and livelihoods in a context of competing power for access to and control over resources, there is need to x-ray the nexus between drivers of change of environmental resources and conflict.

ENVIRONMENTAL RESOURCES, DRIVERS OF CHANGE, AND CONFLICT

The nexus between ecosystems, environmental resources, social systems, and the fulfilment of human needs and aspirations is complex. While environmental resources provide the ecosystem services, there exist some drivers or causalities of change of processes that affect ecosystem services and human well-being and in turn affect livelihoods. Ochola et al.[20] identify two types of drivers of change as direct and indirect and list the direct drivers as (i) changes in local land use and cover; (ii) species introduction and removal; (iii) technological adaptation and use; (iv) external inputs (e.g., fertilizer use, pest control, and irrigation); (v) harvest and resource consumption; (vi) climate change; and (vii) natural, physical, and biological disturbances. The indirect drivers of change include: (i) demographic; (ii) economic; (iii) sociopolitical (governance, institutional, and legal frameworks); (iv) science and technology; and (v) cultural and religious (e.g., beliefs). These drivers of change have over time degraded the ecosystem services and rendered them incapable of delivering the expected goods and services to humanity, in terms of basic material for a good life (health, security, freedom of choice, and action). Effort should, therefore, be to bridge this gap and to understand the vital links between ecosystems and social systems that finally determine the overall well-being of humanity.

One of the challenges in Africa's development is related to the rapid rate of degradation of environmental resources due to a complex combination of these factors, usually called drivers. Such degradation reduces the environmental resources both quantitatively and qualitatively, thereby compromising development activities based on these resources.[21] For example, loss of or declining biodiversity can have a negative impact on the functioning and resilience of ecosystems, decrease the species diversity of plants and soil organisms, and reduce structural diversity of vegetation, which in turn will cause loss of nutrients and affect soil structure. This, again, can lead to reduced nutrient cycling and cause land degradation and soil erosion, and hence productivity decline and food insecurity. Soil erosion also reduces carbon sequestration; hence, it increases carbon dioxide emissions and accelerates climate change.

The impacts of climate change are now inevitable and affect people in developing countries the most. Climate change particularly affects ecosystems, environmental resources, and health and livelihoods of people. According to IPCC,[22] Africa's vulnerability arises from a contribution of many factors, including extreme poverty, a high rate of population increase, frequent natural and cultural disasters such as drought and floods, and agricultural systems (both crop and livestock production) that depend heavily on rainfall. Literature on climate change and environmental resource degradation shows two integrated processes.[23] First, increasing temperature, precipitation irregularities, and adverse weather are expected to worsen the existing environmental resource degradation. Secondly, rising sea levels and additional adverse weather conditions will cause forced migrations that are likely to lead to higher pressure on existing environmental resources. The key consequences of climate change are likely to be sea level rise; changes in the intensity, timing, and spatial distribution of precipitation; and changes in temperature and greater vulnerability in the frequency, magnitude, and duration of extreme climate events such as droughts, floods, and tropical storms.[24] All these factors could influence the availability and distribution of environmental resources and thereby further aggravate or exacerbate scarcity of supply. This can further compound the drivers of scarcity by directly damaging environmental resources, triggering migration, or increasing demand for environmental resources.[25] Climate change is recognized as a driver of environmental resource conflict and hence is considered as presenting a security challenge. Poverty, as a driver of change, is generally recognized as one of the most important causes of vulnerability to environmental threats, on the basis that the poor tend to have fewer coping capacities and thus they bear a disproportionate burden of the impact of disasters, conflict, drought, desertification, and pollution. According to Homer-Dixon,[26] these result in social violence, particularly in developing societies that are more dependent on environmental resources and less able to buffer themselves from the social crisis that environmental scarcities cause.

Another driver of change is the evolving governance challenges and approaches. Governance systems at all levels (global, national, regional, and local) are under increased pressure because of mounting challenges due to persisting social inequalities. For instance, groups of lower socioeconomic status tend to be more negatively affected by environmental hazards. A large part of rural population has more limited access to and control over environmental resources and services than the urban few who are rich. Governance policies on land tenure could create problems for the poor in livelihood diversification. Poor rural people face a series of interconnected environmental resource management challenges, are often disempowered, and thus are unable to sustainably manage environmental resources.

The challenge therefore is to sustain the resilience of ecosystems—their capacity to cope with disturbances and maintain an adequate supply of goods and services. This is especially important in the face of global environmental change, which Ochola et al.[27] say has grave implications for human health, wealth, livelihoods, food security, social cohesion, and even democracy. Preventing conflicts that may be triggered or affected by climate change and natural hazards depends on the identification of vulnerable livelihoods and the provision of dedicated support for adaptation and vulnerability. Nevertheless, the ability of people and societies to cope with environmental change is varied and is a combination of all the natural and social characteristics and resources available in a particular location.[28] These include factors such as wealth, technology, education, information, skills, infrastructure, and access to resources and management capabilities.

ACCESS TO ENVIRONMENTAL RESOURCES AND ENVIRONMENTAL CONFLICT

The world has a considerable abundance of environmental resource endowments, yet its distribution is not even across countries. Even where they exist, they differ in quantity and quality, and hence the concept of environmental resource scarcity.[29] For instance, water scarcity is defined as a situation where the volumes of water withdrawn from lakes, rivers, or groundwater are so large that water supplies are no longer adequate to satisfy all human or ecosystem requirements, resulting in increased competition between users and demands.[30] Situations of water scarcity greatly compromise people's entitlement to water security. Water security entails every person having reliable access to enough safe water at an affordable price to lead to a healthy, dignified, and productive life while maintaining the ecological systems that provide water and also depend on water.[31]

Environmental resource security, therefore, entails the reliable access to and utilization of resources for sustainable livelihoods and is a core component of the broader notion of human security. This largely explains why, when access to environmental resources is constrained or disrupted, people become exposed to serious human security risks such as loss of livelihoods, increased vulnerability, poverty, migration, and conflict.[32] For instance, lack of access to vital environmental resources (due to harsh environmental trends) has made the seasonal migration of Fulani herders to the southern part of Nigeria for fresh water and arable or grazing land more permanent and has contributed to resource conflicts in these areas with the potential to spill over into ethnic clashes.[33]

Another dimension of access is its nature, with respect to the quality and quantity of environmental resources. Significant changes in the quality and quantity of environmental resources can give rise to conflict between different users. Such conflicts can occur between users in the same place or between users in different places. A typical example of this kind of conflict is the conflict between livelihood systems. For instance, water access and use conflict may be caused by competition between groups of people practicing different livelihood activities, such as livestock keepers, crop cultivators, dry cleaners, and so forth, as all of them may tend to use the same surface water source at the same time. In most cases, water access and use conflicts are more pronounced during the dry season, when water is scarcer than during the wet season. This type of water use conflict is also manifested between upstream and downstream users of water, regardless of whether they practice the same or different livelihood activities.[34]

For Blaikie,[35] perhaps the most critical set of factors determining patterns of environmental resource use, such as land, are those that structure the ways in which different environmental resource managers (households, governments, corporate actors, etc.) gain access to resources. These discussions of resource access have been deepened in important ways.[36] While much could be said about the way in which resource access might feature in an agenda for research, several themes emerge such as a notion of resource access that goes well beyond purely environmental resources. Access to these diverse resources is in turn structured (and frequently constrained) by local and national policies and institutions. The relationship between environmental resource use and access has been explored, demonstrating the significance of a range of formal and informal institutions in governing such access.[37]

Discussions of access to environmental resources often focus on property rights, ownership, and the terms of physical entry to resources, such as land. While ownership status and regulations that set the conditions for physical entry are important components of access to natural resources, other factors also affect access, such as structural and relational mechanisms, as well as biophysical factors, that is, environmental conditions and spatial proximity. A conceptualization of access, therefore, includes consideration of those mechanisms that are relevant to an array of resource users.[38]

However, we agree with Ribot and Peluso's argument to expand conceptualizations of access beyond rights-based approaches, to consider "a larger array of institutions, social and political-economic relations and discursive strategies that shape benefit flows".[39] For instance, land management decisions and ecological processes create environmental conditions that influence access to environmental resources, such as pollution or toxicity, changing habitat, loss of habitat and changing land use, and spatial proximity. This provides a fuller understanding of access and can be more useful in explaining and identifying potential responses to resource management issues.

A notion of conflict is at the core of political ecology, which, according to Philippe,[40] aims to address the narratives of environmental conflict driven by scarcity and access to environmental resources. More broadly, political ecology is about politics and about recognizing the political character of environmental and resource issues. Robins[41] identifies environmental conflict as one of the four major theses of political ecology, through which he sees an effort to demonstrate that the actors and causes of conflicts over environmental resources are "part of larger gendered, classed and race struggles and vice versa." These conflicts not only take place over the environment but within the context of economic, ecological, and cultural differences.[42]

Environmental conflict can be broadly understood as a social conflict relating to the environment, and this relation can take several forms and directions. According to Ribot and Peluso,[43] it can be a conflict over the environment, most notably in terms of access to and control over environmental resources. Turner[44] sees environmental conflict as social conflict associated with both struggles to gain access to environmental resources and struggles resulting from the use of environmental resources. Conflicts are at the center of many political ecology studies, generally considered over the defense of the environment as a source of livelihoods for indigenous and marginalized communities, constituting what Martin-Alier[45] defines as the "environmentalism of the poor."

Several studies[46] have found a positive correlation between environmental resources endowment and the onset of civil conflicts. The so-called resource curse argument, which has been widely discussed in the economic literature, hinges on the idea that environmental resource abundance has the likelihood to impact negatively on economic growth rates.[47] A common point of view, according to Mohammed,[48] is that the discovery of environmental resources has played a key role in the conflicts that have plagued few African countries over the last decade, both motivating and fueling armed conflicts. According to Peter,[49] environmental resource scarcity, such as water scarcity, causes and sustains conflict in many parts of the globe. Violent conflict over water resources has broken out in countries as diverse as China, particularly in the Shandong and Guangdong Provinces in 2000; Ethiopia in 2006; India in 2004; Kenya in 2005; and Yemen in 1999. Also, the U. S. Institute of Peace[50] observed that, traditionally, most wars have been fought over control of land (along with other issues); for example, Ecuador and Peru have fought several wars over their disputed border. Violent conflicts over land also occurred in China, East Timor, Kosovo, Rwanda, and Tajikistan.

Conflicts over environmental resources arise when parties disagree about the ownership, allocation, use, management, and protection of environmental resources and related ecosystems. Increasing scarcity of environmental resources or grievances over the governance can drive, reinforce, or compound existing stress factors and play a contributing role in the decision to resort to violence. Understanding both the dynamics of environmental resources and how it can contribute to conflict entails addressing questions of environmental resource ownership, access, control, and management. Environmental resource management is a form of conflict prevention. Traditions, customs, rules, laws, and policies regulating access to use and management of environmental resources all aim to bring order and predictability to situations where competition and conflicting interests are present.

Conflicts over environmental resources are essentially political issues concerning who should have access to and control over resources.[51] This key political question can become a source of tension, based on the competing interests of different individuals, groups, or countries. Weak institutions, fragile political systems, and divisive social relations can be drawn into cycles of conflict and violence. Access to environmental resources (such as water) became an official basic human right following the resolution introduced by Bolivia and adopted by the UN. Although the decision does not make the right to water legally enforceable, it is symbolically important and places more political obligations on national governments.[52] The main and general sources of conflict over environmental resources, such as water, according to UNEP[53] include: (i) competition between different environmental resource sectors (agriculture, industrial, and domestic); (ii) competition between different livelihood groups (farming, livestock, fishing, etc.); (iii) degradation of environmental resource quality caused by pollution (industrial, agriculture, and urban); (iv) reduction of environmental resource supply caused by development/infrastructure projects; (v) lost access to environmental resource supplies and/or exhaustion of supply; (vi) exclusion from control of environmental resource and access; and (vii) unclear environmental resource use and access rights.

Beyond these sources of conflicts over environmental resources, UNEP[54] identified three main drivers of conflict over environmental resources, based on existing academic theory, combined with UN and European Union (EU)'s field experiences, assessments, and case studies. These drivers, working alone or in combination, have been important factors in contributing to violent conflicts when acting alongside other stresses.

The first driver is competition over increasing scarce environmental resources. The concept of resource scarcity describes a situation where the supply of environmental resources such as water, forests, rangeland, and crop lands is not sufficient to meet the local demand. Increasing competition between user groups or between economic sectors can result from increasing scarcity of environmental resources needed to sustain livelihoods. Several social responses to rising competition over environmental resources exist, and they include migration, technological innovation, cooperation, and violent conflict. There are three main causes of increasing environmental resource scarcity, working separately or in combination.[55] The first is demand-induced scarcity, which arises when demand for a specific environmental resource increases and cannot be met by the existing supply. The second is supply-induced scarcity, which occurs when environmental degradation, natural variation, or a breakdown in delivery infrastructure constrains or reduces the total supply of a specific environmental resource. The third is structural factor, which occurs when different groups in a society face unequal environmental resource access. FAO[56] identifies five key conditions that influence the livelihoods that increasingly scarce environmental resources will contribute to conflict. These are: (i) the degree of absolute physical environmental resource scarcity; (ii) the extent to which the scarce supply is shared by two or more groups/sectors; (iii) the relative power of those groups/sectors; (iv) the ease of access to alternative environmental resources; and (v) the capacity to deploy coping mechanisms together with their expected duration. When the supply of an environmental resource cannot meet local demand, the resource will simply be depleted or degraded by competing user groups, as each group struggles to maintain its livelihood.

The second driver is poor governance of environmental resources and the environment. Governance of environmental resources and the environment refers to the institutions, policies, and processes that are established to regulate their management, ownership, allocation, use, and protection. According to EU–UN Partnership on Natural Resources and Conflict,[57] resource rights and related laws determine who can use what resources, for how long, and under what conditions.

According to UNEP,[58] in general terms, there are four types of grievances that can be generated by poor resource and environmental governance. These are: (i) unclear, overlapping, or poor enforcement of resource rights and laws and (ii) discriminatory policies, rights, and laws that marginalize specific groups. When one group controls access to environmental resources to the detriment of others, environmental-dependent communities are often marginalized: (iii) unequal distribution of benefits and burdens from development projects. Major development projects can provide multiple benefits to local communities as well as seriously degrade or pollute environmental resources and become a major source of grievance; and (iv) lack of public participation and transparency in decision-making in environmental resource policies and interventions.

The third driver is trans-boundary environmental resource dynamics and pressures. The challenge of managing environmental resources often extends beyond national borders. UNEP[59] identifies four main types of trans-boundary dynamics and pressure that can cause conflicts over environmental resources: First, when trans-boundary environmental resources, such as water or fisheries, are shared between countries, conflicts can arise when one country consumes the resources at a higher rate than another or violates agreed allocations or demonstrates inflexibility when faced with natural variation. Second, when the quality or quantity of trans-boundary environmental resources, such as water, fisheries, wildlife, and air, are negatively impacted in one country by infrastructure, industrial development, or changes in land use, in the other country. Third, while national borders define the sovereign boundary of states, these are often not respected by pastoral livelihood groups that migrate on a seasonal basis along traditional routes, based on the availability of environmental resources, such as water and grazing land.

Finally, one of the emerging threats to the national resource base of countries comes from illicit activities and criminal groups operating on a global and trans-boundary basis. Illicit extraction and trade of environmental resources deprive local communities of resource benefits and can lead to conflict.

A CRITICAL THEORY OF ENVIRONMENTAL SECURITY, SCARCITY, AND ACCESS TO ENVIRONMENTAL RESOURCES

Attempts to think about environmental problem in terms of national security, according to Eli,[60] began in the late 1970s, motivated by concerns to address the specter of resource wars that arose after the oil crises.[61] During the 1980s, the environmental security movement picked upstream, with the writings of environmental advocates[62] and with the support of foundations, conferences, and the initiation of several research efforts. The largest research project was the Toronto-based Environmental Change and Acute Conflict Project, led by Thomas Homer-Dixon.[63] Thomas Homer-Dixon had been the foremost academic advocate of environment and conflict linkages and the coordinator of the three-year Toronto-based project. He developed models to demonstrate the linkages between environmental resource scarcity and conflict.[64] These models of environmental scarcity leading to conflict have generated deep concern for the environmental underpinnings of conflict. The foremost finding of the project is that the degradation or depletion of agricultural land, forests, water, and fish will make a greater contribution to social turmoil in the coming decades than will climate change or ozone depletion.[65] The project addressed the question of how scarcity of environmental resources is related to violent conflict. The independent variable for the project is environmental scarcity, which includes scarcity from three sources, namely: (i) supply-induced scarcity, caused by degradation or depletion of environmental resources; (ii) demand-induced scarcity, caused by population growth or an increase in per capita consumption of environmental resources; and (iii) structural scarcity, arising from the unequal distribution of environmental resources.

The dependent variable is violent conflict, with intervening variables, such as social effects like migration, expulsion, and so forth. Thus, the two main causal steps show how environmental scarcity gives rise to social effects that result in violent conflict. The research revealed that different interactions often exist among the three sources of scarcity and identified that the most frequent interactions are resource capture and ecological marginalization. In the first interaction, resource capture is seen as a cause of environmental scarcity, whereby resource capture, demand, and supply-induced scarcities interact to produce structural (increasing) scarcities. Restricted or unequal access to environmental resources by different livelihood groups is a driver of resource scarcity and a source of conflict, especially when linked to grievances around equity, fairness, and justice. As Homer-Dixon and Blitt[66] explain, powerful groups within the society, anticipating future shortages due to increased population growth and degradation of resources, shift resource distribution in their favor, which subjects the remaining population to scarcity. This is also amplified by Brown et al.[67] in their conceptualization of the link between climate change, environmental resource scarcity, and conflict and established that environmental resource scarcity leads to competition and conflict.

The second interaction shows ecological marginalization as a cause of environmental scarcity. This, therefore, means that the fundamental basis of environment and conflict linkages is that environmental factors, such as natural resource scarcity and environmental degradation, leading to a decrease in quantity and quality of environmental resources, are significant sources of conflict. In other words, demand-induced and structural scarcities interact to produce supply-induced scarcities. As Homer-Dixon[68] notes, different livelihood groups begin a process of resource capture, whereby each attempt to secure access to and control over key services, to the exclusion of others.

It is, therefore, revealed that the three sources of environmental resource scarcity interact through two common patterns (resource capture and ecological marginalization) to give rise to the pattern of access to environmental resources, namely unequal availability (access to environmental resources), and the status of environmental resources (quality and quantity). This explains the finding by Homer-Dixon and Blitt[69] that lack of access to resources caused by inequitable distribution forces population migration from regions where resources are scarce to regions that are ecologically buoyant, hence justifying conflict as a core concept of political ecology. Also, significant changes in the nature of access (quality and quantity) to environmental resources can give rise to conflict between different users—users in the same place or users in different places, especially between livelihood systems. It, therefore, means that the environment, encapsulated in the notions of lack of access to environmental resources, is a ready threat on which to hinge new conceptualizations of conflict.

In the preceding section, the three main drivers responsible for environmental damage and scarcity were examined. These are competition over scarce environmental resources (internal), poor governance of environmental resources, and environmental resource dynamics and pressure (external). These internal and external stress factors combine with weak institutions and governance to also create the nature of access to environmental resources.

These drivers and sources of environmental scarcity, which can be regarded as socio-ecological dynamics, create the nature and pattern of access to environmental resources, which exacerbate environmental conflict in developing countries. This gives credence to Rakodi's view[70] on the need to capture the dynamics of changing access to environmental resources. According to her, the assets (environmental resources) that people possess or have access to, the livelihoods they derive, and the strategies they adopt are influenced by the context within which they live and are conceptualized as having two broad dimensions—factors that influence their vulnerability and policies, institutions, and processes (as discussed in the SLF). While vulnerability refers to environmental changes, policies refer to laws, social norms, and so forth and institutions (structures or organizations) refer to public (e.g., political, legislative, and governmental) and private (e.g., commercial, civil, and NGOs). All these influence entitlements and constrain access to environmental resources.

CONCLUSION

Environmental resources provide ecosystem services that are categorized into provisioning, cultural, regulatory, and supporting services. The world has a considerable abundance of natural resources endowment, yet its distribution is uneven across countries. There is a growing concern about environmental resources and their availability, access, and utilization and conflict. Access is a major indicator of deprivation, which is the most important dimension of poverty, and the key determinant of other aspects of welfare. Livelihood activities can often destroy the natural resource base, which in turn contributes to environmental damage, in addition to other stressors, causing environmental scarcity. There are three main drivers responsible for environmental damage and scarcity. They are competition over scarce environmental resources (internal), poor governance of environmental resources, and environmental resource dynamics and pressure (external). These internal and external stress factors combine with weak institutions and governance to create the nature of access to environmental resources. Also, environmental scarcity stems from three main sources—supply-induced (environmental degradation), demand-induced (population growth), and structural (unequal distribution). These three sources create two common patterns of access to environmental resources, namely, unequal availability and status (quantity and quality) of environmental resources. These drivers and sources of environmental scarcity create the nature and patterns of access to environmental resources, which exacerbate environmental conflict. Environmental conflicts have emerged as key issues challenging environmental and social security. Many environmental conflicts are related directly to the provision and utilization of goods and services from ecological processes, as people attempt to obtain access and secure control over environmental resources. Thomas Homer-Dixon models of environmental scarcity leading to environmental conflict have generated deep concern for the environmental underpinnings of conflict. This new way of explaining conflict, by underpinning it to access to environmental resources, is increasing the pathways for managing conflict.

In agreement with UNEP,[71] and conscious of the socio-ecological dynamics that give rise to the nature and patterns of access to environmental resources, we make the following recommendations, which can help human beings to cooperate over scarce resources and move a step closer to peace and a sustainable development for all, both now and in the future. These recommendations are as follows:

(i) The reduction of competition over scarce resources between livelihood groups, by supporting sustainable livelihoods, reducing vulnerability to resource scarcity, and increasing the availability of renewable resources through protection, restoration, infrastructure, and efficient use;
(ii) Improving resource governance, accountability, and dispute resolution capacity by establishing the governance framework for natural resources, strengthening implementation capacity, and recognizing resource rights, as well as building the capacity of stakeholders and civil society to participate in decision-making, in order to monitor compliance with the governance frameworks and access to justice mechanisms;
(iii) Improving trans-boundary management institutions and cooperation by establishing or strengthening trans-boundary information, resource-sharing agreements, joint institutions, and dispute resolution processes;

(iv) Designing and implementing cross-cutting measures and programs for the identification of environmental conflict-sensitive hotspots and for the resource management, adaptation, assessment, and analysis in those hotspots;
(v) Conducting and relating the environmental impact assessment of the various conflicts by evaluating the direct and indirect impacts of conflict on the environment; and
(vi) Focusing on proper land use planning, implementation, and monitoring.

NOTES

1. Bucknall, J. *Poverty/Environment Background Paper.* Washington, DC: World Bank, 2000.
2. Ogwuche, J. A. "Assessment of Physical Capital as a Sustainable Livelihoods Approach to Poverty Reduction in Otukpo, Benue State, Nigeria." PhD diss., Enugu State University of Science and Technology, ESUT, Enugu, 2005.
3. Ochola, Sanginga, and Bekalo, eds. *Managing Natural Resources for Development in Africa: A Resource Book.* International Development Research Centre, International Institute of Rural Reconstruction, Regional Universities Forum for Capacity Building in Agriculture, Nairobi: University of Nairobi Press, 2010.
4. Ibid, Ogwuche, 2005.
5. NEPAD, *Action Plan;* Khan Poverty, Environment; Comim, Poverty and Environment; IAASED, Summary; Chowdhury and Ahmed, Poverty-Environment.
6. Ibid, Ochola *et al, 2010.*
7. Ibid.
8. Tansley, A. G. "The Use and Misuse of Vegetational Terms and Concepts." *Ecology* 16, (1935): 284–307.
9. Chapin, Stuart F., Gary Kofinas, and Carl Folke. *Principles of Ecosystem Stewardship: Resilience-Based Natural Resource Management in a Changing World.* New York: Springer, 2009.
10. Ibid, Ochola *et al, 2010.*
11. Ibid.
12. Forsyth, T., Leach, M., and Scoones, I. *Poverty and Environment: Priorities for Research and Policy.* Prepared for the United Nations Development Programme and European Commission. London: Institute of Development Studies, 1998.
13. Ibid, Ochola *et al*, 2010.
14. Bishnu, R. A., and Ulrike, M. *Livelihood in Insecurity and Social Conflicts in Nepal.* Kathmandu: South Asia Regional Coordination Office, Swiss National Centre of Competence in Research, North-South Kathmandu. 2010.
15. Ibid, Ochola *et al*, 2010.
16. Chambers, R. "Vulnerability, Coping and Policy." *IDS Bulletin* 20, no. 2 (1989): 1–7.
17. DFID. *Sustainable Livelihoods Guidance Sheet.* London: *DFID* 445 1999.

18. Netra, P. T. "Natural Resources Governance and Livelihoods in Nepal: Political Economy and Political Ecology Perspectives." In *Livelihood in Insecurity and Social Conflicts in Nepal*, edited by Bishnu and Ulrike, (add page numbers). Kathmandu: South Asia Regional Coordination Office, Swiss National Centre of Competence in Research, North-South Kathmandu, 2010.

19. Paudel N. S. "Conservation and Livelihoods: Exploration of Local Responses to Conservation Interventions in Royal Chitwan National Park, Nepal." PhD diss.,Reading: The University of Reading, 2005.

20. Ibid, Ochola *et al*, 2010.

21. Ibid.

22. IPCC. *Climate Change 2007: Synthesis Report. Geneva, Intergovernmental Panel on Climate Change,* 2007.

23. Raleigh, C. and Urdal, H. *Climate Change, Democracy, Environmental Degradation and Armed Conflict.* Environmental Change and Security Program (ECSP) Report, Issue 13, 2008–2009. The Woodrow Wilson International Center for Scholars, One Woodrow Wilson Plaza, 1300 Pennsylvania Avenue, NW Washington, DC 20004–3027, 2009.

24. Oscar, G. M. "Climate Change Hydropolitics and Security in Lesotho." In *Climate Change and Food Security: A Framework Document,* Ogwuche, J. A. ed. 45–62. Rome: FAO, 2008.

25. Ibid.

26. Homer-Dixon, Thomas. "Thresholds of Turmoil: Environmental Scarcities and Violent Conflict." In *Contested Grounds: Security and Conflict in the New Environmental Politics*, edited by Daniel H. Deudny and Richard A. Matthew, 61–90. Albany, New York: State University of New York, 1999a.

27. Ibid, Ochola *et al*, 2010.

28. IATFDR. *Framework for Action for the Implementation of the International Strategy for Disaster Reduction (ISDR)*, 2001.

29. Onuoha, F. C. "Climate Change, Population Surge and Resource Overuse in the Lake Chad Area: Implications for Human Security in the North-East Zone of Nigeria." In *Climate Change and Food Security: A Framework Document,* edited by Donald Anthony Mwiturubani and Jo-Ansie van Wyk, 23–43. Rome: FAO, 2008.

30. UNECOSOC. *Comprehensive Assessment of the Fresh Water Resources of the World.* Report of the Secretary-General, 1997.

31. Ibid, Onuoha, 2008.

32. James, I. "Lake Chad as an Instrument of International Cooperation." In *Borderlands in Africa: A Multidisciplinary and Comparative Focus on Nigeria and West Africa,* edited by A. I. Asiwaju and P. O. Adeniyi, (add page numbers). Lagos: University of Lagos Press, 1989.

33. Ibid, Onuoha, 2008.

34. Donald, A. M. "Climate Change and Access to Water Resources in the Lake Victoria Basin." *In Climate Change and Food Security: A Framework Document.* Donald, A. and Jo-Ansie eds. 45–62. Rome: FAO, 2008.

35. Blaikie, P. "Explanation and Policy in Land Degradation and Rehabilitation for Developing Countries." *Land Degradation & Rehabilitation* 1, (1989a): 23–37.

36. Carney, J. "Struggles over Crop Rights within Contract Farming Households in a Gambian Irrigated Rice Farming Project." *Journal of Peasant Studies* 15, (1988): 334–349.

37. Berry, S. "Social Institutions and Access to Resources." *Africa* 59, (1989): 41–55.

38. Ribot, J. C., and N. L. Peluso. "A Theory of Access." *Rural Sociology* 68, 2 (2003): 153–181.

39. Ibid.

40. Philippe, L. *Environmental Conflict*. Chapter 46. The Routeledge Handbook of Political Ecology, 598–608, 2015.

41. Robbins, P. *Political Ecology: A Critical Introduction*. Oxford: Blackwell, 2004.

42. Escobar, A. "Difference and Conflict in the Struggle over Natural Resources: A Political Ecology Framework." *Development* 49, no. 3 (2006): 6–13.

43. Ibid, Ribot and Peluso, 2003.

44. Turner, M. D. "Political Ecology and the Moral Dimensions of 'Resource Conflicts': The Case of Farmer–Herder Conflicts in the Sahel." *Political Geography* 23, no. 7 (2004): 863–889.

45. Martinez-Alier, J. *The Environmentalism of the Poor: A Study of Ecological Conflicts and Valuation*. Cheltenham: Edward Elgar Publishing, 2003.

46. Fearon, J. D. and Laitin, D. D. "Ethnicity, Insurgency, and Civil War." *The American Political Science Review* 97, no. 1 (2003): 75–90.

47. Auty, R. *Sustaining Development in the Mineral Economies: The Resource Curse Thesis*. London: Routledge, 1993.

48. Mohammed, J. "Natural Resources Endowment and Economic Growth: The West African Experience." *Journal of Natural Resources and Development* 03, (2013): 66–84.

49. Peter, G. *Environment and Security Water Conflict Chronology*. Pacific Institute, 2006. http://www.worldwater.org/chronology.html.

50. United States Institute of Peace. *Natural Resources, Conflict, and Conflict Resolution. A Study Guide Series on Peace and Conflict.* Washington, DC: United States Institute of Peace, (no date).

51. UNEP. *Toolkit and Guidance for Preventing and Managing Land and Natural Resources Conflict*, UNEP, Paris, 2012.

52. Ibid.

53. Ibid.

54. Ibid.

55. Homer-Dixon, T. *Environment, Scarcity, and Violence*. Princeton, NJ: Princeton University Press, 1999b.

56. FAO. *Negotiation and Mediation Techniques for Natural Resource Management*. Rome: FAO, 2005.

57. EU-UN Partnership on Natural Resources and Conflict. *Guidance Note on Land and Conflict.* New York: United Nations, 2012.

58. Ibid, UNEP, *2012*.

59. Ibid.

60. Eli, M. *A Critical Theory of Environmental Security: Denaturalizing Scarcity-Conflict Research with Marcuse, Habermas, and Bourdieu.* Minneapolis: University of Minnesota (nd).

61. Ullman, R. H. "Redefining Security." *International Security* 8, no. 1 (1983): 129–153.

62. Dalby, Simon. *Environmental Security.* Vol. 20. Minneapolis: U of Minnesota Press, 2002.

63. Ibid, Homer-Dixon, 1999a.

64. Homer-Dixon, Thomas, and Jessica Blitt, eds. *Ecoviolence: Links among Environment, Population, and Security.* Lanham, MD: Rowman and Littlefield, 1998.

65. Ibid, Homer-Dixon, 1999b.

66. Ibid, Homer-Dixon and Blitt, 1998.

67. Brown, O. Hammill, A. and McLeman, R. "Climate Change as the New Security Threat: Implications for Africa." *International Affairs* 83, no. 6 (2007): 1141–1154.

68. Ibid, Homer-Dixon, 1999b.

69. Ibid, Homer-Dixon and Blitt, 1998.

70. Rakodi, C. "A Livelihoods Approach: Conceptual Issues and definitions." In *Urban Livelihoods: A People-Centred Approach to Reducing Poverty,* edited by Rakodi and Lloyd-Jones, 3–22. London: Earthscan Publications, 2002.

71. Ibid, UNEP, 2012.

REFERENCES

Auty, R. *Sustaining Development in the Mineral Economies: The Resource Curse Thesis.* London: Routledge, 1993.

Bebbington, A. J., and Perreault, T. "Social Capital and Political Ecological Change in Highland Ecuador." *Economic Geography* 75, no. 4 (Oct., 1999): 395–418. https://www.jstor.org/stable/144478.

Berry, S. "Social Institutions and Access to Resources." *Africa* 59, (1989): 41–55.

Bishnu, R. A. and Ulrike, M. *Livelihood in Insecurity and Social Conflicts in Nepal.* Kathmandu: South Asia Regional Coordination Office, Swiss National Centre of Competence in Research, North-South Kathmandu, 2010.

Blaikie, P. "Explanation and Policy in Land Degradation and Rehabilitation for Developing Countries." *Land Degradation & Rehabilitation* 1, (1989a): 23–37.

Blaikie, P. "Environment and Access to Resources in Africa." *Africa* 59, (1989b): 18–40.

Brown, Lester. "Redefining National Security." *Worldwatch Paper* 14, Washington, Worldwatch Institute, 42 . (1977).

Brown, O. Hammill, A., and Mcleman, R. "Climate Change as the New Security Threat: Implications for Africa." *International Affairs* 83, no. 6 (2007): 1141–1154.

Bucknall, J. *Poverty/Environment Background Paper.* Washington, DC: World Bank, 2000.

Carney, J. "Struggles over Crop Rights within Contract Farming Households in a Gambian Irrigated Rice Farming Project." *Journal of Peasant Studies* 15, (1988): 334–349.

Chambers, R. "Vulnerability, Coping and Policy." *IDS Bulletin* 20, no. 2 (1989): 1–7.

Chambers, R., and Conway, G. "Sustainable Rural Livelihoods: Practical Concepts for the 21st Century." Institute of Development Studies (IDS) Discussion Paper 296. Brighton (1992).

Chapin, Stuart F., Gary Kofinas, and Carl Folke. *Principles of Ecosystem Stewardship: Resilience-Based Natural Resource Management in a Changing World.* New York: Springer, 2009.

Chowdhury, M., and Ahmed S. U. "Poverty-Environment Nexus: An Investigation of Linkage Using Survey Data." *International Journal of Environment and Sustainable Development* 9, no. 1–3 (2010): 91–113.

Comim, F., Kumar, P., and Sirven, N. "Poverty and Environment Links: An Illustration from Africa." *Journal of International Development* 21, (2009): 447–469.

DFID. *Sustainable Livelihoods Guidance Sheet.* London: Institute of Development Studies, Sussex. 1999.

Mwiturubani, Donald Anthony. "Climate Change and Access to Water Resources in the Lake Victoria Basin." *ATE Change and Natural Resources Conflicts in Africa*, (2010): 63.

Eli, M. *A Critical Theory of Environmental Security: Denaturalizing Scarcity-Conflict Research with Marcuse, Habermas, and Bourdieu.* Minneapolis: University of Minnesota, (n.d.).

Escobar, A. "Difference and Conflict in the Struggle over Natural Resources: A Political Ecology Framework." *Development* 49, no. 3 (2006): 6–13.

EU-UN Partnership on Natural Resources and Conflict. *Guidance Note on Land and Conflict.* New York: United Nations, 2012.

FAO. *Negotiation and Mediation Techniques for Natural Resource Management.* Rome: FAO, 2005.

Fearon, J. D., and Laitin, D. D. "Ethnicity, Insurgency, and Civil War." *The American Political Science Review* 97, no. 1 (2003): 75–90.

Forsyth, T., Leach, M., and Scoones, I. *Poverty and Environment: Priorities for Research and Policy.* Prepared for the United Nations Development Programme and European Commission. London: Institute of Development Studies, 1998.

Gelb, A. H. *Windfall Gains: Blessing or Curse?* New York: Oxford University Press, 1988.

Homer-Dixon, T. "Environmental Scarcities and Violent Conflict: Evidence from Cases." *International Security* 19, no. 1 (1994): 5–40.

Homer-Dixon, Thomas, and Jessica Blitt, eds. *Ecoviolence: Links among Environment, Population, and Security.* Lanham, MD: Rowman and Littlefield, 1998.

Homer-Dixon, Thomas. "Thresholds of Turmoil: Environmental Scarcities and Violent Conflict." In Daniel H. Deudny and Richard A. Matthew (eds.) pp. 61–90, *Contested Grounds: Security and Conflict in the New Environmental Politics*, Albany, New York: State University of New York, 1999a.

Homer-Dixon, T. *Environment, Scarcity, and Violence*. Princeton, NJ: Princeton University Press, 1999b.

Humphreys, M. "Natural Resources, Conflict, and Conflict Resolution: Uncovering the Mechanisms." *Journal of Conflict Resolution* 49 (2005): 508–537.

IATFDR. *Framework for Action for the Implementation of the International Strategy for Disaster Reduction (ISDR)*, UN International Strategy for Disaster Reduction (UNISDR) | Department of Economic and Social Affairs, 2001.

Intergovernmental Panel on Climate Change (IPCC). "Mitigation of Climate Change." *IPCC Fourth Assessment Report – Working Group 111 Report* (2007). http://www1.ipcc.ch/ipccreports/ar4-wg3.htm.

International Assessment of Agricultural Knowledge, Science, and Technology for Development (Project), and Beverly D. McIntyre. *International Assessment of Agricultural Knowledge, Science and Technology for Development (IAASTD): Synthesis Report: a Synthesis of the Global and Sub-global IAASTD Reports*. Island Press, 2008.

IPCC. *Climate Change 2007: Synthesis Report. Geneva, Intergovernmental Panel on Climate Change.* AR4 Climate Change 2007: Synthesis Report—IPCC, 2007.

James, I. "Lake Chad as an Instrument of International Cooperation." In A. I. Asiwaju and P. O. Adeniyi (eds.), *Borderlands in Africa: A Multidisciplinary and Comparative Focus on Nigeria and West Africa,* (add page numbers). Lagos: University of Lagos Press, 1989.

Khan, H. "Poverty, Environment and Economic Growth: Exploring the Links among Three Complex Issues with Specific Focus on the Pakistan's Case." *Environment, Development and Sustainability* 10, no. 6 (2008): 913–929.

Maloney, W. F. *Missed Opportunities: Innovation and Resource-Based Growth in Latin America. Working paper* No. 2935. Office of the Chief Economist Latin America and Caribbean Region; The World Bank (Dec. 2002); JSTOR; New York University Bobst? 2008.

Malthus, T. R. *An Essay on the Principle of Population*. Cambridge: Cambridge University Press, 1798.

Martinez-Alier, J. *The Environmentalism of the Poor: A Study of Ecological Conflicts and Valuation.* Cheltenham: Edward Elgar Publishing, 2003.

Millennium Ecosystem Assessment. *Ecosystems and Human Wellbeing: Synthesis*. Washington D.C.: Island Press, 2005.

Mohammed, J. "Natural Resources Endowment and Economic Growth: The West African Experience." *Journal of Natural Resources and Development* 03, (2013): 66–84.

New Partnerships for Africa Development (NEPAD). *Action Plan for Environmental Initiative.* NEPAD, Sponsored by UNEP, AMCED, and GEF, (In collaboration with UNEP, AMCEN, GEF), 2003.

Netra, P. T. "Natural Resources Governance and Livelihoods in Nepal: Political Economy and Political Ecology Perspectives." In Bishnu and Ulrike (eds.), *Livelihood in Insecurity and Social Conflicts in Nepal*, (add page numbers). Kathmandu: South Asia Regional Coordination Office, Swiss National Centre of Competence in Research, North-South Kathmandu, 2010.

Ochola, Sanginga, and Bekalo, eds. *Managing Natural Resources for Development in Africa: A Resource Book.* International Development Research Centre, International Institute of Rural Reconstruction, Regional Universities Forum for Capacity Building in Agriculture, Nairobi: University of Nairobi Press, 2010.

Ogwuche, J. A. "Assessment of Physical Capital as a Sustainable Livelihoods Approach to Poverty Reduction in Otukpo, Benue State, Nigeria." PhD diss., Enugu State University of Science and Technology, ESUT, Enugu, 2005.

Okali, C. "Issues of Resource Access and Control: A Comment." *Africa* 59, (1989): 56–60.

Onuoha, F. C. "Climate Change, Population Surge and Resource Overuse in the Lake Chad Area: Implications for Human Security in the North-East Zone of Nigeria." In Donald Anthony Mwiturubani and Jo-Ansie van Wyk (eds.), *Climate Change and Food Security: A Framework Document*, Rome: FAO, 23–43, 2008.

Oscar, G. M. "Climate Change Hydropolitics and Security in Lesotho." In Ogwuche, J. A. ed. 45–62.*Climate Change and Food Security: A Framework Document,* Rome: FAO, 2008.

Paudel NS. "Conservation and Livelihoods: Exploration of Local Responses to Conservation Interventions in Royal Chitwan National Park, Nepal." PhD diss., Reading: The University of Reading, 2005.

Peter, G. *Environment and Security Water Conflict Chronology.* Pacific Institute 2006. http://www.worldwater.org/chronology.html. (Assessed on January 9, 2022).

Philippe, L. "Environmental Conflict." In *The Routledge Handbook of Political Ecology,* 2015 June 12 (pp. 598–608). Routledge.

Rakodi, C. "A Livelihoods Approach: Conceptual Issues and definitions." In Rakodi ānd Lloyd-Jones (eds.), *Urban Livelihoods: A People-Centred Approach to Reducing Poverty,* London: Earthscan Publications, 2002.

Raleigh, C. and Urdal, H. *Climate Change, Democracy, Environmental Degradation and Armed Conflict.* Environmental Change and Security Program (ECSP) Report, Issue 13, 2008–2009. The Woodrow Wilson International Center for Scholars, One Woodrow Wilson Plaza, 1300 Pennsylvania Avenue, NW Washington, DC 20004–3027, 2009.

Ribot, J. "Theorizing Access: Forest Profits along Senegal's Charcoal Commodity Chain." *Development and Change* 29, (1998): 307–342.

Ribot, J. C., and Peluso, N. L. "A Theory of Access." *Rural Sociology* 68, no. 2 (2003): 153–181.

Robbins, P. "Authority and Environment: Institutional Landscapes in Rajasthan." *India', Annals of the Association of American Geographers* 88 (1998): 410–435.

Robbins, P. *Political Ecology: A Critical Introduction.* Oxford: Blackwell, 2004.

Ross, M. "Does Oil Hinder Democracy?" *World Politics* 53, (2001): 325–361.

Ross, Michael. "A Closer Look at Oil, Diamonds, and Civil War." *Annual Review of Political Science* 9, no. 1 (2006): 265–300.

Rukuni, M., Tawonezvi, P., Eicher, C. with M. Munyuki-Hungwe and P. Matondi, eds. *Zimbabwe's Agricultural Revolution.* Second edition, Harare: University of Zimbabwe Publications, 2006.

Sachs, J. D., and Warner, A. M. *Natural Resource Abundance and Economic Growth.* Center for International Development and Harvard Institute for International Development. Cambridge, Massachusetts: Harvard University, 1997.

Sachs, J. D., and Warner, A. M. *Natural Resource Abundance and Economic Growth.* NBER Working Paper No. 5398, Cambridge, Massachusetts: National Bureau of Economic Research, 1995.

Scoones, I. "Sustainable Rural Livelihoods: A Framework for Analyses." Institute of Development Studies (IDS) Working Paper 72. Brighton (1998).

Tansley, A. G. "The Use and Misuse of Vegetational Terms and Concepts." *Ecology* 16, (1935): 284–307.

Turner, M. D. "Political Ecology and the Moral Dimensions of "Resource Conflicts": The Case of Farmer–Herder Conflicts in the Sahel." *Political Geography* 23, no. 7 (2004): 863–889.

Ullman, R. H. "Redefining Security." *International Security* 8, no. 1 (1983): 129–153.

UNECOSOC. *Comprehensive Assessment of the Fresh Water Resources of the World.* Report of the Secretary-General, 1997.

UNEP. *Toolkit and Guidance for Preventing and Managing Land and Natural Resources Conflict*, 2012.

UNEP. *Global Environment Outlook 3.* London: Earthscan, 2002.

United States Institute of Peace. *Natural Resources, Conflict, and Conflict Resolution. A Study Guide Series on Peace and Conflict.* Washington, DC: United States Institute of Peace, (no date).

Chapter 4

Environmental Conflicts, Politics, and Development in Africa

A Review

Fagdéba Bakoyéma

The incidence of environmentally triggered conflict is not new. Krakowka, Heimel, and Galgano[1] state that history and contemporary events have demonstrated that environmental stress often results in disputes. There is an ongoing discussion within the body of current and emerging literature on the complex linkages between environmental conflicts and governance and policy decision-making in Africa and the role played by international entities in causing or instigating conflicts. This chapter catalogs critical research findings on local and global environmental conflicts due to resource scarcity, large-scale land acquisition (LASLA), poor governance, and environmental policies. Climate change was first an environmental issue, then an energy issue, and now a security threat, as recognized by the UN Security Council.[2] Consequently, climate change is considered a "security" issue that could make violent conflicts more likely.

Literature offers many views on the causes of environmental conflicts. This chapter focuses on environmental changes, politics, and socioeconomic development disputes. There is a theoretical framework in which policies are being developed to address African environmental issues. The review highlights research on the relationship between people, governance, and environmental conflicts. Concepts such as environment, resources, security, and Africa are inextricably interwoven in human interactions. Politicians, policymakers, and researchers have debated attempts to grasp the impact of environmental issues on populations and living entities. History shows that poor environmental deterioration consequences for humans and the world, especially from an over-consumption standpoint, has a proud seventy-five-year history[3] since the Cold War; state-centric ideas have dominated security debates, with critics calling them "Malthusian" or "neo-Malthusian."[4] Scholars in conflict studies, international law, and environmental security have explored direct and indirect links between armed conflicts and the environment. Scholars in geography, planning, and urban studies have highlighted the destruction of infrastructure in urban settings. In addition to disputes, researchers posit that deforestation, for survival, has been viewed as negatively impacting the environment. The chapter also stresses the role of "land grabbing" by international entities and their role in instigating new environmental conflicts.

At the core of environmental conflicts in Africa is the issue of environmental resources, which refers to anything people find helpful in their environments or surroundings. The etymological definition of environmental resources is rooted in the Middle French preposition "environ," "around," environment. In its most basic meaning, environ is that which surrounds us. When preceded by the definite article, "the" environment usually refers to the natural world, both abiotic (physical or nonliving) and biotic (living) environment. Ecological resources include food from plants and animals, wood for cooking, heating, and building, metals, coal, and crude oil. Clean land, air, and water are environmental resources, as are the abilities of land, air, and water to absorb society's waste products.

This chapter explores three research areas covering Africa's environmental conflicts, politics, and development. First, it reviews the vulnerability of local places and social groups to climate change, livelihoods, and violent clashes and the role of policy and politics in the continent's development. Second, discourses and debates highlighting new insights into the intersectionality of the environment and disputes, politics, and development initiatives in Africa are examined. Additionally, global and regional public policies debated and implemented to address environmental conflicts are discussed.

THEORETICAL FRAMEWORK

Several theoretical frameworks inform the literature review on African environmental conflict, politics, and development. These theories are challenged by realism, neo-realism, structuralism, post-structuralism, modernism, and institutionalism, enhancing the discourse and debate on environmental conflicts' impacts on African politics and development.[5] The conventional approach to security studies, known as realism, focuses on the causes of war and the conditions of peace between and among states. "Security" may be defined as protecting a state (and, by logical extension, its citizens). Given the systemic situation of anarchy, wherein sovereign states pursue self-help policies, states threaten each other by their existence. Proponents of realism hold that environmental scarcity does not inevitably or deterministically lead to social disruption and violent conflict. They argue that in the face of environmental changes such as global warming and deforestation, stakeholders should provide explanations that will ensure stability and bring order. Yet, answers regarding causality diverge significantly. Critical theorists, including Peterson,[6] argue that the forms and practices giving rise to environmental degradation cannot simultaneously lead to its salvation.

The 1990s saw two interrelated discussions: one about the redefinition of security and the other about how environmental change threatens (individual, state, and global) security.[7] There are hypotheses linked to the conflict of environmental change: (1) declining supplies of physically controllable resources like water and agricultural land would produce "simple scarcity" conflicts; (2) massive population movements driven by environmental stress would cause "group identity" conflicts; and (3) extreme environmental scarcity would raise economic deprivation, upset essential social institutions (especially the state), and trigger "deprivation" disputes.[8] Environmental change alone would not cause African domestic wars. A body of literature points to other causes of environmental conflicts in which state and non-state actors are implicated in environmental conflicts.[9] In light of these, legal approaches to protecting the environment should take environmental consciousness as a moral duty, work to improve the economy and make society relatively prosperous, and, most importantly, choose nonviolent solutions to resource scarcity.[10]

RESOURCE SCARCITY AND CONFLICTS IN AFRICA

Scarcity can come from an increase in demand (demand-induced), such as population growth or increased per capita consumption; decreased supply (supply-induced), such as cropland erosion; or unequal access to and distribution of a resource (structural), which concentrates it in the hands of relatively few people while the remaining population suffers from severe shortages.[11] Some research findings posit that environmental conflicts generally originate from barriers to access, poor management, and scarcity of resources. Claims of ownership of environmental resources can lead to all types of disputes.[12]

Environment, conflict, and development interact in various ways. In their apparent forms, environmental conflicts are fights among people or wars over a few resources. Nomadic pastoralists and sedentary farmers compete over grazing pastures, arable land, water wells, and wood for the fire. Desertification and resource scarcity push humans against each other for survival. Indeed, radical climate changes like desertification, drought, shrinking water reserves, and general environmental deterioration threaten the viability of societies and sometimes pit groups against one another, such as pastoralists against agriculturalists. Linkages among environmental resource scarcity, insecurity, poverty, and vulnerability hinder sustainable development.

The scarcity of environmental resources affects human populations in various ways. Bereketeab[13] examines the Horn of Africa's environmental changes, conflicts, and sustainable development and suggests the need for a dialectical correlation between environment, conflict, and development. Environmental deterioration and conflicts hinder sustainable development; conversely, environmental deterioration caused by unsustainable development can also cause conflicts. Conflict, in turn, degrades the environment, slowing progress. Social structures and socioeconomic lifestyles affect the environment, which either exacerbates or mitigates conflicts, influencing sustainable development. These social structures and institutions influence environmental norms, values, and behaviors.

Violence between farmers and herders in northern Niger is an example of conflict due to the scarcity of resources. Like other Sahel countries, Niger suffers from rising temperatures, reduced rainfall, and dwindling pasturelands. Herders and farmers fight for scarce natural resources, often violently. Natural resource constraints have violently impacted Sabo Gida, Tountoube, Katanga, and Bokki communities in Niger. Herders' agricultural damage and unlicensed grazing of crop wastes after harvest have caused conflict. In the past twenty-five years, herder–farmer confrontations have increased, especially in northern Niger, and some are resolved locally by community traditions. These migrating pastoralists often encroached on the farm and, in the process, caused strife in Niger as they looked for greener pastures.

Similarly, farmers and Fulani herders sometimes fight over pastures, crop residues, water, and grazing land. Researchers examined the farmer–herder disputes to stress how complex these conflicts can be. Cluster sampling was selected from four local government areas[14] for the study. Overall, 235 randomly selected farmers and herders were questioned using standardized questionnaires. Most farmers (90.7 percent) and herders (82.7 percent) in the research area had witnessed farmer–herder conflict. Most farmers (84.4 percent) also raise animals, while most herders (65.3 percent) are sedentary. Herders (57.3 percent) and farmers (56.3 percent) reported oscillating farmer–herder relations. Nomads and semi-settled herders fought most.

Economic Communities of West African States (ECOWAS) has bolstered cross-border pastoralist mobility in West Africa.[15] Migratory pastoralists look for pasture freely under the free and fair movement convention. The policy benefits the region's socioeconomic growth, and the organization quickly downplays the cross-border conflict and its spillover effects throughout the continent. For instance, in northern and central Togo, herders from Niger are entangled in a confrontation with Togolese farmers over crop destruction. These conflicts, so far, have often been resolved amicably, with herders offering a few cows to compensate for destroyed crops. Violent clashes are largely avoided.

Disputes stemming from environmental changes do not only impact human-to-human relations; they also occur between humans and wildlife. Indeed, climate change worsens environmental degradation and destroys wildlife habitats in vulnerable areas, forcing animals to migrate or die. The situation forces wildlife into densely populated human areas and causes human–wildlife conflict. For example, climate change has intensified natural disasters and changes to terrestrial ecosystems, making human–wildlife conflicts inevitable in some East African countries.[16] Also, many West African countries are considered more vulnerable to the impacts of environmental change due to a history of desertification, drought, and inadequate adaptation and mitigation, as seen in the Lake Chad region.

Environmental change impacts in Africa have caused new conflicts and exacerbated preexisting forms of violence in recent decades. Ethiane and Moyo[17] argue that while the causes and drivers of environmental conflict could be contextual and multifaceted, the impacts of the disputes could lead to marginalization, inequality, exclusion, and poor governance in Africa.[18]

ENVIRONMENTAL CONFLICTS AND DEVELOPMENT IN AFRICA

Environmental resources play a critical role in the development of the African continent. Farmers–herders disputes, intercommunal conflicts, and international tensions over scarce environmental resources are common in Africa. Environmental security politics should limit these conflicts to allow local, national, and regional growth. However, current literature on the ownership and use of environmental resources highlights conflicting and disappointing approaches that adversely impact efforts to develop Africa. Nkansah-Dwamena and Yoon[19] posit that SSA's development depends on foreign land-grabbing investments. They argue that land grabbing has positive aspects, such as technology transfer, employment creation, infrastructure improvement, and global market integration. It also has negative aspects, including environmental degradation, food insecurity, displacement, and loss of livelihoods.[20]

A land grab[21] is the purchase or leasing of enormous expanses of land from poor, developing countries by wealthier, food-insecure governments and private investors. According to Shepard Daniel,[22] land grabbing has led to the acquisition of nearly 50 million hectares of farmland. While rising food costs, demand for biofuels, and investors wanting quick returns have been emphasized as the main drivers of this trend, the World Bank has gone largely unnoticed.

The Oakland Report adds that following the 2008 food and financial crises, the World Bank was supposed to play a crucial role in what was expected to be a massive reform in international food policy and a vast increase in food security in the poor world. Research shows that World Bank Group (WBG)'s policies and initiatives are doing just the opposite. IFC has helped international investors buy land in underdeveloped countries. It promotes "products" like "Access to Land" and "Land Market for Investment" to give investors land access. Moreover, investment promotion agencies and national legislation rewrite have institutionalized such investments. It has ignored the pressing food problem in client nations and lost sight of its primary objective: alleviate poverty.[23]

LASLA is the acquisition (by transfer of rights, purchase, ownership, or lease) of massive tracts (>200 ha) of land, usually for twenty-five to ninety-nine years in developing countries by foreign investors looking to produce food, biofuel, and other agricultural commodities for export to their home country or the international market. LASLA projects require water, increasing water demand in countries where LASLA operates.[24] LASLA investors in SSA are from rising economies such as China and Brazil, the Gulf States, the Arab Emirates, and the Global North, specifically the United States and EU countries.[25] Foreign governments, individual investors, and multinational corporations (MNCs) are the three main types of joint ventures. Governments provide direct financial, technical, and policy support to private investors. Land governance regimes and agroclimatic circumstances lead to LASLA in SSA. Investors choose countries with poor land tenure where investment regulations favor opaque, top-down, politically brokered transactions. In nations with uncertain and overlapping land tenure rules,[26] LASLA negotiations may be private.

International finance and development agencies have driven LASLA in SSA.[27] In 2009, the World Bank and Food Agriculture Organization (FAO) advised SSA to use its 202 million hectares of agricultural land to decrease poverty.[28] Although these groups have called for excellent land governance and natural resources management, they have intentionally supported LASLA by giving technical assistance to investors and influencing investment legislation.[29] The promotion of land leasing had profound development implications: The World Bank presented leasing at an expert conference on managing risk in financing agriculture in Johannesburg in April 2009. With its growing focus on Africa, IFC advised leasing facilities in Ghana, Tanzania, Rwanda, Madagascar, Senegal, Cameroon, DRC, Mali, and Ethiopia. IFC invested in the first African access leasing company, Ethiopia Access Leasing Company. In 2009, similar initiatives were to be created in the DRC, Madagascar, and other West African states. Leasing development and IFC promotion of leasing businesses must be critically scrutinized, considering the rapid acquisition of leased land.

Several governments promote "idle" or "underutilized" land for foreign investment. They say that residents would not lose access to land and that FDI and the production of "unproductive" land will benefit the country. One of the significant implications of the WBG's privatization efforts and IFC and FIAS' land law reforms endangers African communal property ownership and deprives farmers and pastoralists of the lands they have worked for millennia.[30] LASLA increases environmental scarcity and negatively impacts the socioeconomic and cultural communities in the countries affected. The involvement of prominent organizations must be reflected more in environmental conflict research.

ENVIRONMENTAL CONFLICT AND PUBLIC POLICY

Global governance shifts typically stem from paradigmatic concerns over the reconfiguration of corporate food regimes, international capitalist accumulation structures, and neocolonialism, as evidenced by the rising global influence of middle-income countries, the increasing exploitation of frontier resources to accommodate external food and energy security agendas, and financialization. Academics have critiqued land-grabbing activities in SSA for anticipating "win-win" outcomes. The critics say that such projects promote unsustainable corporate practices.[31] Findings from studies on the abundance of farmland and water, inexpensive initial costs, streamlined legal procedures, government effectiveness, and corruption support LASLA in SSA affect policy and falsely lead to the belief that good governance improves and sustains LASLA transactions.

Despite international commitments to small farmers, numerous reports show that food production programs have targeted better-off and more productive farmers than marginal farmers and non-farming rural groups like small farmholders and pastoralists. On the other hand, high corruption, low transparency and accountability, and governments' incapacity to negotiate better arrangements that advance their development agenda cause much of LASLA's harm to host nations. Governments should modify LASLA processes and enforce LASLA contract regulations to address these issues.[32]

GLOBAL ENVIRONMENTAL POLICY

Residents often own "idle" land without a title. The WBG's privatization activities and IFC and FIAS's land law reforms endanger African communal property ownership. Farmers and pastoralists have worked on this land for ages. Governments suggest that selling or leasing this land to private investors is ideal.[33] Investors would o't hire many people and build up manufacturing. They may create low-paying jobs and lousy working conditions if they pick the former. These have been the experiences based on colonialism's exploitation of natural resources and people. The label "unused" land is particularly disturbing. One of the justifications given for LASLA is that the land being grabbed is not used; it is idle. The fact is that these lands are pastoralist corridors, fallow space for soil regeneration, and provide access to scarce water sources, future generations, or local farmers' produce. We cannot call land unused because prosperous and growing economies do not have pastoralists or maintain age-old customs.[34]

Since mid-2008, UN agencies, countries, and civil society organizations have voiced concerns that land grabs jeopardize food security and land access for poor, vulnerable populations.[35] Land grabbing will continue threatening hunger and poverty unless IFC is held accountable for ensuring local people's rights in investment promotion operations. The Oakland report shows how IFC's guidance to governments enhances investor access to land markets, threatening local communities' land rights and food security.[36]

INVOLVEMENT OF INTERNATIONAL DEVELOPMENT AGENCIES

IFC and FIAS are WBG entities that provide developing nations with technical assistance and advice. IFC and FIAS promote foreign investors' land grabs to strengthen developing countries' "investment climates" and "business-enabling environments." IFC/FIAS facilitates land grabs by analyzing its "products" and providing governments with technical assistance and advisory services. The Oakland Report describes three case studies—Sierra Leone, Liberia, and Ethiopia—where IFC/FIAS advice services have led to regulatory and legislative reforms and expanded investor participation in land markets. The report indicates that investor access to developing nations' land markets affects local food security, displaces local populations, and violates IFC's Performance Standards and many UN Human Rights agreements. The Oakland Institute recommends investigating IFC/FIAS technical support and advice services and their "products." IFC must be held accountable for the land grabs its technical support and consulting services promote to defend the food security and livelihoods of the world's most vulnerable population, those in sub-Saharan African countries.

CONCLUSION

Recent research has focused on the social and environmental impact of increased investor demand for Africa's farms. The studies have spawned a vast body of literature that has shed light on drivers, trends, social and environmental repercussions, discursive implications, and global governance choices. Host nation's government factors, which enable and legitimize unsustainable farming investments, have been neglected even though large-scale farmland investments are associated with displacement, dispossession, and environmental degradation.[37] The apparent ease with which legislative safeguards are bypassed indicates more serious institutional concerns than land, environment, and investment law shortcomings. Conflicts of interest, co-optation, elite capture, insufficient inter-institutional coordination, capacity, and pro-investment ideology are examples. Sustainability impacts are similar, but institutional issues vary by country. Legislative approaches to environmental protection should consider ecological consciousness as a moral duty, enhance the economy, make society relatively affluent, and pick nonviolent solutions to resource shortages.[38]

Although concise and limited, the literature review in this chapter reveals the scope and depth of research on environmental conflicts, politics, and development in Africa. Some researchers use the human security theory to establish the interface between climate impacts, human insecurity, and local conflicts.[39] Additionally, bad governance, poor environmental policy on resource management, and local and international agencies' large-scale land grabbing increase the gap between the haves and the have-nots, leading to socioeconomic challenges and a recipe for violent conflicts. Research on environmental conflict, politics, and development in Africa suggests many approaches to address the root cause of the so-called environmentally induced conflicts by improving social stratification and reducing the gap between the poor and the rich.

NOTES

1. Krakowka, Amy Richmond, Natalie Heimel, and Francis A. Galgano. "Modeling Environmental Security in Sub-Saharan Africa." *Geographical Bulletin* 53, no. 1 (May 2012): 21–38. https://search.ebscohost.com/login.aspx?direct=true&db=a9h&AN=74447846&site=ehost-live.
2. Raleigh, Clionadh. "Political Marginalization, Climate Change, and Conflict in the African Sahel States." *OUP Academic*, March 1, 2010. https://academic.oup.com/isr/article/12/1/69/1796517.
3. Mumford, L. *Technics and Civilization*. New York: Harcourt Brace and World, 1934.
4. Dalby, S. *Environmental Security*. Minneapolis: University of Minnesota Press, 2002a.
5. Swatuk, Larry A., and A. Omari. "Environmental Security in Practice: Transboundary Natural Resources Management in Southern Africa." presentation in Section 31 (2004).

6. Peterson, V. Spike. *A Critical Rewriting of Global Political Economy: Integrating Reproductive, Productive and Virtual Economies*. London, and New York: Routledge, 2003.
7. Ibid.
8. Ibid.
9. Ibid.
10. Theisen, Ole Magnus, Helge Holtermann, and Halvard Buhaug. "Climate Wars? Assessing the Claim That Drought Breeds Conflict." *International Security* 36, no. 3 (2011): 79–106. http://www.jstor.org/stable/41428110.
11. Homer-Dixon, T., and J. Blitt, eds. *Ecoviolence: Links Among Environment, Population and Security*. Lanham, MD: Rowman and Littlefield, 1998.
12. Ibid.
13. Bereketeab, Redie. "Environmental Change, Conflicts and Problems of Sustainable Development in the Horn of Africa." *African & Asian Studies* 13, no. 3 (2014): 291–314. https://doi.org/10.1163/15692108-12341300.
14. Ibid.
15. Ibid.
16. Otiang'a-Owiti, G. E., S. Nyamasyo, E. EMalel, and R. Onyuro. "Impact of Climate Change on Human-Wildlife Conflicts in East Africa." *Kenya Veterinarian* 35, no. 2 (2011): 103–110.
17. Ibid.
18. Ibid.
19. Nkansah, Dwamena, Ernest, and Hyunsoo Yoon. "Why Is Sub-Saharan Africa an Attractive Destination to Foreign Land Grabbers? Evidence from Country Characteristics." *African Development Review* 34, no. 2 (2022): 280–292. https://doi.org/10.1111/1467-8268.12632.
20. Ibid.
21. Daniel, Shepard, and Anuradha Mittal. "The Great Land Grab: Rush for World's Farmland Threatens Food Security for the Poor." *The Oakland Institute,* 2009. https://www.oaklandinstitute.org/sites/oaklandinstitute.org/files/great-land-grab.pdf.
22. Ibid.
23. Ibid.
24. Ibid.
25. Anseeuw, W., Boche, M., Breu, T., Giger, M., Lay, J., Messerli, P., and Nolte, K.*Transnational Land Deals for Agriculture in the Global South: Analytical Report Based on the Land Matrix Database*. Rome, Italy: International Land Coalition, 2012.
26. Ibid.
27. Ibid.
28. Ibid.
29. Ibid.
30. Daniel, Shepard, and Anuradha Mittal. "The Great Land Grab: Rush for World's Farmland Threatens Food Security for the Poor." https://www.oaklandinstitute.org/sites/oaklandinstitute.org/files/great-land-grab.pdf.

31. Borras, S. M., Franco, J. C., Wang, C. "The Challenge of Global Governance of Land Grabbing: Changing International Agricultural Context and Competing Political Views and Strategies." *Globalizations* 10, no. 1 (2013): 161–179.
32. Daniel, Shepard, and Anuradha Mittal. "The Great Land Grab: Rush for World's Farmland Threatens Food Security for the Poor." https://www.oaklandinstitute.org/sites/oaklandinstitute.org/files/great-land-grab.pdf.
33. Ibid.
34. Ibid.
35. Ibid.
36. Ibid.
37. Schoneveld, George C. "Host Country Governance and the African Land Rush: 7 Reasons Why Large-scale Farmland Investments Fail to Contribute to Sustainable Development." *Geoforum* 83, (2017): 119–132.
38. Theisen, Ole Magnus, Helge Holtermann, and Halvard Buhaug. "Climate Wars? Assessing the Claim That Drought Breeds Conflict." *International Security* 36, no. 3 (2011): 79–106. http://www.jstor.org/stable/41428110.
39. Ethiane, Stanley, and Philani Moyo. "Climate Change, Human Insecurity and Conflict Dynamics in the Lake Chad Region." *Journal of Asian & African Studies* (Sage Publications, Ltd.) 57, no. 8 (2022): 1677–1689. https://doi.org/10.1177/00219096211063817.

REFERENCES

Ahmed, J. "The Farmer/Herder Conflict: What Way Forward?" *Daily Trust*, September 8, 2019. https://www.dailytrust.com.ng/the-herder-farmer-conflict-what-way-forward.html.

Anseeuw, W., Boche, M., Breu, T., Giger, M., Lay, J., Messerli, P., and Nolte, K. *Transnational Land Deals for Agriculture in the Global South: Analytical Report Based on the Land Matrix Database*. Rome, Italy: International Land Coalition, 2012.

Bereketeab, Redie. "Environmental Change, Conflicts and Problems of Sustainable Development in the Horn of Africa." *African & Asian Studies* 13, no. 3 (2014): 291–314. https://doi.org/10.1163/15692108-12341300.

Bello Farouk Umar1, and Kankiya, HAI. "Analysis of Farmer-Herder Conflicts in Zamfara State of Nigeria." *Journal of Agriculture and Environment* 7, no. 2 (Dec. 2011: 105–115): 105–115.

Borras, S.M., Franco, J.C., and Wang, C., 2013. "The Challenge of Global Governance of Land Grabbing: Changing International Agricultural Context and Competing Political Views and Strategies." *Globalizations* 10, no. 1, 161–179.

Dalby, S., *Environmental Security*. Minneapolis: University of Minnesota Press, 2002a.

Environmental Resources. "Environmental Encyclopedia." Encyclopedia.com. (December 20, 2022). https://www.encyclopedia.com/environment/encyclopedias-almanacs-transcripts-and-maps/environmental-resources.

Ethiane, Stanley, and Philani Moyo. 2022. "Climate Change, Human Insecurity and Conflict Dynamics in the Lake Chad Region." *Journal of Asian & African Studies* (Sage Publications, Ltd.) 57, no. 8: 1677–1689. https://doi.org/10.1177/00219096211063817.Krakowka.

Richmond, A., Heimel, N., and Galgano, F. A. "Modeling Environmental Security in Sub-Saharan Africa." *Geographical Bulletin* 53, no. 1 (May 2012): 21–38. https://search.ebscohost.com/login.aspx?direct=true&db=a9h&AN=74447846&site=ehost-live.

Homer-Dixon, T., and J. Blitt, eds. *Ecoviolence: Links Among Environment, Population and Security*. Lanham, MD: Rowman and Littlefield, 1998.

Mumford, L. *Technics and Civilization*. New York: Harcourt Brace and World, 1934.

Nkansah, Dwamena, Ernest, and Hyunsoo Yoon. "Why Is Sub-Saharan Africa an Attractive Destination to Foreign Land Grabbers? Evidence from Country Characteristics." *African Development Review* 34, no. 2 (2022): 280–292. https://doi.org/10.1111/1467-8268.12632.

Otiang'a-Owiti, G. E., S. Nyamasyo, E. EMalel, and R. Onyuro. "Impact of climate change on human-wildlife conflicts in East Africa." *Kenya Veterinarian* 35, no. 2 (2011): 103–110.

Peterson, V. Spike. *A Critical Rewriting of Global Political Economy: Integrating Reproductive, Productive and Virtual Economies*. London, and New York: Routledge, 2003.

Raleigh, Clionadh. "Political Marginalization, Climate Change, and Conflict in the African Sahel States." *OUP Academic*, March 1, 2010. https://academic.oup.com/isr/article/12/1/69/1796517.

Daniel, Shepard, and Anuradha Mittal. "The Great Land Grab: Rush for World's Farmland Threatens Food Security for the Poor." *The Oakland Institute,* (2009). https://www.oaklandinstitute.org/sites/oaklandinstitute.org/files/great-land-grab.pdf.

Swatuk, Larry A. "Environmental Security in Practice: Transboundary Natural Resources Management in Southern Africa." *Presentation in Section* 31, (2004). http://www.afes-press.de/pdf/Hague/Swatuk_environmental_security.pdf.

Schoneveld, George C. "Host Country Governance and the African Land Rush: 7 Reasons why Large-scale Farmland Investments Fail to Contribute to Sustainable Development." *Geoforum* 83, (2017): 119–132.

Theisen, Ole Magnus, Helge Holtermann, and Halvard Buhaug. "Climate Wars? Assessing the Claim That Drought Breeds Conflict." *International Security* 36, no. 3 (2011): 79–106. http://www.jstor.org/stable/41428110.

Chapter 5

Climate and Environmental Changes

A Nexus of Environmental Changes in Sub-Saharan Africa

Ahmad Hamza Abdullahi and Abdulkadir Maharazu

Throughout human history, states and nations aspire for the protection of their borders against foreign aggression. However, after the Cold War, the concept of national security across the universe was expanded to include protection of surrounding environments, including natural and human resources. The rate at which people struggle for scarce environmental resources and have divergent beliefs and attitudes and other institutional factors lead to what is now known as environmental conflicts. Environmental conflicts generally become one of the major issues that challenge local, regional, national, and global coexistence of humanity. Climate change has become one of the major factors that threaten to deter the progress made in reducing environmental and other related conflicts. It is estimated that every year, about 25 million people temporarily flee their homes due to weather-related catastrophes; given the rising population in the developing world from 6 billion to 9.5 billion by 2050,[1] the crisis is expected to increase tremendously. In fact, climate change and its negative consequences on the available environmental resources such as agriculture, vegetation, land and water resources, economic activities, and other natural disasters invariably add fuel to environmental conflict and instability.

Generally speaking, climate change is having more impact on the people and environment in small islands, landlocked countries, and arid and semiarid areas of the world such as SSA. The rate at which climate has been affecting the region could add risks to and render people prone to violent conflict over environmental resources. Climate change in the region is expected to push millions of people into poverty if no concrete developmental action is taken because it affects the quantity and quality of environmental resources. SSA experiences considerable environmental changes due to climate change and human activities. For instance, the rate at which land is degraded due to natural hazards such as floods and desertification in the region is on the increase. These environmental changes may be one of the reasons that brought about more conflict in the region, such as the conflict between Fulani herders and farmers. These, therefore, imply that climate and environmental changes can reverse the progress made by governments and various organizations such as the Intergovernmental Panel on Climate Change (IPCC) in controlling environmental conflicts in SSA. Climate and environmental change are among the drivers of poverty and inadequate environmental resources in SSA, which consequently becomes a threat to people's survival in their original environment. This chapter argues that climate and environmental changes in SSA are, in part, responsible for environmental conflicts in the region.

Conceptual frameworks have previously been developed by researchers in contexts of the influence of climate and environmental change to the existence of environmental conflict, such as the one developed by Aaron in Climate Change and Conflict in Nigeria: A Basic Casual Mechanism.[2] The framework was basically developed by looking into how climate

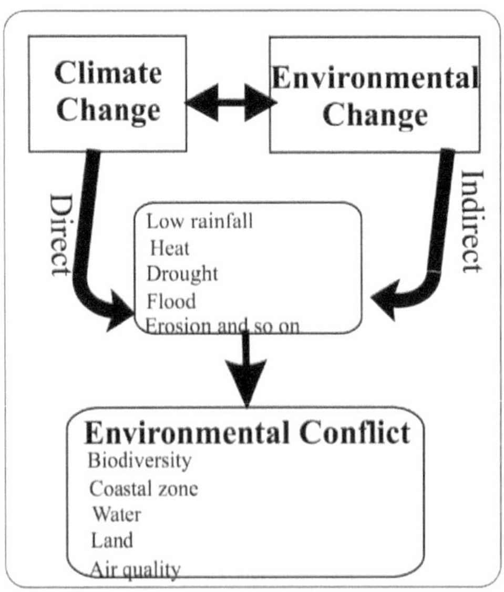

Figure 5.1 The Nexus between Climate Change, Environmental Change, and Environmental Conflict in SSA. *Source*: Created by the author based on information from Sayne Aaron, "Climate Change Adaptation and Conflict in Nigeria." United States Institute of Peace, special report, 2011: 1–274, 2301 Constitution Ave., NW, Washington, DC 20037 202.457.1700, fax. 202.429.6063 http://www.usip.org.

change contributes to resource shortage, the possible secondary impacts of the shortages, and how these effects increase the rate of conflicts in Nigeria. This framework is, therefore, adapted by this chapter because it explicitly integrates major factors that relate climate change and the existence of conflict. The modified framework (from the work of Aaron) presents the pathways by which climate and environmental change influence the existence of environmental conflict in SSA (figure 5.1). It was also conceptualized based on the argument that climate and environmental change are linked with each other; while climate change directly resulted in various environmental stress such as shortage of rainfall, heat, drought, and flooding, environmental change on the other hand is indirectly related to them. The framework further indicates that the resultant environmental stress could fuel environmental conflict such as conflict on biodiversity and water in SSA.

THE REGION OF THE SUB-SAHARAN AFRICA

Sub-Saharan African region can be demarcated based on the UN Children's Fund (UNICEF)'s definition as consisting of the whole of the African continent but with the exclusion of North Africa.[3] The region is bordered with Northern African countries to the north, the Red Sea in the northeast, the Indian Ocean in the east and the Atlantic Ocean in the southern and western parts of the continent (figure 5.2). SSA is one of the areas of the world reportedly with fast-growing population. The region is projected by the UN Department of Economic and Social Affairs (UNDESA) to reach 2 billion people by 2050 and nearly 4 billion by 2100.[4] Network of African Science Academies (NASAC) also reports that the region is highly dynamic in terms of great climatic, ecological, and cultural diversity.[5] Away from the Equator, north and south in the region, there is high variability in temperature and rainfall. IPCC reported multiple and compound climate-related risks in the region, which increases global warming between 1.5°C and 2°C and raises the frequency and intensity of natural disasters.[6]

Agriculture is the major source of economic activities in the SSA; however, climate and environmental changes highly affect the system, which will, most likely, increase the rate of conflict in the region. The nature of the SSA environment shows that most of its parts have generally low-to-moderate aquifer productivity, with some very high systems located in the part of Congo, western Guinea, and Gambia. These negatively affect water supply in the region and, therefore, increase the rate of potentials to environmental conflict.

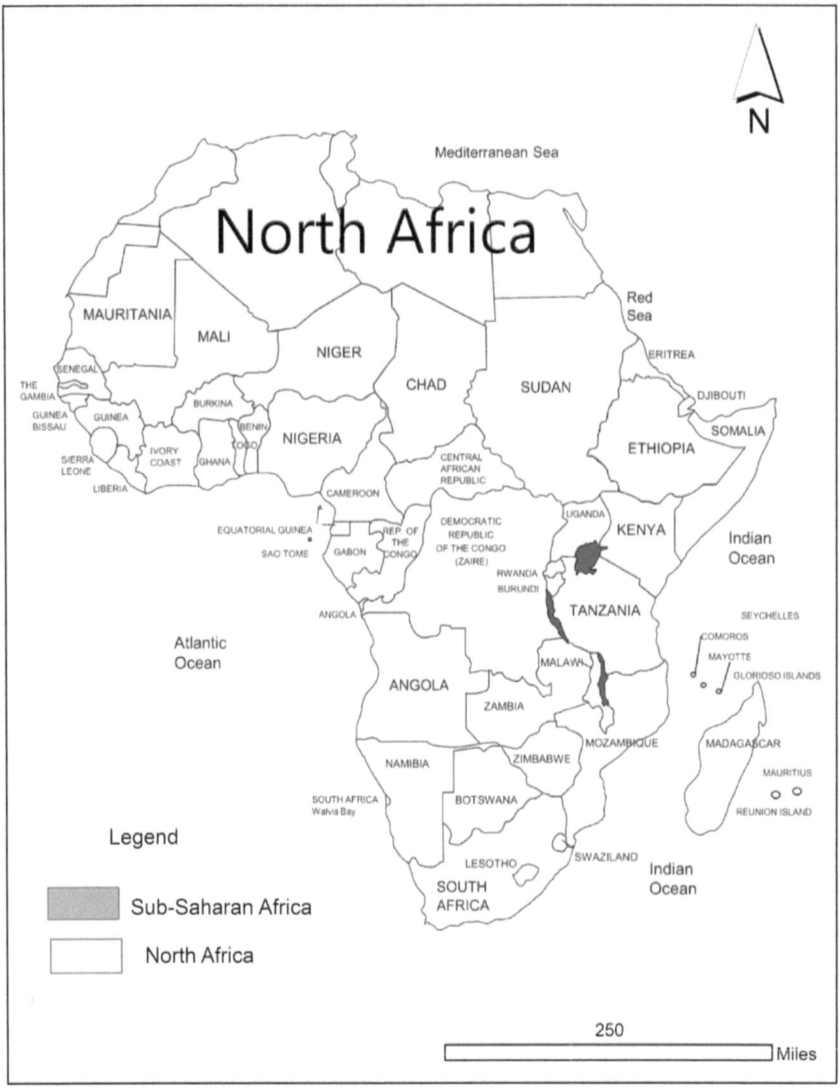

Figure 5.2 Sub-Saharan Africa. *Source*: Created by the author based on information from UNICEF Child Alert (2021, cover page). https://www.unicef.org/esa/media/8486/file/UNICEF-Sub-Saharan-Africa-Child-Alert-2021.

CLIMATE CHANGE IN SUB-SAHARAN AFRICA

Climate simply refers to the description of the mean and variability of relevant atmospheric elements, which include temperature, precipitation, atmospheric pressure, wind, and solar radiation. Basically, climate is determined by average values of the climatic elements over several years and has been explained by climatologists based on their various characteristics. These atmospheric parameters are of primary importance for the life of all ecosystems on our planet; however, people and environment do experience their negative consequences.

It is the changes in the natural characteristics of the elements at various temporal and spatial scales that are called climate change. As a more precise form, climate change may simply be referred to as the variation in the statistical distribution of average weather conditions over a prolonged period within a particular area.[7] It usually occurs due to natural processes and human activities such as deforestation and rising fossil fuel use that lead to the increase in the level of GHGs. The GHGs as contained in the UN Framework Convention on Climate Change include carbon dioxide (CO_2), methane (CH_4), and nitrogen dioxide (N_2O). An increase in the level of these gases results in the increase in the amount of heat from the sun withheld in the Earth's atmosphere, which consequently brings about global warming and higher changes in climatic conditions after a long period of time.

Currently, the pace and scale of this global warming and climate change, generated by the emission of GHGs, have become a worldwide phenomenon of concern. Although African countries produce fewer GHGs when compared with developed countries, the region, especially countries referred to as fragile states, recorded a higher index of climate change more than the Western world. This is especially in the light of the fact that many researchers see the region as the most vulnerable to the rate and impacts of climate change, mainly owing to the nature of and inadequate adaptive capacity. The region's vulnerability to climate change and other natural disasters may also be linked to the increasing level of temperature, structural weaknesses, poor government support, and lack of institutional basic functions. The months of April, May, and June in SSA, for instance, are predicted to be extensively hotter, amounting to more than five sigma, with especially strong increases in tropical West Africa between 2071 and 2099.[8] Central and Eastern Africa, according to NASAC, are expected to experience major impacts that will be driven by changes in seasonal rainfall and extreme rainfall and drought events. On the other hand, large parts of western, northern, and southern Africa will likely experience significant reductions in overall rainfall amounts, becoming hotter and drier. As a whole, for decades the entire region of SSA has been disproportionately impacted by climate change through an increase or decrease in rainfall, an increase in temperature, and extended droughts.

Climate change, through rising temperatures and changes in precipitation, causes droughts, desertification, rising sea levels, and higher vapor pressure, which increase the rate of flooding and storms such as hurricanes and tropical cyclones. The African continent, especially SSA, has been reported as a *hotspot* for climate change and is exposed to various climate change challenges, including extreme weather events, drought, diseases, and water scarcity, and has low coping capacity.[9] These impacts are greater, especially within the most vulnerable populations with high poverty and low coping ability. More efforts are, therefore, required to investigate the situation, which include the need to reconcile the culture of the community with other mechanisms that limit the progress of climate change impacts in the region.

Generally, climate change has been connected to changes in the nature and structure of the SSA as it has upset physical, chemical, and biological characteristics of the environment. The impacts of climate change on the environment may be categorized into biophysical and socioeconomic. Based on these, the focus of this chapter is on biophysical impact, which has to do with effects of changes on the physical environment such as drought, vegetation resources, water stress, and flooding.

ENVIRONMENTAL CHANGE IN SUB-SAHARAN AFRICA

The environment may simply be considered as our surroundings, which include weather and climate, water system, and the rich diversity of life in all its forms. Thus, the environment as an important component of human life has always been undergoing many challenges and change, and hence, the term environmental change. Environmental change may be referred to as destabilizing interference in the ecosystem's equilibrium. The world as a larger human environment experiences numerous changes through the influence of various factors including human-induced climate change. This prompted the call for environmental sustainability, which appears as the key theme in the post-2015 sustainable development agenda.[10] In this regard, governments and organizations in Africa, especially the SSA, are expected to be in the forefront for the campaign for environmental sustainability since they are most likely affected by the negative consequences of environmental changes. Although SSA is blessed with a wide range of ecosystems, the region's land resources continue to be degraded through natural and anthropogenic activities such as overgrazing, deforestation, agricultural mismanagement, fuelwood consumption, and urbanization. Numerous empirical studies show that these activities produce various environmental changes like flooding, erosion, pollution, as well as drought. According to the African Development Bank, European Commission, and the Food and Agriculture Organization (ADB/EC/FAO), for example, Sudan, Zambia, and the DRC alone account for about 44 percent of Africa's deforestation.[11]

Other environmental changes related to climate disturbances in SSA include flood such as was experienced by many coastal areas of African countries on the Indian Ocean coastline during the tsunami of 2008, where 138,000 people lost their lives and properties worth millions of dollars were lost.[12] This and other subsequent occurrences snowballed into various forms of environmental challenges across many SSA countries. For example, the coastal areas of southwestern Nigeria are vulnerable to floods, having a weighted impact index of 3.1 to 4.4 out of the maximum possible score of 5.0.[13] A similar situation is lamented by Grote and Warner,[14] who found a reduction of forest cover, land degradation, coastal erosion, and a reduction in the volume of Lake Volta in Ghana. These researchers also reported outbreaks of severe floods in Mozambique during the years 1997, 1998, 2000, and 2003.

Another key environmental challenge in the region is drought, which attracts the use of drought-tolerant crops for positive environmental impacts. In 1977, for instance, Senegal was affected by land degradation and drought that affected many people. Others include the droughts of northern Sudan and western Sahel; the droughts of Mali and Ethiopia; and the severe drought of Niger in the mid-1970s up to the 1980s, which affected more than 3.5 million people.[15] Other related environmental challenges across SSA countries include: significant increase in the rate of soil degradation along Lake Chad, shrinking of the water body, and the reductions in soil quality. In fact, flooding, water stress, and reduction in the available vegetation resources in several parts of SSA such as Republic of Benin, Burkina Faso, Ethiopia, Ghana, Niger Republic, Mali, western Botswana, and western Sahara are the major symbols of environmental changes in the region. Therefore, the scarcity of these environmental resources in conjunction with other economic stress contributes to the region experiencing high numbers of environmental conflicts.

ENVIRONMENTAL CONFLICT IN SUB-SAHARAN AFRICA

The global environment is evidently rich and resource endowed. Thus, the incorporation of resources in a definition of environmental conflicts appears inevitable. The relationship between the key concepts of environment and conflicts can be defined in terms of the availability or otherwise of these environmental resources. Accordingly, environmental conflict may basically be referred to as disputes between two individuals, social groups, or communities linked with claims over environmental resources such as land, water, and vegetation, which may tend to or turn into an open fight. It may also include disputes over the contamination of the environment through the discharge of unwanted or harmful substances into the environment. For a considerable amount of time now, SSA faces some severe environmental difficulties like deforestation, soil erosion, pollution, desertification, wetland, and general ecosystem depletion, which result in environmental conflicts between individuals or groups.

Environmental conflicts in the Republic of Niger, for instance, often occur in the process of land acquisition as land in the country increasingly becomes scarce and expensive to acquire especially among rural farmers. In contrast, this gives opportunity to rich, urban-based merchants to possess land. This prompts the outbreak of conflicts between the incoming rich people who buy land and the local inhabitants, who need land for their economic activities, more importantly agricultural activities. Similarly, an environmental conflict occurs between Senegalese and Mauritanians and between Mauritanian social groups as a result of local tensions over landed resources associated with irrigation farming. In the Niger Republic, droughts challenge social and political coexistence between communities, whereby pastoralists and farmers among the Fullas and Toubou communities have engaged in armed conflicts over the limited natural resources for several years.[16] In Mozambique, environmental change such as environmental deterioration and soil erosion increases the rate of environmental conflicts over the access to natural resources.[17]

EFFECTS OF CLIMATE AND ENVIRONMENTAL CHANGES ON ENVIRONMENTAL CONFLICTS IN SSA

Climate change leads to various forms of natural hazards and environmental stresses that bring about environmental change and subsequently lead to social and political impacts including conflicts, within an environment. In other words, climate change influences the occurrence of environmental change and vice versa. Similarly, many researchers establish a direct relationship between climate change, environmental change, and environmental conflict. An example can be seen in the research conducted by Khan, Ozturk, Gul, and Ahmed,[18] who report a positive correlation between community conflict and climate change. Conflict, according to ecologists, may be defined as a phenomenon that brings about competition among organisms on scarce but essential resources, normally within an environment. Thus, conflict occurs when people interact with each other to attain scarcely limited resources. This, therefore, implies that environmental events such as climate and environmental changes result in conflicts due to the need to leave disaster-affected locations or contest for the available limited resources as an adaptive mechanism.

Desertification, drought, and flood are among the major outcomes of climate and environmental changes in SSA. These disasters often increase poverty, human suffering, and the dislocation of people through migration and create deficits in the availability of natural resources in the region. Hence, there are outbreaks of conflict on biodiversity, coastal zones, water resources, land resources, air quality, and many more.

Biodiversity conflicts may be regarded as conflicts that exist between humans, wildlife, vegetation resources, and/or other components of the biosphere. Broad and ample evidence such as the study conducted by Agyemang[19] shows that whenever there is a conservation process or interaction between humans and the environment such as wildlife or vegetation resources, conflicts might be inevitable. This type of conflict is mainly triggered by climate and environmental change and these often affect not only disputing parties, but snowballs to produce collateral damages. Indeed, the complex nature and delays in dealing with conservation and restitution reform in SSA might be among the leading factors to conflict in the region. The Fulani and farmers' conflicts in various parts of the region such as those in the southwestern and southeastern Nigeria are common examples. Others include the 1997 conflict between Fulani herders and farmers in Niger; the clashes between Fulani herders and Djerma farmers in the arrondissements of Téra and Birni N'Gaouré in Niger; and similar crises where the Fulanis fled Nigeria to Cameroon were all associated with climate and environmental changes.

Most of SSA countries have coastlines. The issues of coastal ecosystem change and coastal development are undeniably two key causes of coastal zone conflicts. Coastal zones are a foremost attraction to many sectors: commerce, seafood, industry, offshore waters, oil exploitation, and ocean dumping. Others include mineral mining, fish harvesting, holiday making and tourism, home seekers, the military, and a variety of other industries. These ultimately make the regions fertile conflict grounds between many coastline users. The coastal zones and oceans of SSA constitute a vast storehouse of food, energy, and mineral resources, which could provide for sustainable development. The prospects for economic fortune and better life in this zone could, therefore, lead to skirmishes and sometimes avoidable conflicts in these areas. This includes impacts on water security in the region, which create unavoidable pressure on people's life and food insecurity that may lead to conflict.

The riverine or coastal zones of SSA experience frequent conflict partly due to climate change, which increases the risk of food insecurity in the region. Climate and other extreme weather events in SSA are found to have affected the rate at which pastoralists migrate to the coastal zone of Africa in search of green pastures. The host communities in the coasts of Ethiopia,

Somalia, and Eritrea often entered into conflict due to encroachments onto their land. The Government of Finland, in collaboration with UN Environmental Protection and Priority Actions Programme/Regional Activity Centre also reported nasty experiences and increasing pressures in the coasts of Africa over the use of resources that often resulted in unwarranted conflicts.[20]

Similarly, climate and environmental change are important factors that determine outbreaks of conflict related to water resources in many parts of the world, including the Sub-Saharan African region. Inadequate water resources lead to higher risk of conflict in river-sharing dyads, for instance due to the fact that many people come together. Therefore, the higher the number of people congregate in one place over a given resource such as water, the greater the risk of violence. However, upstream and downstream communities are more at risk in this situation. Whatever the case may be, this risk may be reduced with the increase in the size of basins. The Mbororo community, for instance, experiences higher intercommunal conflicts due to the reduction of water in Lake Chad. In Nigeria, shortage of water also increases the rate of conflict, especially in the Sahel region of the country. It was also observed that in Sudan, inadequate water increases the risk of communal conflict. In fact, in the 1990s, the World Bank identified the region of South Africa as one in which water scarcity could lead to potential violent conflicts.[21]

Conflicts over natural resources such as land are bound to happen especially in agrarian societies of SSA. The risk to occurrences of conflict becomes higher with increase in climate and environmental changes through the scarcity of renewable resources as arable land, water, and vegetation resources; this is more evident especially between migrants and residents. Although SSA is endowed with many natural resources, climate and environmental changes affect ways through which these resources can be useful to people. On the contrary, they serve as the causes of numerous and complex conflicts in the region. Land resource has multiple functions in SSA: a commodity, a factor of production, as family or community property, a capital asset, and a source of cultural identity and/or citizenship. However, scarcity of this resource endangers the life of the people through the outbreak of tension and conflict. Perhaps, two major factors are responsible for land conflicts in SSA, and these are high population growth rate and natural disaster. SSA has a population of over one billion, and this will double in the foreseeable future. These countries (SSA) would account for more than half of the growth of the world's population between 2019 and 2050.[22]

Positive variation in land use for agricultural and urban purposes significantly increases both the probability and the number of communal conflicts in SSA. For example, a persistent drought smashed the Sahel in the late 1970s, North–East Africa in the 1970s and 1980s, and Southern Africa in early 1980s and 1990s, leading to severe conflict between migrant farmers and inhabitants of various communities over land resources. The region experiences high reduction in the available land resource for agricultural and rising in the urban population due to climate and environmental change, which significantly increases both the probability and the number of communal conflicts. In Kenya, for instance, herders move beyond their territory in search for pasture and water for their livestock, leading to conflict between herders and farmers. There are similar incidences in Nigeria because communities and individuals encounter conflicts and other social violence owing largely to the shortage of land for agriculture.

Air quality is usually determined by the level of pollutants in the atmosphere such as SOx, NOx, PM2.5, or PM10. Studies reveal how climate and environmental change affect the quality of air through changes in the speed of chemical processes in the atmosphere, the altitude where pollutants mix, and changes in air flow which affect the transport of pollutants. For example, an analysis on chemical transport models (CTMs) driven by General Circulation Model show that "climate change alone will increase summertime surface ozone in polluted regions by 1–10 ppb over the coming decades, with the largest effects in urban areas and during pollution episodes."[23] Another study reported that "the presence of the soot constitutes hazards of air pollution and thus has notable effects on the air quality, visibility and contributes immensely to the changing climate impacts in Port Harcourt."[24] There are six major ways through which war and other human activities affect the quality of air, dust from reconstruction sites, increased emissions from transport, emissions from burning fossil fuel, forest fires, damaged industrial facilities, and emissions from exploded weapons and ammunitions.

The SSA is characterized by rapid growth of the population, urbanization, traditional system of agriculture, strives for industrialization, and overall economic growth. These processes usually result in environmental challenges, particularly air quality, which most often lead to environmental conflict. However, many factors, including poverty, lackadaisical attitude, and lower level of education among others, may affect the rate at which people dispute or take their cases to the appropriate authority on the contamination of the atmosphere due to the discharge of pollutants. Hence, there is a lack of sufficient studies that evaluate the relationship between air quality and environmental conflict in the region. As a whole, social problems, which include poor working conditions, lack of social security, disregard for the rule of law, poor freedom of expression guarantees, widespread corruption, and poverty, diminish the concern that citizens and politicians show toward air quality in SSA. Therefore, many studies related to air quality in SSA are epidemiological research on indoor air pollution from household use of cooking fuels.

CONCLUSION

The chapter reviews studies and establishes the relationship between climate and environmental changes as evidently found in many parts of SSA. While climate change has been directly related with the occurrence of various environmental challenges such as drought, flooding, and desertification, environmental change has been indirectly related to most of these issues. The study further indicates that these twin phenomena to some degree influence the occurrence of environmental conflict in the region such as conflict on biodiversity, water, and land resources. The study finally calls for increasing public awareness of climate and environmental changes, implications, and adaptation strategies. It also recommends for the inclusion of knowledge of climate and environmental changes in the region's education policy.

NOTES

1. Federal Ministry for the Environment, Nature Conservation and Nuclear Safety, "Climate Change and Conflict." *Environmental Policy* 6(2002): 2. Public Relations Division, Alexanderplatz, Berlin, German. http://www.bmu.de.

2. Sayne Aaron, "Climate Change Adaptation and Conflict in Nigeria." United States Institute of Peace, special report 274(2011): 3. Constitution Ave., NW, Washington, DC 20037 202.457.1700, fax. 202.429.6063. available at http;//www.usip.org.

3. United Nations Children's Fund Child Alert, "Sub-Saharan Africa: Growing Up in Crisis in a World of Opportunities." *The Lasting Impact of Covid-19 on Children,* 2021: Cover page. https://www.unicef.org/esa/media/8486/file/UNICEF-Sub-Saharan-Africa-Child-Alert-2021.pdf.

4. United Nations, Department of Economic and Social Affairs (UNDES), Population Division, "World Population Prospects: The 2012 Revision." *Comprehensive Tables* I(2013.): 97. ST/ESA/SER.A/336.

5. Network of African Science Academies, "Climate Change Adaptation and Resilience in Africa." *Recommendations to policymakers,* 2015, 6–7. https://nasaconline .org/index.php/2016/09/12/climate-change-adaptation-and-resilience-in-africa-recommendations-to-policymakers-2

6. Intergovernmental Panel on Climate Change, "Summary for Policymakers." In *Global Warming of 1.5°C. An IPCC Special Report on the Impacts of Global Warming of 1.5°C above Pre-industrial Levels and Related Global Greenhouse Gas Emission Pathways, in the Context of Strengthening the Global Response to the Threat of Climate Change, Sustainable Development, and Efforts to Eradicate Poverty,* edited by V. Masson-Delmotte, P. Zhai, H. O. Pörtner, D. Roberts, J. Skea, P. R. Shukla, A. Pirani, W. Moufouma-Okia, C. Péan, R. Pidcock, S. Connors, J. B. R. Matthews, Y. Chen, X. Zhou, M. I. Gomis, E. Lonnoy, T. Maycock, M. Tignor, and T. Waterfield, 2018: 12. World Meteorological Organization, Geneva, Switzerland, ISBN 978-92-9169-151-7.

7. Michael Eagles Ikehi, "Impacts of Climate Change on Animal and Crop Production in the Niger Delta Region of Nigeria." *Research Journal of Agriculture and Environmental Management* 3(10), 2014: 528, http://www.apexjournal.org.

8. Olivia Serdeczny et al., "Climate Change Impacts in Sub-Saharan Africa: From Physical Changes to their Social Repercussions." *Regional Environmental Change* 2(63), 2016: 3. Springer. ·https://doi.org/10.1007/s10113-015-0910-2.

9. Kheira Tarif, "Climate Change and Violent Conflict in West Africa: Assessing the Evidence." *SIPRI Insights on Peace and Security* 3(2022): 1. www.sipri.org.

10. Uduak Akpan et al., "Effect of Technology Change on CO2 Emissions in Japan's Industrial Sectors in the period 1995–2005: An Input-Output Structural Decomposition Analysis." *Environmental and Resources Economics* 61(2), 2015: 173. dx.doi.org/10.1007/s10640-014-9787-1.

11. African Development Bank, European Commission, and the Food and Agriculture Organization of the United Nations, "Forestry Outlook Study for Africa: Sub Regional Report – Southern Africa." 2003: 4. Rome, ftp://ftp.fao.org/ docrep/ fao/005/y 8672e/y8672e00.pdf.

12. Andrew Collins, et al, "World Disasters Report 2009-Focus on Early Warning, Early Action." *Project Report*. 2009: 12. International Federation of Red Cross and Red Crescent Societies. http://www.ifrc.org/Global/WDR2009-full.pdf.

13. Ibidun Adelekan and Fregene Tosan, "Vulnerability of Artisanal Fishing Communities to Flood Risks in Coastal Southwest Nigeria." *Climate and Development* 7(4), 2015: 327. https://doi.org/10.1080/17565529.2014.951011.

14. Ulrike Grote and Koko Warner, "Environmental Change and Migration in Sub-Saharan Africa." *International Journal of Global Warming* 2(1), 2010: 29, 32.

15. Ulrike Grote and Koko Warner.

16. *Brigitte*, Thébaud. 2002. "Sécheresse et conflit pastoral au Niger oriental." *Foncier pastoral et gestion de l'espace au Sahel* 103–172. Paris: Karthala/Nordic Consulting Group.

17. Bowen, M. L, Chilundo, A. and Tique, C. A. 2003. "Social Differentiation Farming Practices and Environmental Change in Mozambique." In *African Savannas. Global Narratives and Local Knowledge of Environmental Change,* edited by Bassett T. J. and Crummey D., 225–247. Portsmouth: James Currey/Heinemann.

18. Muhammad Ajmal Khan, Munir Ozturk, Bilquees Gul, and Muhammad Zaheer Ahmed, "Halophytes for Food security in Dry lands." *The Boulevard*, Langford Lane, Kidlington, 2016: 113, Oxford OX5 1GB 225 Wyman Street, Waltham MA 02451 Elsevier Inc.

19. Agyemang, Emmanuel, 2017. Farmer-Herder Conflict in Africa: An Assessment of the Causes and Effects of the Sedentary Farmers-Fulani Herdsmen Conflict. A Case Study of the Agogo Traditional Area, Ashanti Region of Ghana. Unpublished Master's thesis, Department of Education, University of Agder. 1–201.

20. Government of Finland, United Nations Environmental Protection and Priority Actions Programme/Regional Activity Centre, "Assessment of Integrated Coastal Management in Africa." *Split, Croatia, Priority Actions Programme*, 1998: 5 Regional Activity Centre, UNEP/MAP.

21. Larry A. Swatuk, "Environmental Security in Practice: Transboundary Natural Resources Management in Southern Africa." *Paper prepared for presentation in Section 31 of the Pan-European Conference on International Relations* 2004: 16. The Hague.

22. United Nations, "World Population Prospects 2019 Highlights." (ST/ESA/SER.A /423). Department of Economic and Social Affairs, Population Division (2019): 1. www.un.population.org.

23. Daniel J. Jacob, and Darrell A. Winner, "Effect of Climate Change on Air Quality." *Atmospheric Environment* 43, 2009: 51–63. https://doi.org/10.1016/j.atmosenv.2008.09.051.

24. Ajiere, Susan, and Nwaerema, Peace. "Climate Change and Air Pollution: Implication for Human Health and Environment in Rivers State." *International Journal of Research and Innovation in Social Science* (IJRISS) IV(V), 2020: 276–279. https://www.researchgate.net/publication/342000441.

REFERENCES

Aaron, Sayne. *Climate Change Adaptation and Conflict in Nigeria.* United States Institute of Peace, special report, 2011: 1–274, 2301 Constitution Ave., NW, Washington, DC 20037 202.457.1700, fax. 202.429.6063 http://www.usip.org.

Abera, Asmamaw, Johan Friberg, Christina Isaxon, Michael Jerrett, Ebba Malmqvist, Cheryl Sjöström, Tahir Taj, and Ana Maria Vargas8. "Air Quality in Africa: Public Health Implications." *Annual Review of Public Health* 42, 2021: 193–210. doi:10.1146/annurev-publhealth-100119-113802.

Adelekan, Ibidun, and Fregene Tosan. "Vulnerability of Artisanal Fishing Communities to Flood Risks in Coastal Southwest Nigeria." *Climate and Development* 7(4), 2015: 322–338. doi:10.1080/17565529.2014.951011.

Afifi, Tamer. "Economic or Environmental Migration? The push factors in Niger." *International Migration* 49(s1), 2011: e95–e124. doi:10.1111/j.1468-2435.2010.00644.x.

African Development Bannk, European Commission and the Food and Agriculture Organization of the United Nations. *Forestry Outlook Study for Africa: Sub Regional Report – Southern Africa.* Rome, 2003, ftp://ftp.fao.org/ docrep/fao/005/y8672e/y8672e00.pdf.

Ahmad, Nafees, S. K. Shahnawaz, Muzafar Husain, Sajid Qamar, and Zaid Alam. "Food Insecurity: Concept, Causes, Effects and Possible Solutions." *IAR Journal of Humanities and Social Science* 2(1), 2021: 105–113. doi:10.47310/jiarjhss.v02i01.016

Akpan, Uduak, Ovunda Green, Subhes Bhattacharyya, and Salisu Isihak. "Effect of Technology Change on CO2 Emissions in Japan's Industrial Sectors in the period 1995–2005: An Input-Output Structural Decomposition Analysis." *Environmental and Resources Economics* 61(2), 2015: 165–189. doi:10.1007/s10640-014-9787-1.

Anderson, Weston, Charles Taylor, Sonali McDermid, Elisabeth Ilboudo-Nébié, Richard Seager, Wolfram Schlenker, Fabien Cottier, Alex de Sherbinin, Dara

Mendeloff, and Kelsey Markey. "Violent Conflict Exacerbated Drought-related Food Insecurity Between 2009 and 2019 in sub-Saharan Africa." *Nature Food,* 2021: 1–16. Springer. doi:10.1038/s43016 021 00327 4.

Asongu, Simplice, Ghassen El Montasser, and Hassen Toumi. "Testing the Relationships between Energy Consumption, CO2 Emissions, and Economic Growth in 24 African Countries: A Panel ARDL Approach." *Environmental Science and Pollution Research* 23(7), 2016: 6563–6573. doi:10.1007/s11356-015-5883-7.

Batterbury, Simon. "Adapting to Drought in the West African Sahel." In *Natural Disasters and Adaptation to Climate Change,* edited by Sarah Boulter, Jean Palutikof, David Karoly, and Daniela Guitart, 1–14. Cambridge University Press, 2013. doi:10.1017/CBO9780511845710.020.

Coker, Eric and Samuel Kizito. "A Narrative Review on the Human Health Effects of Ambient Air Pollution in Sub-Saharan Africa: An Urgent Need for Health Effects Studies." *International Journal of Environmental Research and Public Health* 15(427), 2018: 1–15. doi:10.3390/ijerph15030427.

Collins, Andrew, Nick Maunder, Michele McNabb, Anne Moorhead, and Maarten van Aalst. "World Disasters Report 2009-Focus on Early Warning, Early Action." *Project Report.* 2009. International Federation of Red Cross and Red Crescent Societies. http://www.ifrc.org/Global/WDR2009-full.pdf.

Eagles, Ikehi Michael. "Impacts of Climate Change on Animal and Crop Production in the Niger Delta Region of Nigeria." *Research Journal of Agriculture and Environmental Management* 3(10), 2014: 528–537, http://www.apexjournal.org.

Federal Ministry for the Environment, Nature Conservation and Nuclear Safety. "Climate Change and Conflict." *Environmental Policy* 6(11055), 2002: 1–146. Public Relations Division, Alexanderplatz, Berlin, German, http://www.bmu.de.

Gleditsch, Nils Petter, Kathryn Furlong, Håvard Hegre, Bethany Ann Lacina, and Taylor Owen. "Conflicts over Shared Rivers: Resource Wars or Fuzzy boundaries?" *Political Geography* 25(4), 2006: 361–382.

Government of Finland, UNEP and PAP/RAC. "Assessment of Integrated Coastal Management in Africa." *Split, Croatia, Priority Actions Programme* 1998: viii + 72. Regional Activity Centre, UNEP/MAP.

Grote, Ulrike, and Koko Warner. "Environmental change and migration in Sub-Saharan Africa." *International Journal of Global Warming* 2(1), 2010: 17–47.

Gunn, Lee. "National Security and the Accelerating Risk of Climate Change." Elementa: Science of the Anthropocene, 5(2017): 1–30. doi:10.1525/elementa.227.

Ibe, A. Chidi. "The Coastal Zone and Oceanic Problems of Sub-Saharan Africa." In *Economic, Social, and Environmental Change in Sub-Saharan Africa: Sustaining the Future. Economic, Social, and Environmental Change in SubSaharan Africa,* edited by J. Benneh, W.B. Morgan, and J.I. Uitto. Tokyo: United Nations University, 1998. http://hdl.handle.net/1834/273.

Intergovernmental Panel on Climate Change (IPCC). "Summary for Policymakers." In *Managing the Risks of Extreme Events and Disasters to Advance Climate Change Adaptation,* edited by C.B. Field, V. Barros, T.F. Stocker, D. Qin, D.J. Dokken, K.L. Ebi, M.D. Mastrandrea, K.J. Mach, G.-K. Plattner, S.K. Allen, M. Tignor, and P.M. Midgley, 3–21. A Special Report of Working Groups I and II of the Intergovernmental Panel on Climate Change. Cambridge and New York: Cambridge University Press, 2012.

IPCC. "Summary for Policymakers." In *Global warming of 1.5°C. An IPCC Special Report on the Impacts of Global Warming of 1.5°C above Pre-industrial Levels and Related Global Greenhouse Gas Emission Pathways, in the Context of Strengthening the Global Response to the Threat of Climate Change, Sustainable Development, and Efforts to Eradicate Poverty,* edited by V. Masson-Delmotte, P. Zhai, H. O. Pörtner, D. Roberts, J. Skea, P. R. Shukla, A. Pirani, W. Moufouma-Okia, C. Péan, R. Pidcock, S. Connors, J. B. R. Matthews, Y. Chen, X. Zhou, M. I. Gomis, E. Lonnoy, T. Maycock, M. Tignor, T. Waterfield, 1–26. Geneva, Switzerland: World Meteorological Organization, 2018. ISBN 978-92-9169-151-7.

Japino, Bramka Arga, Brian Walsh, Julie Rozenberg, and Stephane Hallegatte. "Revised Estimates of the Impact of ClimateChange on Extreme Poverty by 2030." *Policy Research Working Paper* No. 9417. World Bank, 2020. Washington, DC. https://open knowle dge.worldbank.org/handle/10986/34555.

Khan, Muhammad Ajmal, Munir Ozturk, Bilquees Gul, and Muhammad Zaheer Ahmed. "Halophytes for Food Security in Dry Lands." In *The Boulevard*, 1–338, Kidlington: Langford Lane, 2016. Oxford OX5 1GB 225 Wyman Street, Waltham MA 02451 Elsevier Inc.

Koubi, Vally. "Climate Change and Conflict." *Annual Review of Political Science* 22, 2019: 343–360. doi:10.1146/annurev-polisci-050317-070830.

Koubi, Vally, Mohamed Behnassi, Antoniett Elia, Manolis Grillakis, and Ethemcan Turhan. "Human Security." In *Climate and Environmental Change in the Mediterranean Basin – Current Situation and Risks for the Future. First Mediterranean Assessment Report*, edited by W. Cramer , J. Guiot, and K. Marini, 1–23. Plan Bleu: Union for the Mediterranean, 2020, UNEP/MAP, Marseille, France, in Press.

Krampe, Florian, Luc Van De Goor, Anniek Barnhoorn, Elizabeth Smith, and Dan Smith. "Water Security and Governance in the Horn of Africa." *Stockholm International Peace Research Institute* 2020: 1–13, https://www.jstor.org/stable/resrep 24527.8.

Leroy, Marcel and Fana Gebresenbet. "Climate Conflicts in the Horn of Africa?" *Vasu Gounden, Conflict Trends, Environment, Climate Change and Conflict* 2, 2011: 1–57. The African Centre for the Constructive Resolution of Disputes.

MacDonald, A.M., H.C. Bonsor, B. Dochartaigh, and R.G. Taylor. "Quantitative Maps of Groundwater Resources in Africa." *Environmental Research Letters* 7(024009), 2012: 1–7.doi:10.1088/1748-9326/7/2/024009.

Maino, Rodolfo and Drilona Emrullahu. "Climate Change in sub-Saharan Africa's Fragile States: Evidence from Panel Estimations." *International Monetary Fund,* Working paper, WP/22/54. 2022: 1–31.

Nagabhatla, Nidhi, Martha Cassidy-Neumiller, N. Ntugulo Francine, and Neil Maatta. "Water, Conflicts and Migration and the Role of Regional Diplomacy: Lake Chad, Congo Basin and the Mbororo Pastoralist." *Environmental Science and Policy* 122, 2021: 35–48. doi:10.1016/j.envsci2021.03.019.

Network of African Science Academies (NASAC). "Climate Change Adaptation and Resilience in Africa." *Recommendations to Policymakers,* 2015: 1–43. https://www.researchgate.net/publication/342396497.

Owonikoko, Babajide, and Jude A. Momodu. "Environmental Degradation, Livelihood, and the Stability of Chad Basin Region." *Small Wars and Insurgencies* 31(6), 2020: 1295–1776322. doi: 10.1080/0959 2318.2020.1 092.

Patana, Pindi, Herman Mawengkang, and Maya Silvi Lydia. "Conceptual Model for Mitigating Human – Wildlife Conflict based on System Thinking." *IOP Conf. Series: Materials Science and Engineering* 300, 2018: 1–6, doi: 10.1088/1757-899X/300/1/012052.

Rigaud, Kumari, Alex de Sherbinin Kanta, Bryan Jones, Jonas Bergmann, Viviane Clement, Kayly Ober, Jacob Schewe, Susana Adamo, Brent McCusker, Silke Heuser, and Amelia Midgley. "Groundswell: Preparing for Internal Climate Migration." *International Bank for Reconstruction and Development* 2018: 1—222. The World Bank, Washington.

Sayan, Ramazan Caner, and Nidhi Nagabhatla. "The Co-constitution of Regional Politics and Massive Infrastructures in the Transaqua Water Project." *Territory, Politics, Governance* 2022: 1–23. doi: 10.1080/21622671.2022.2043178.

Schmitz, J. "Anthropologie des conflits fonciers et hydropolitiques du fleuve Sénégal (75–91)." *Cahiers de sciences humaines* 29(4), 1993: 591–623.

Serdeczny, Olivia, Sophie Adams, Florent Baarsch, Dim Coumou, Alexander Robinson, William Hare, Michiel Schaeffer, Mahé Perrette, and Julia Reinhardt. "Climate Change Impacts in Sub-Saharan Africa: From Physical Changes to their Social Repercussions." *Regional Environmental Change* 2(63), 2016: 1–19. Springer ·doi: 10.1007/s10113 015 0910 2 https://www.researchgate.net/publication/290194107.

Swatuk, Larry A. "Environmental Security in Practice: Transboundary Natural Resources Management in Southern Africa." *paper prepared for presentation in Section 31 of the Pan-European Conference on International Relations* 2004: 9–11. The Hague.

Tarif, Kheira "Climate Change and Violent Conflict in West Africa: Assessing the Evidence." *SIPRI Insights on Peace and Security* 3, 2022: 1–24. www.sipri.org.

United Nations. "World Population Prospects 2019 Highlights" (ST/ESA/SER.A/423). Department of Economic and Social Affairs, Population Division (2019): 1–39. www.unpopulation.org.

United Nations Children's Fund (UNICEF) Child Alert. "Sub-Saharan Africa: Growing Up in Crisis in a World of Opportunities." *The Lasting Impact of Covid-19 on Children* 2021: 1–28. https://www.unicef.org/esa/media/8486/file/UNICEF-Sub-Saharan-Africa-Child-Alert-2021.

United Nations, Department of Economic and Social Affairs, Population Division. "World Population Prospects: The 2012 Revision." *Comprehensive Tables* I, 2013: 1–439. ST/ESA/SER.A/336.

United Nations Framework Convention on Climate Change. "Climate Change: Impacts, Vulnerabilities and Adaptation in Developing Countries." 2007: 1–68. Climate Change Secretariat Martin-Luther-King-Strasse 853175 Bonn, Germany. Up to 2060*?* Paper prepared for the Foresight project: Migration and Global Environmental.

Wolf. Tanja, and BettinaMenne. "Environment and Health Risks from Climate Change and Variability in Italy." *WHO Regional Office for Europe*, 2007: vi + 114. Scherfigsvej 8 DK-2100 Copenhagen Ø, Denmark, http://www.euro.who.int/pubrequest.

Part II

CONFLICTS AND POLITICS OF ENVIRONMENTAL SECURITY

Chapter 6

Threats, Triggers, and Tipping Points of Environmental Conflicts

A Case Study of North Central Nigeria, 2002–2022

Zainab Nuhu and Tasi'u Yalwa Rilwanu

Managing the environment is one of the challenges facing the global community in the twenty-first century. These challenges are especially true concerning population expansion, urbanization, and climate change. The challenges lead to dwindling resources and low skills, sometimes leading to conflicts or contestations over natural resources and affecting the environment. In numerous global regions, the intersection of limited resources, overpopulation, and environmental degradation has spawned severe poverty and income inequality, leading to conflicts. Policymakers, private organizations, and scholars employ various concepts to manage these conflicts, viewing them as outcomes of competition for scarce resources[1] and in managing environmental conflicts.[2] Conflicts are seen as a resultant effect or consequence of competition for scarce resources in human society. The access to natural resources in an environment is often contested and, in most cases, characterized by all manner of conflicts. The need for people to legitimize their rights and access to natural resources has remained central in competition and contestation for natural resources.[3] This can be seen in the works of many scholars who have shown these contestations using models or theories such as the basic livelihood approach[4] and the human basic needs theory.[5]

A secure environment provides a safe ground for humans to thrive; therefore, identifying the threats, triggers, and tipping points of conflict is critical, especially when mitigation measures are to be devised. However, the complex dynamics of conflict demand careful and well-thought-out strategies for building sustainable peace and ensuring human security. This chapter synthesizes official statistics and studies on environmentally related conflicts in North Central Nigeria (NCN) to identify and explain their threats, triggers, and tipping points.

THEORETICAL/CONCEPTUAL FRAMEWORK

This section explains the relationship between conflicts and what triggers them, situating it using the basic needs theory. The theory, having its roots in psychology, posits that the basic need of man is food, and then other needs emerge when the primary need is satisfied. Maslow's hierarchy of needs notes the following:

Human beings are motivated by a hierarchy of needs, which are organized in a hierarchy of prepotency in which more basic needs must be met (rather than all or none) before higher needs; the order of needs is not rigid but may be flexible based on external circumstances or individual differences and most behavior is multi-motivated, that is, simultaneously determined by more than one basic need.[6]

Conflict can be triggered by competition for limited resources (physiological needs), such as water, arable land, or food, when these resources are insufficient to meet the population's basic survival requirements. A lack of safety (safety needs) due to threats, violence, or political instability may drive individuals or groups to engage in conflicts to establish or restore security.[7] When social needs are unmet, individuals or communities feel marginalized or discriminated against; conflicts may arise as they seek to address perceived injustices and secure a sense of belonging. A lack of esteem may trigger conflicts and need fulfillment as individuals or groups strive for recognition, respect, or the acknowledgment of their identity, culture, or contributions. Societal structures that limit opportunities for education, personal development, or self-actualization can lead to frustration and may contribute to conflicts as individuals aspire to fulfill their potential. Maslow's hierarchy of needs provides a lens through which we can understand the underlying motivations contributing to conflict triggers. Addressing these fundamental needs makes it possible to implement strategies that promote peace, stability, and the well-being of individuals and communities.

Contextualizing the basic needs theory about livelihoods and development, environmental conflicts hinder these developments and threaten livelihood. However, Hans[8] argues for the rise of another type of livelihood studies, which aims through meta-analysis of the multitude of livelihood studies available and comparative research at broader generalizations that may challenge existing theories because the generation of livelihood studies managed to come to grip with that shortcoming and how it developed an understanding of the operation of power in livelihood strategies that can effectively contribute to livelihood enhancement. Nevertheless, this notwithstanding, it is worth noting that the provision of land, which is essential for food production, grazing of animals, and space for human habitation for different actors in conflict, utilizes the concept of the basic needs theory to explain their needs as threats, triggers, and tipping points of conflicts. The basic livelihood needs theory is rooted in many disciplines, and contextualizing the theories for this study will have its application in environmental resource utilization and its servitude as a means of livelihood. Even though the concept evolved within the context of the intentional development approach as posited by Morse, McNamara, and Acholo,[9] interventions on environmental development and what will bring about positive progress are advocated for.

ENVIRONMENTAL RESOURCES CONFLICTS IN NIGERIA

Conflicts or contestations over resource access and use are common in most parts of Nigeria and are characterized by violence primarily because the claim and ownership of land and agricultural resources, the proliferation of small arms and light weapons, and the absence of neutrality in the intervention by the Nigerian state in conflicts have been blamed for the persistence and escalation of resource conflicts in the country. Similarly, the Crisis Group highlighted drought, desertification, loss of grazing reserves, changes in pastoralism and farming practices, rural banditry, and cattle rustling as factors escalating environmental insecurity.[10]

Land as a resource for contestation has been managed customarily. Nigeria's Land Use Act of 1978 vested all lands for grazing, farming, or human habitation in the hands of the state's governor. Several land categorizations have emerged. However, since the incorporation of the Grazing Reserve Law of 1965, some grazing reserves were gazetted for the strategic objectives of settlement of nomadic pastoralists and sedentarization through the provision of land for grazing and demarcation of permanent water points. However, settling nomadic pastoralists tend to be difficult, especially when structural infrastructure is not established.

Environmental insecurity hinders growth and development and (un)availability of resources, such as land, trigger conflicts. This assertion concurs with Colier and Hoeffler's thesis that countries with abundant natural resources are more prone to violent conflicts than those without such resources. In addition, insurgent groups are more likely to be motivated by control over resources than by actual political differences with government authorities, ethnic divisions, or other factors typically viewed as root causes of civil war.[11] Ostrom and Slager advocate the human basic needs theory and experimented how actors involved in the use of the common property (mostly natural resources) bring into play different choices and behaviors due to human–nature couplings that vary across space, time, and organizational units and found out that institutional restructuring will broker peace.[12] Dimelu, Salifu, Enwelu, and Igbokwe are of the opinion that land use conflict in Nigeria is, in general, an annual phenomenon and suggest that traditional authorities should be given the formal mandate to resolve the conflicts.[13] Chukwuma and Atehle explore the herder/farmer conflicts situating its nature, causes, dynamics, effects, and implications, viewing the conflict through a political ecology perspective, and concluded that these conflicts hinder sustainable societal progress and national security.[14]

ENVIRONMENTAL RESOURCES CONFLICT IN NORTH CENTRAL NIGERIA

The NCN is a region home to multiethnic, multireligious people with a rich cultural setup having different tribes and languages even among the same ethnic groups. The region has abundant agricultural land, with some areas having rich, fertile soils that facilitate the growth of different crops and fruits. The region is witnessing an intractable and unprecedented conflict, especially between nomadic Fulani herdsmen and sedentary farmers, over access to and control of water points and grazing/farming lands.[15] With the escalation of conflicts occasioned by weak security enforcement, the region has witnessed a rapid increase in conflicts which over time is changing patterns and expanding in scope and intensity.

Even though other conflicts, including ethnic or religious conflicts, exist in the NCN, the most common conflicts experienced are environmentally related (resource access, use, and control). The National Bureau of Statistics (NBS), Nigeria's official office of statistics, reported that land or resource conflicts accounted for 55 percent of conflicts in NCN in 2017 alone.[16] As identified by different studies, the critical drivers of conflict in northern Nigeria are associated with access to natural resources, especially land. Unless the drivers are understood and managed, they threaten lives and property.

The region is vulnerable to tenure insecurity, which is a great challenge even in post-conflict situations because of fear of attack. Some conflicts spill into other areas where cases of reprisal attacks happen. For example, Nuhu found that most conflicts in Benue and Nasarawa states are because of spillovers from neighboring communities and that these disrupt the lives of members of other communities.[17] Similarly, the Nigeria Conflict Report discoursed that the incidences of conflict and violence in NCN remained steady from 2013 to 2016; of all the reported cases, more than half of the conflict events were caused by disputes over access to land or resources.[18]

CAUSES/TRIGGERS OF ENVIRONMENTAL RESOURCE CONFLICTS

Single or multiple reasons can trigger conflicts. Factors such as population pressure, environmental constraints, the proliferation of small and light weapons, and the absence of neutrality in the intervention by the Nigerian state in these conflicts have been blamed for the persistence and escalation of resource conflicts in the country.[19] Ungoverned spaces that allow for insecurity, institutional failures, and the incapacity of government employees to manage conflicts between herders and farmers without bias are the causes of these conflicts. Social inequities, ethnic and cultural discrimination, and erroneous land tenure regulations are also responsible for eco-violence. Similarly, overstay by herders also breeds conflicts because of shrinking grazing lands and sedentarization.

Analyzing the Grazing Reserve Act of Nigeria, 1965,[20] the causes and triggers of conflicts associated with the environment and their corresponding

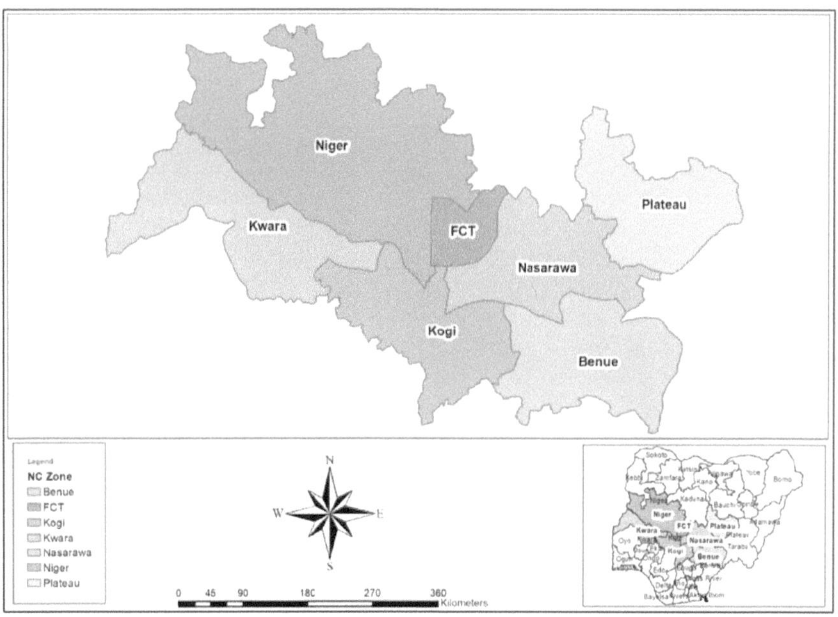

Figure 6.1 Nigeria Showing the North Central State. *Source*: National Bureau of Statistics, Nigeria, 2023.

features and manifestations are complex and dynamic. Increases in population, contestations over access, and the use of limited resources are some of the threats associated with these conflicts. Since the environment provides a safe ground for humans to thrive, identifying the threats, triggers, and tipping points of conflict is critical, especially in devising mitigation measures. Most reasons are socioeconomic and are because of access to resources or insecurity about resource utilization. Other causes of conflicts include the destruction of farms, spillover of conflicts from other places, resource ownership claims, and even poverty in some cases.

The findings from data analysis on conflicts in the specified region (NCN) suggest that land-related issues are a significant driver of conflict, either directly or indirectly. Land encroachment emerges as the primary cause, contributing to approximately 23.6 percent of environmental conflicts. Other significant factors include spillover effects (19.4 percent), destruction of farmlands (11.1 percent), ownership disputes (9.7 percent), and old grievances (8.3 percent). Collectively, these land-related issues account for a substantial portion of the identified causes. Additionally, data indicate that a considerable portion (27.9 percent) of environmental conflicts in NCN are attributable to factors beyond land issues. These include politics, poverty, reprisal attacks, proliferation of arms, and selfish interests. This argument implies a diverse range of factors contributing to the overall conflict landscape in the region,

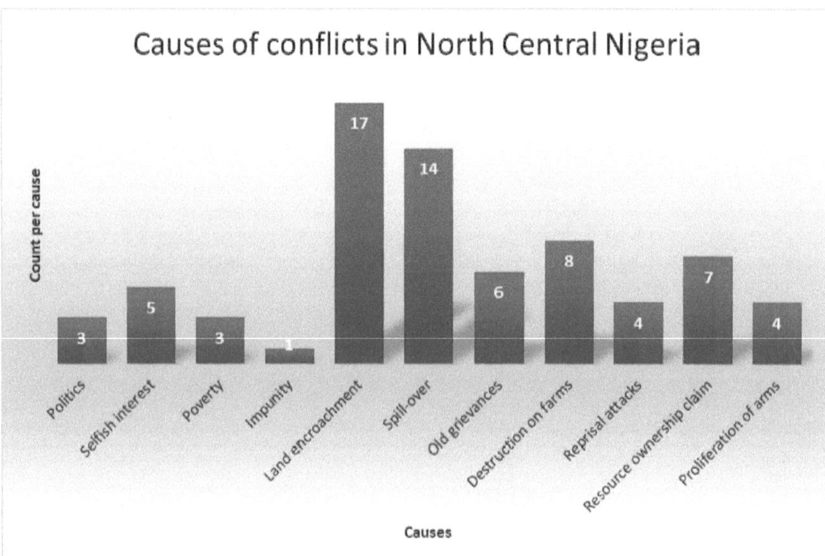

Figure 6.2 Causes of Conflicts in North Central Nigeria. *Source*: Created by the author based on information from Armed Conflict Location and Event Dataset (ACLED), 2021, at www.acleddata.com.

with non-land-related issues also playing a significant role. In summary, the analysis highlights the prominent role of land-related conflicts in the region. However, it also underscores the complexity of the overall situation, with various other sociopolitical and economic factors contributing to environmental conflicts in NCN.

MANIFESTATIONS AND EFFECTS OF ENVIRONMENTAL CONFLICTS

The manifestations and effects of conflict in the NCN are visible in the absence or interruption of tenure security because of conflicts, which affect livelihoods, displaced persons, and damage the environment. Loss of lives and destruction of properties are common in NCN, especially in areas with abundant land resources, whether used for subsistence or on a large scale. These losses breed poverty, giving rise to vulnerable populations, threatening food security, and rendering the environment unproductive. A study conducted by Mohammed, Ekhamheye, and Meeqaeel in Nasarawa state, Nigeria. concluded that farmers–herders' conflicts significantly negatively impact economic and agricultural activities.[21] Grazing reserves and farmlands are lost due to environmental insecurity, defeating the aims of grazing reserves, including getting and protecting pasture space for the national herds and removing discord between agronomists and pastoralists living in the same geographic area. Separating the herders from the cultivators will create a potential peaceful coexistence between them, which can be fostered by making the grazing reserve a no-conflict zone. Aside from the loss of lives and destruction of property, incessant resource conflicts have forced farmers to abandon their fertile farmlands, rendering several farming communities deserted and have consequently reduced crop farmers to tenants in IDPs camps.

MANAGEMENT OF ENVIRONMENTAL CONFLICTS

Several attempts have been made to foster peace among resource users identified as significant conflict actors. Community-based initiatives from community-based organizations (CBOs), local and international NGOs, and traditional institutions have made a series of peacebuilding initiatives with conflicting groups, particularly in the most conflict-prone areas and areas identified as vulnerable to conflicts. Many programs, policies, and projects aim to address conflicts; their threats, triggers, and environmental impacts have emerged over time. Both the public and the private sectors' involvement is visible, as seen in some local, state, or national intervention programs. Some of these include NGOs' management of conflicts, localized agreements and treaties, and State Intervention Dialogue and Mediation programs.

Even though some trajectories of the conflicts in the region are not promising, for example, the ACLED, in their midyear report, predicted that in the NCN, there is a limited prospect of diminishing intercommunal conflict in the absence of adequate measures adopted by the government to address the underlying issues causing hostilities, such as changes in land use resulting from drought and displacement.[22] The short-term implication of this, if not addressed, will give way to the conflict escalating and in the long term, transforming them into wars, especially taking into cognizance the multidimensionality of the critical elements involved (actors, resources, ethnicity, religion, and space). Security of lives and land will be threatened; therefore, there will be a threat to food, livelihood, and safe spaces.

Although eco-violence decreased by over 46 percent in 2021 compared with the peaks of 2018, it risks escalating and further undermining Nigeria's security in both the near and long terms. To achieve sustainable environmental conflict management, there is a need for a holistic approach that includes balanced representation, accountability, openness, and fairness, as well as employing monitoring and evaluation as a continuous process.

CONCLUSION

This chapter drew explanations on the links between conflict and the extraction of a given resource, how they trigger and serve as a threat to peaceful coexistence, and the roles natural resources play in conflicts, and the effect of the conflict itself implies that if the conflicts are not managed properly, there is a threat to food security and poor development. Strengthening intergroup relations, advancing resource use tolerance, and ensuring tenure security are some of the measures that will reduce conflicts. It, therefore, suggests using a robust approach to conflict transformations, especially in post-conflict situations and areas where trajectories show the likelihood of conflicts if not checked. There is a need for a paradigm shift in putting measures for developing conflict trajectories and corresponding response mechanisms. This paradigm shift is where systems thinking approaches in conflict management, which can inform strategy for decision-making, come into play. Creating safe spaces using unarmed security protection to minimize conflicts and ensure environmental security is another approach that can aid in managing environmental conflicts.

Minimal attention has been given to the need for an all-inclusive approach in conflict-related problem-solving strategies for environmental security and sustained peace; hence, actions need to identify capacities and support local or indigenous strategies in managing threats and triggers before they reach tipping points of conflicts.

There is an urgent need to explore the option of predicting conflicts and linking them to early warning response mechanisms while taking into cognizance that these predictions are based on assumptions. Hence, there is a need for a holistic approach by stakeholders and predictors to have a robust collaboration for positive outcomes to fit all cases. Identifying the trends of conflicts, actors, causation, history of previous events, and management strategies employed before will aid in arriving at a decisive conclusion in making the predictions. Some key indices include the vulnerability index and environmental variations and their triggers that result in conflicts. Apart from identifying the indices, the need for action in all sectors will ensure minimal conflict and a security-safe environment.

NOTES

1. Birkmann, J., E. Liwenga, R. Pandey, E. Boyd, R. Djalante, F. Gemenne, W. Leal Filho, P.F. Pinho, L. Stringer, and D. Wrathall. 2022: Poverty, Livelihoods and Sustainable Development. In *Climate Change 2022: Impacts, Adaptation and Vulnerability*. Contribution of Working Group II to the Sixth Assessment Report of the Intergovernmental Panel on Climate Change. Cambridge and New York, NY: Cambridge University Press, pp. 1171–1274. https://doi.org/10.1017/9781009325844.010. https://www.mercycorps.org/sites/default/files/Mercy_Corps_2016_Annual_Report.pdf.

2. Blench, R. *Natural Resource Conflicts in North-Central Nigeria: A handbook of Case Study*. Boulder: Lynne Reiner Publishers, 2004.

3. Liman, B. M. Conflict, and Identity in Nigeria: An Emerging Culture of Conflict in Northern Nigeria. PhD thesis. SOAS University of London, 2015. https://doi.org/10.25501/SOAS.00023674.

4. Ellis, F. and Allison, E. "Livelihood Diversification and Natural Resource Access." *Livelihood Support Programme,* 2004. ftp://ftp.fao.org/docrep/fao/006/AD689E00.pdf.

5. Ostrom Eleanor and Schlager. "Property Rights Regimes and Natural Resources: A Conceptual Analysis." *Land Economics*, University of Winscousin Press, 68, (1992): 49–262.

6. Maslow, A. H. A Theory of Human Motivation. *Psychological Review* 50, no. 4 (1943): 370–396.

7. Trivedi A. J. and Mehta A. A. Maslow's Hierarchy of Needs - Theory of Human Motivation. *International Journal of Research in all Subjects in Multi Languages* 7, no. 6 (2019): *Academy for International Journals of Multidisciplinary Research* (RAIJMR).

8. Haan, L. The Livelihood Approach: A Critical Exploration. *Erdkunde*, 2012. Accessed on April 13, 2022. https://www.academia.edu/69631691/The_livelihood_approach_a_critical_exploration.

9. Morse, Stephen, McNamara, Nora, and Acholo, Moses. Sustainable Livelihood Approach: A Critique of Theory and Practice. *Geographical Paper* 189, (2009): 1–68. https://doi.org/10.1007/978-94-007-6268-8.

10. Watch List 2017 – Second Update the Crisis Group. Entries on Nigeria, Qatar, Thailand, and Venezuela. EU Watch List/From Early Warning to Early Action, July 20, 2017.

11. A., Collier P., and Hoeffler. Greed and Grievance in Civil War. *World Bank.*, 2001. Accessed on March 20, 2018. https://economics.ouls.ox.ac.uk/12055/1/2002-01text.pdf.

12. Ostrom Eleanor, and Schlager. "Property Rights Regimes and Natural Resources: A Conceptual Analysis." *Land Economics* 68, (1992): 49–262. University of Wisconsin Press.

13. Dimelu M. U., Salifu E. D, Enwelu A. I., and Igbokwe E. M. "Resource Uses Conflict in Agrarian Communities, Management and Challenges. A Case of Farmers-Herdsmen Conflict in Kogi State- Nigeria." *African Journal of Agricultural Research* 46, (2016): 147–154. https://doi.org/10.1016/j.jrurstud.2016.06.011.

14. Chukwuma O. A., and Atehle G. Nomads against Natives: A Political Ecology of Herder/Farmer Conflicts in Nasarawa State. *American International Journal of Contemporary Research* 4, no. 2 (2014): 82.

15. Ibid.

16. National Bureau of Statistics. *Nigeria Statistics Report.,* 2018.

17. Nuhu, Z., Idris, H. A., Mohammed, U. K., Muhammad, U. M. and Nuhu, H. "Conflict, Tenure Security and Implication on Agriculture in Nasarawa State, Nigeria." *2nd International Conference on Dryland Africa.,* 2016.

18. Dada, J. P., Zhema, S., and Ochoga, E. O. Conflicts in Nigeria. In *HandBook on Conflict Prevention in Nigeria. Wukari International Studies Group,* (2018): 22–43. Wukari, Taraba State: Department of History and Diplomatic Studies, Federal University. https://www.researchgate.net/publication/340628835.

19. Awogbade, M. O. Grazing Reserves in Nigeria. *Nomadic Peoples* 23, (1987): 19–30. http://www.jstor.org/stable/43123287.

20. Azad A., Crawford E., and Kaila H. *Conflict and Violence in Nigeria Results from the Northeast, North Central, and South-South zones Preliminary Draft Report.* Nigeria: World Bank and National Bureau of Statistics, 2018.

21. Musa, Mohammed, Sylvester Ekhamheye, and I. G. Meeqhaeel. "Causes of Herders and Farmers Conflicts on Economic Activities in Nasarawa State, Nigeria." *International Journal of Research and Innovation in Social Science (IJRISS)* 5, no. 11 (2021): 508–517.

22. ACLED. *The Armed Conflict Location & Event Data Project (ACLED)*, 2022. www.acleddata.com.

REFERENCES

A, Chukwuma O. A and Atehle G. "Nomads against Natives: A Political Ecology of Herder/Farmer Conflicts in Nasarawa State." *American International Journal of Contemporary Research* 4, no. 2 (2014): 82.

A., Collier P. and Hoeffler. "Greed and Grievance in Civil War." *World Bank.*, 2001. Accessed on March 20, 2018. https://economics.ouls.ox.ac.uk/12055/1/2002-01text.pdf.

Akinrinde, Olawale Olufemi. *Confictis in Nigeria.*, 2018. Accessed on October 11, 2022. https://www.researchgate.net/publication/340628835.

Armed Conflict Location and Event. *Armed Conflict Location and Event Data.* July 20, 2021.

Awogbade, M. O. "Grazing Reserves in Nigeria." *Nomadic Peoples* 23, (1987): 19–30. http://www.jstor.org/stable/43123287.

Azad, M Abul Kalam, Emily Elizabeth Crawford and Heidi Kristiina. Kaila. *Conflict and Violence in Nigeria: Results from the North East , North Central and South South Zones (English).* Washington DC: World Bank Group., 2018. Accessed on July 5, 2022. http://documents.worldbank.org/curated/en/111851538025875054/conflcit-and-Violence-in-Nigeria-Results-from-the-North-East-North-Central-South-South-Zones.

Blench, R. *Natural Resource Conflicts in North-Central Nigeria: A Handbook of Case Study.* Boulder: Lynne Rienner Publishers, 2004.

Dimelu M. U., Salifu E. D, Enwelu A. I and Igbokwe E. M. "Resource Ues Conflict in Agrarian Communities, Management and Challenges. A Case of Farmers-Herdsmen Conflict in Kogi State- Nigeria." *African Journal of Agricultural Research* 46, (2016): 147–154. https://doi.org/10.1016/j.jrurstud.2016.06.011.

Egobueze Anthony, Ajunwa Daniel and Ogele Eziho Promise. "Farmers/Herdsmen Crisis and Sustainable Food Production in Nigeria." *The International Journal of Humanities and Social Studies* 8, no. 4 (2020): 66–74.

Ellis, F. and Allison, E. "Livelihood Diversification and Natural Resource Access." *Livelihood Support Programme,* 2004. ftp://ftp.fao.org/docrep/fao/006/AD689E00.pdf.

Institute for Economics & Peace. *Global Terrorism Index 2019: Measuring the Impact of Terrorism*, Sydney, November 2019. Available from: http://visionofhumanity.org/reports (accessed 21/03/2021).

International Crisis Group. N°168 – December 20, 2010. "Northern Nigeria: Background to Conflict." *Africa Report*, 2010.

International Peace Institute. "Toward an Environmental and Climate-Sensitive Approach to UN Peacekeeping Operations." https://www.ipinst.org/2022/10/toward-an-environmental-and-climate-sensitive-approach-to-un-peacekeeping-operations. (accessed 22/12/2022).

Leo. J. De Haan. *The Livelihood Approach: A Critical Exploration* Bd. 66, H. 4 (October - December 2012), pp. 345–357 (13 pages) Published By: Erdkund.

Mercy Corps. "Annual Report." 2016. https://www.mercycorps.org/sites/default/files/Mercy_Corps_2016_Annual_Report.pdf.

Mohammed, L. M. "Conflict and Identity in Nigeria: An Emerging Culture of Conflict in Northern Nigeria." *SOAS Research Online*, 2015. https://eprints.soas.ac.uk/23674.

Morse, Stephen, McNamara, Nora and Acholo, Moses. "Sustainable Livelihood Approach: A Critique of Theory and Practice." *Geographical Paper* 189, (2009): 1–68. https://doi.org/10.1007/978-94-007-6268-8.

Musa, Mohammed, Sylvester Ekhamheye and I. G. Meeqhaeel. "Causes of Herders and Farmers Conflicts on Economic Activities in Nasarawa State, Nigeria." *International Journal of Research and Innovation in Social Science (IJRISS)* 5, no. 11 (2021): 508–517.

National Bureau of Statistics. *Nigeria Statistics Report.,* 2018. https://nigerianstat.gov.ng/elibrary (accessed 22/11/2022).

Nigeria Stability and Reconciliation Programmed. "Nigeria Watch." *Report on Violence.*, 2016.

Nuhu, Z., Idris, H. A., Mohammed, U. K., Muhammad, U. M. and Nuhu, H. "Conflict, Tenure Security and Implication on Agriculture in Nasarawa State, Nigeria." *2nd International Conference on Dryland Africa.*, 2016.

Ostrom Eleanor and Schlager. "Property Rights Regimes and Natural Resources: A Conceptual Analysis." *Land Economics*, University of Winscousin Press 68, (1992): 49–262.

P., Le Billion. "Buying Peace or Fueling War: Natural Resources and Armed Conflict." *Journal of International Development* 4, no. 15 (2003): 413–426. https://doi.org/10.1002/jid.993.

Trivedi A. J. and Mehta A. A. "Maslow's Hierarchy of Needs—Theory of Human Motivation." *International Journal of Research in all Subjects in Multi* Languages 7, no. 6 (2019). Academy for International Journals of Multidisciplinary Research (RAIJMR).

United States Institute of Peace. Managing Natural Resources Wealth. Part of Stabilization and Reconstruction Series by Jill Shankleman., 2006. Accessed on January 9, 2022. https://www.usip.org/pubs/specialreports/srs/srs4.html.

Chapter 7

The Impact of Ethiopian Dams on the Environment in Northeast Africa

Domagoj Krpan

The concept of environmental security has many definitions, and it is still contested. According to one of the definitions, environmental security defines the relationship between individuals, communities, or nations and threats of environmental events. Also, it examines how international or internal relationships are affected by environmental problems. The changes in the environment can start a chain reaction in society. For example, the primary sector or mining can be devastated, leading to social unrest and internal conflict. The country can lose its capability to defend itself, leaving itself vulnerable to external threats.[1]

Contemporary authors claim that future wars will be fought over water resources.[2] In many cases, rivers do not follow a country's borders, so more than a couple of countries may share water from the same river. This creates tension between the countries since a society's well-being depends on water. Water affects many different spheres of society; it is vital for the everyday survival of human beings and a county's economy. Farming and cattle breeding partly depend on proper water supply. Furthermore, many industries, such as textile manufacturing, paper mills, oil and gas, and dairy industries, depend on water supply. Finally, hydropower plants use water as their fuel for electricity production.

Hydropower plants need a dam for proper function. However, dam construction is a complicated process that seriously affects the environment and communities living downstream. That is why engineers must consider how the construction will affect not only surrounding areas but also communities living a thousand kilometers away. Sometimes, the state governments consider what is, as a whole, in their home country's best interest, so life in some regions may be altered beyond recognition. At the same time, they can alienate other countries, which could lead to friction and conflict since a peaceful solution, such as a treaty, cannot be reached.

In recent decades, Ethiopia has invested a lot of resources in dam construction. The country is rich in water resources. The Blue Nile is probably the most famous river, which has a source in Ethiopia. Putting the water resource under control would make Ethiopia the powerhouse in Africa since the dams would not be used only for electricity but also for irrigation, drinking water, and flood control. However, there are many concerns about this program. Sudan and Egypt disapprove of the GERD construction on the Blue Nile since they believe it would hurt their everyday life. A similar situation is with the Gilgel Gibe III Dam on the Omo River, which causes tensions between Kenya and Ethiopia. Furthermore, there is a serious concern regarding the impact on local communities inside Ethiopia, and the environmental issues should not be overlooked.

This chapter examines how Ethiopia's dam construction influences its relationship with Kenya, Sudan, and Egypt. In addition, it presents an example of how West African countries sharing the Senegal River have found a peaceful solution for their water supply. Furthermore, it analyzes how dams impact the environment and the local population.

THE GRAND ETHIOPIAN RENAISSANCE DAM

The Nile is probably one of the most famous rivers in the world. Many people would automatically connect the river with Egypt, but although the Nile delta is situated in Egypt, its sources are much further south. Two rivers, the White Nile, and the Blue Nile, comprise the Nile. The source of the White Nile is still disputed between a couple of countries in equatorial Africa, while the source of the Blue Nile is in Ethiopia.

Lake Tana is the source of the Blue Nile.[3] It flows from the southern part of the lake and through the Ethiopian Highlands, where it receives water from its many tributaries.[4] The Blue Nile and the White Nile confluence at Khartoum. Overall, the Blue Nile can be considered much more important than its counterpart since about 85 percent of all water in the Nile comes from Ethiopia.[5]

The Ethiopian government started the construction of the GERD near the border with Sudan.[6] It is believed that the construction cost will be close to five billion USD, and it is primarily financed by crowdfunding since it is assumed that Egypt is campaigning against the GERD's construction.[7] The main reason why Egypt looks unfavorably upon GERD is that Egypt's society is dependent on the River Nile. For example, Egypt's agriculture uses 80 percent of the Nile's freshwater and contributes 11.3 percent to the national GDP. Furthermore, industries connected with agriculture are responsible for

16 percent of GDP, 50 percent of the rural population is dependent on the river, and around 30 percent of the labor force is employed in agriculture.[8] In Sudan, the agriculture sector is responsible for over 30 percent of the country's GDP.[9] Around 80 percent of the labor force is employed in agriculture, and more than 80 percent of exports are agricultural products.[10] In other words, the well-being of Egypt and Sudan is heavily dependent on the Nile. Therefore, the possibility of losing the water from the Blue Nile would seriously undermine the stability of both countries.

On the other hand, Ethiopia is looking to solve many domestic problems by building the GERD. It is expected that the fully operational GERD will have a capacity of 6,000 MW, and the reservoir will hold twice the amount of water than the biggest lake in Ethiopia, Lake Tana, which can hold up to 32 billion m3.[11] As a result, it is expected that a surplus of electricity, especially during the rainy season, will be sold to other countries in the region.[12] In addition, the new reservoir will create new agricultural land around the dam. However, to create this newly arable land, some 20,000 people are displaced, and the government will create a five-kilometer-long buffer zone for malaria control.[13] It should also be noted that electric energy from the GERD would help lower the number of trees cut for firewood, which would positively impact climate change.

Some reports suggest that during the filling of the reservoir, it should be filled at annual rates from 8 percent to 15 percent since this is the optimal rate for hydroelectric power generation, and it would have a minimum impact on the rest of the river.[14] However, there is fear that the delicate balance between the freshwater and seawater could be lost in Nile's delta since the seawater will pierce inland.[15] It should be noted that losing the groundwater will increase the number of illegal wells drilled and will also impact the spreading of seawater.[16] The result will be salt accumulation, and the process will be irreversible, so the area of agricultural land in the Nile's delta will decrease. According to the Sudanese reports from 2021, the reservoir filling did not impact the annual Nile flow.[17] However, even if the reservoir filling would not impact the environment seriously, there are still worries about GERD's future effects. The main issue is multiyear droughts. A non-cooperation from the Ethiopia side, which would only look upon their hydropower generation, would cut the amount of water flowing downstream, hurting Egyptian agriculture and water resource.[18]

Because of all that, Egypt is very vocal about its dissatisfaction with the GERD construction. From Egypt's perspective, there are some historical legal binding agreements under which the Nile water resource is divided between the countries. According to this interpretation, Ethiopia is breaking these agreements. Although there are many agreements concerning the water use of the Nile, the most famous and controversial one is the 1929 Nile Waters Agreement reached between Great Britain and Egypt in which Great Britain recognized Egypt's historical and natural rights and promised that the quantities of water flowing through Egypt would not be changed.[19] At that time, all riparian states, except for Ethiopia, were British colonies, so they forfeited the rights to make any changes to the river Nile that could endanger the river flow through Egypt. As a result, Egypt received exclusive rights on the river and maintained that any new agreement must be based on this agreement. After Sudan gained independence, the 1959 Nile Waters Agreement was reached between Sudan and Egypt. It was agreed that Egypt has the rights to 55.5 billion cubic meters (bcm) and Sudan for 18.5 bcm annually, while 10 bcm is lost due to evaporation.[20] This agreement was necessary for Egypt since they planned to construct the Aswan Dam, and the 1959 Agreement was a steppingstone for their future irrigation plans. Today, one of the consequences of this agreement is 17,000 kilometers of public drainage networks and 80,000 kilometers of private canals and farm drainage systems.Okbazghi Yohannes, (Albany: State University of New York Press, 2008), 38.[21] Again Ethiopia was not included in the negotiation or final agreement. Egypt is basing its claim on these two agreements that the GERD should not be built since it will reduce the quantity of water flowing down. However, these agreements are not legally binding on Ethiopia since it was not included in the negotiation processes, although it was an independent state.

Egypt's population is highly dependent on the Nile, but so are the other countries through which the Nile flows, and Egypt has not shown any concern for their needs in decades. In 2010, the Nile Basin Initiative (NBI) was signed between Uganda, Rwanda, Tanzania, and Ethiopia, while Kenya, Burundi, and the DRC sent officials to the meeting. The goal of the NBI is to determine how much water each country should receive for its needs, but Egypt and Sudan are strongly opposing any initiative which would change their share of water quantity.[22] Egypt is keeping all the options about GERD on the table. In 2013, the Egyptian government discussed sabotaging or destroying the dam or supporting Ethiopian anti-government groups in an attempt to stop GERD construction.[23] Sudan is less reluctant about GERD since it hopes to gain from its provision of cheap electricity in the future. Sudan could also benefit from increased summer flows, reduced flooding, and sedimentation.[24] Furthermore, climate change is an issue that needs to get resolved since there is a prediction that by 2040 more than 80 million people living in the Nile Basin could face water scarcity.[25] For now, it does not seem that this is much of a concern since, in August 2022, the Ethiopian government announced it would be completing GERD's third phase of filling.[26]

The other potentially dangerous factor is internal instability in Ethiopia. The current Ethiopian political system has its roots in the early 1990s. The Ethiopian People's Revolutionary Democratic Front, or the EPRDF, ousted the Derg from power and formed a new government composed of different ethnic groups. A new constitution in which the country was administratively divided between ethnic groups with a significant level of autonomy was influenced by this political diversity.[27] As a result, the country became a federation based on ethnic division. This arrangement helped stabilize the country, and, in 2019, there were talks about the possibility of Ethiopia becoming a new powerhouse. Prime Minister Abiy Ahmed won the Nobel Peace Prize for his involvement in de-escalating the tension between Ethiopia and Eritrea.[28] From 2004 to 2019, the Ethiopian economic growth rate was 7 percent per capita on average, and it was the fastest-growing economy in the world.[29] It all seemed like an incredible story from the region famous for instability. However, the stability in Ethiopia, a country with over 100 million people, started to crumble after 2019.

Prime Minister Abiy began distancing himself from ethnic federalism, which led him to a rift with the Tigray People's Liberation Front. The Tigray People's Liberation Front, or the TPLF, is a left-wing ethnic nationalist party formed in 1975 among the Tigray ethnic group.[30] The TPLF was a member of the EPRDF until Abiy removed it from power in 2018, which sparked the conflict between the TPLF and the central government.[31] The Tigray War started in 2020, and after some initial success, the TPLF retreated to their home province while the government forces kept them "under siege." As a result, the humanitarian situation continues to deteriorate by the day.[32] There are other anti-government military groups, but the TPLF is currently the strong and most influential one. Ethiopia's ethnic diversity is one of the internal obstacles to stability since the central government is expected not to allow any ethnic group to have domination over others. After Abiy assumed the office of prime minister, the period of stability ended since he abandoned the concept of ethnic federalism. Discarding this policy is returning Ethiopia to ethnic violence, and this internal stability is also problematic for the central government since it has a strained relationship with its neighbors.

One of the solutions to this deadlock could be joint water management of the Nile River basin. The proposed joint water management of the Nile River would help the countries since it would help them to get better control over river floods, and in times of need, such as during the drought, they could use the water from the reservoir. However, probably the best consequence of joint management is avoiding water conflict since all parties can resolve their water supply concerns through joint management. If we looked across the continent, we could find a successful story of cooperation between the riparian states. The Organisation pour la Mise en Valeur du fleuve Sénégal (OMVS; Eng. Senegal River Basin Development Authority) is the cooperation between Senegal, Mauritania, Mali, and Guinea for joint management of the Senegal River and its tributers.[33] The OMVS was founded in 1972, but Guinea joined only in 2006 since it had a tense relationship with Senegal for a few decades.[34] Overall, the OMVS has had mixed results. The positive impacts could be considered a supply of drinking water and energy to Dakar, the capital of Senegal, and Nouakchott, the capital of Mauritania, preventing the intrusion of saltwater in the lower parts of the basin and reducing the immigration from the area. On the other hand, the negative impacts could be considered the weakness of the irrigable crop potential, the proliferation of harmful aquatic plants, and the increasing spread of waterborne diseases such as malaria, bilharzia, and cholera.[35] Although the results of the OMVS are not at the level the member countries had expected, it should be noted that their cooperation showed the framework for other countries worldwide on how to resolve disputes over river management.

However, in northeast Africa, the countries are stubbornly entrenched in their positions, and none wants to take a step back. Considering that discussing military options was on the table, it is uncertain what might happen in the future, especially if any of the participants were led by a hotheaded leader. The main concern for all parties involved is the well-being of their population, so the balance between developing their own country while at the same time not hurting other countries' societies should be achieved. Ethiopian geographical position as a landlocked country is dictating its effort to develop a self-sustained economy, and energy independence is a steppingstone on that path. The environmental change is causing a security threat to Egypt and Sudan because the consequence of losing a large amount of water could bring social unrest among the downstream countries. From the individual to the national level, many will be affected by this change, which could lead to violent conflicts. In other words, there is a security problem common to all states, a threat to national security, and a possible cause of violent conflict. The best course of action could be joint management of the Nile basin since examples, such as the OMVS, showed that working together can ease the tension. The research showed some possible benefits of the GERD construction, so if the governments of all three nations sit together and try to resolve their concerns about their security, the dam could benefit everybody involved.

THE GIBE III DAM

The GERD is still under construction, so many concerns will not be answered until it is finished and fully operational. On the other hand, the Gilgel Gibe III Dam on the Omo River has been operational for a few years, and it might give us insights into the impact on the environment and people. The Omo River flows through south Ethiopia, and its delta is in Lake Turkana. Lake Turkana is divided between Ethiopia and Kenya, but its main tributer, around 80 to 90 percent, is the Omo River. The climate around the Turkana basin is hot and arid, with long periods of unusually intense diurnal winds.[36] Lake Turkana has a mean depth of 35 meters (max. 120 m), a length of 250 kilometers, and an average length of 30 kilometers.[37] The Gilgel Gibe III Dam produces around 6,500 GWh annually, doubling the country's installed capacity.[38] Still, the situation for the local population and environment after the dam had been put into function began to deteriorate.

Different tribes, around 200,000 people, live in the lower Omo as the traditional agro-pastoralist communities that depend on a flood-retreat river system.[39] Furthermore, Lake Turkana supports additional 300,000 people, many of whom live over the border in Kenya.[40] In addition, the fishermen are also endangered since the decline in water level would seriously undermine their livelihood.[41] The smallest decrease would be enough to wipe out major fish reproductive habitats.[42] There is also the threat of conflict between tribal communities from opposite sides of the lake. Also, the Kenyans are angry that the Ethiopian dam will affect their livelihood.[43] In some scenarios, the lake could be split into two smaller lakes if Lake Turkana retreats 40 meters.[44] Also, it should be noted that Lake Turkana is a Salt Lake. So, any decrease in water flow will increase the salinity that will affect the fish population in the lake.

The lake is not the only affected area. The environment of the lower Omo River is also seriously endangered. The last riparian forest in SSA is located along the shores of the Omo River. This forest is very dependent on seasonal flooding and will probably dry out. Furthermore, the Omo River recharges not only Lake Turkana but also the underground water supplies so that less water will impact the erosion of the river banks and loss of soil for crop cultivation.[45] There are also reports of human rights violations. For example, the Ethiopian government is opening a sugar plantation using the water from the reservoir for irrigation. At the same time, the government is using intimidation tactics, including beatings, to stop locals from voicing their dissatisfaction.[46]

The relationship between the Gibe III dam and Lake Turkana could be seen as a classical conflict between economic development and environmental protection. From the economic perspective, the Gibe III dam will boost the Ethiopian and Kenyan economies in the business sector and within households since it will bring electricity to ordinary people. On the other hand, communities living downstream from the dam are threatened with losing their livelihood and will probably have to move away. Their lives are based on regular yearly flooding, which will drastically change. Furthermore, the future of Lake Turkana does not look bright since the Omo River is its main tributary, and the reduction of water that flows from it will significantly affect the lake. The biggest hit will be on the fishing community since the lake's shrinking will destroy the fish's natural habitat. For now, it seems that the central governments do not consider how to help local communities, which could lead to future conflicts since the area does not have much arable land. The local communities will have two options: migration or intracommunity conflict over sparse food sources. The conflict with the central government should not be ruled out since the government does not include local communities in its economic plans. A feeling of exclusion could lead to rebellions, which would not be good news for a country that is already in a state of civil war.

In the case of Lake Turkana, environmental change can be identified as a possible cause of violent conflict and a risk to human security. Many local communities live on the shores of the lake, but they consider the lake as natural protection against the communities on the other shore. In addition, Lake Turkana is their primary source of income, so their personal well-being is in jeopardy. On the other hand, environmental security can be defined as the impacts of human activities on the environment.[47] The last riparian forest in SSA will probably be wiped out, but the main focus is on intercommunity relationships. So, the argument could be that environmental awareness among the people is not so strong since their current main concern is their livelihood and not the environment. It seems that if the forest were destroyed, but their lives remained unchanged, they would not mind. Real progress can be achieved when the communities start to consider environmental security not only as something that can directly impact them but also indirectly through the environment.

CONCLUSION

Ethiopia's growing population and economy are in need of a sustainable power source. The country is rich in water sources, so hydropower energy seems the logical option. However, constructing the dam to fully utilize the water resources is a daunting task. First of all, the lifestyle of local communities is changing forever, especially since some did not change their way of life for hundreds of years. The dam is blocking the natural river flow, so the year cycle of flooding will be broken, which will be devesting for the communities. They will have two options—either to migrate away or stay in their ancestral homeland with fewer resources. The second option will lead to conflicts with other groups because agricultural and pastureland will become a luxury. In that scenario, only the strongest would survive.

Second, the communities living further downstream are also affected by the dams. As a result, the communities, from a few hundred thousand people on Lake Turkana to tens of millions of people on the Nile, are seriously threatened to lose their livelihood. In Egypt, the situation is so severe that the military option is on the table because the future does not seem bright for its economy and social standard. Although there are examples of good cooperation between the countries concerning water management, the countries on the Nile stubbornly refuse to change their position for an inch. Moreover, future conflict could be devastating for the environment since damaging or destroying the dam would release millions of cubic meters of water downstream which would, in its path, destroy everything on the way. Not only external threats but internal instability in Ethiopia could also lead to this scenario.

Furthermore, even a slight lowering of water level would irreparably impact the environment. The natural habitat for some plants and animals would be lost entirely, so that the regional ecosystem would be lost forever. Also, the increase in salinity would contribute to this scenario. The Mediterranean Sea would penetrate further inland, decreasing the arable land area. At the same time, Salt Lake Turkana would destroy the wildlife in its waters if the inflow of fresh water from the Omo River decreased.

To sum up, Ethiopian dams are changing the face of the northeast African region. For many, the future is bringing a life-and-death situation. The reduction of water will impact the environment. Many people will lose their income, especially if they work in agriculture. This will increase social tensions, and the new powder cake may be ready to explode. One could argue that Ethiopian dams are changing more than one environment.

NOTES

1. Braden R. Allenby, "Environmental Security: Concept and Implementation," *International Political Science Review / Revue Internationale de Science Politique* 21, no. 1 (2000): 5–21.

2. Brahma Chellaney, *Water, Peace, and War: Confronting the Global Water Crisis* (Lanham: Rowman & Littlefield, 2013); Vandana Shiva, *Water Wars: Privatization, Pollution, and Profit* (Berkeley: North Atlantic Books, 2016)

3. Jacobus Vijverberg, Ferdinand A. Sibbing, and Eshete Dejen, "Lake Tana: Source of the Blue Nile," in *The Nile: Origin, Environments, Limnology and Human Use*, ed. Henri J. Dumont (Berlin: Springer, 2009), 164.

4. Vijverberg, Sibbing, and Eshete, "Lake Tana: Source of the Blue Nile," 183.

5. John Anthony Allan, "Nile Basin Asymmetries: A Closed Fresh Water Resource, Soil Water Potential, the Political Economy and Nile Transboundary Hydropolitics," in *The Nile: Origin, Environments, Limnology and Human Use*, ed. Henri J. Dumont (Berlin: Springer, 2009), 753.

6. Wossenu Abtew, and Shimelis Behailu Dessu, *The Grand Ethiopian Renaissance Dam on the Blue Nile* (Cham: Springer, 2019), 5.

7. Abtew, and Dessu, *The Grand Ethiopian Renaissance Dam on the Blue Nile*, 161.

8. Dalia M. Gouda, "The Grand Ethiopian Renaissance Dam, Agriculture, and the Rural Poor in Egypt," in *Grand Ethiopian Renaissance Dam Versus Aswan High*

Dam: A View from Egypt, eds. Abdelazim M. Negm and Sommer Abdel-Fattah (Cham: Springer, 2019), 199.

9. Virginia DeLancey, "The Economy," in *Sudan: A Country Study*, ed. Berry LaVerle, 5th edition (Washington, D.C.: Federal Research Division, Library of Congress, 2015), 149.

10. Virginia DeLancey, "The Economy," 171.

11. Agathe Maupin, *Energy Dialogues in Africa: Is the Grand Ethiopian Renaissance Dam Transforming Ethiopia's Regional Role?*. (Johannesburg: South African Institute of International Affairs, 2016), 7.

12. Feilding Cage, Samuel Granados, and Michael Ovaska, "Power Struggle," *Reuters*, July 29, 2021, accessed on August 20, 2023. https://graphics.reuters.com/ETHIOPIA-DAM/movanmkbmpa/index.html.

13. Dalia Abdelhady et al. "The Nile and the Grand Ethiopian Renaissance Dam: Is There a Meeting Point between Nationalism and Hydrosolidarity?," *Journal of Contemporary Water Research & Education* 155, no. 1 (July 2015): 77, accessed on August 20, 2023. https://doi.org/10.1111/j.1936-704X.2015.03197.x.

14. Bruce Keith, David Ford, and Radley Horton. "Considerations in Managing the Fill Rate of the Grand Ethiopian Renaissance Dam Reservoir using a System Dynamics Approach," *The Journal of Defense Modeling and Simulation: Applications, Methodology, Technology* 14, no. 1 (2016): 42, accessed on August 20, 2023. https://doi.org/10.1177/1548512916680780.

15. H. Abd-Elhamid, I. Abdelaty, and M. Sherif, "Evaluation of Potential Impact of Grand Ethiopian Renaissance Dam on Seawater Intrusion in the Nile Delta Aquifer," *International Journal of Environmental Science and Technology* 16, (May 2019): 2331, accessed on August 20, 2023. https://doi.org/10.1007/s13762-018-1851-3.

16. Osama M. Sallam, "Vision for Future Management of Groundwater in the Nile Delta of Egypt After Construction of the Ethiopian Dams," *Hydrology: Current Research* 9, no. 3 (2018). https://doi.org/10.4172/2157-7587.1000302.

17. Reuters, "Sudan says Ethiopian Dam made no Impact on Floods this Year," *Reuters*, August 25, 2021, accessed on August 20, 2023. https://www .reuters .com /world /africa /sudan -says -ethiopian -dam -made -no -impact -floods -this -year -2021 -08 -25/.

18. Stefan Liersch, Hagen Koch, and Fred Fokko Hattermann, "Management Scenarios of the Grand Ethiopian Renaissance Dam and Their Impacts under Recent and Future Climates," *Water* 9, (September 2017): 728, accessed on August 20, 2023. https://doi.org/10.3390/w9100728.

19. Youssef M. Hamada, *The Grand Ethiopian Renaissance Dam, its Impact on Egyptian Agriculture and the Potential for Alleviating Water Scarcity* (Cham: Springer, 2017), 81.

20. Okbazghi Yohannes, *Water Resources and Inter-Riparian Relations in the Nile Basin: The Search for an Integrative Discourse* (Albany: State University of New York Press, 2008), 37.

21. *Water Resources and Inter-Riparian Relations in the Nile Basin: The Search for an Integrative Discourse*

22. BBC, "East Africa seeks more Nile water from Egypt," *BBC*, May 14, 2010, accessed on August 15, 2023. http://news .bbc .co .uk /2 /hi /africa /8682387 .stm.

23. Dina Ezzat, "Egyptian Diplomats Left Red-faced after Nile Dam meeting Broadcast Live," *Ahramonline*, June 4, 2013, accessed on August 15, 2023. https://english.ahram.org.eg/NewsContent/1/64/73144/Egypt/Politics-/Egyptian-diplomats-left-redfaced-after-Nile-dam-me.aspx.

24. Kevin G. Wheeler et al., "Understanding and Managing New Risks on the Nile with the Grand Ethiopian Renaissance Dam," *Nature Communications* 11, (October 2020): 5222, accessed on August 20, 2023, https://doi.org/10.1038/s41467-020-19089-x.

25. Ruth Michaelson, "'It'll cause a Water War': Divisions Run Deep as Filling of Nile Dam Nears," *The Guardian*, April 23, 2020, accessed on August 20, 2023. https://www.theguardian.com/global-development/2020/apr/23/itll-cause-a-water-war-divisions-run-deep-as-filling-of-nile-dam-nears.

26. Giulia Paravicini, "Ethiopia Completes Third Phase of Filling Giant Nile dam," *Reuters*, August 12, 2022, accessed on August 20, 2023. https://www.reuters.com/world/africa/ethiopia-completes-third-phase-filling-giant-nile-dam-2022-08-12/.

27. Mulatu Wubneh, "Ethnic Identity Politics and the Restructuring of Administrative Units in Ethiopia," *International Journal of Ethiopian Studies* 11, no. 1 & 2 (2017): 128, accessed on August 20, 2023, https://www.jstor.org/stable/26586251.

28. Matina Stevis-Gridneff, "Nobel Peace Prize Awarded to Abiy Ahmed, Ethiopian Prime Minister," *New York Times*, October 11, 2019, accessed on August 20, 2023. https://www.nytimes.com/2019/10/11/world/africa/nobel-peace-prize.html.

29. Andres Schipani, and David Pilling, "After the War Ends, can Ethiopia's Economic "Miracle" get Back on Track?," *Financial Times*, June 20, 2022, accessed on August 20, 2023. https://www.ft.com/content/a74b5486-d2fa-4cdd-98d0-20a7eaf4ede2.

30. Tefera Negash Gebregziabher, "Ideology and Power in TPLF's Ethiopia: A Historic Reversal in the Making?," *African Affairs* 118, no. 472 (July 2019): 469, accessed on August 20, 2023, https://doi.org/10.1093/afraf/adz005.

31. Jason Burke, "Rise and Fall of Ethiopia's TPLF – from Rebels to Rulers and Back," *The Guardian*, November 25, 2020, accessed on August 20, 2023. https://www.theguardian.com/world/2020/nov/25/rise-and-fall-of-ethiopias-tplf-tigray-peoples-liberation-front.

32. Alex de Waal, "Ethiopia Civil War: Why Fighting has Resumed in Tigray and Amhara," *BBC*, September 1, 2022, accessed on August 20, 2023. https://www.bbc.com/news/world-africa-62717070.

33. Susanne Schmeier, *Governing International: Watercourses River Basin Organizations and the Sustainable Governance of Internationally shared Rivers and Lakes* (London: Routledge, 2015), 243.

34. Susanne Schmeier, *Governing International.*

35. Cheikh Faye, "Organization for the Development of the Senegal River Basin (omvs) and Integrated Water Resources Management (iwrm): What Benefits and Difficulties of the omvs for iwrm in Senegal?," *Folia Geographica* 61, no. 1 (2019): 23.

36. Thomas C. Johnson, and John O. Malala, "Lake Turkana and Its Link to the Nile," in *The Nile: Origin, Environments, Limnology and Human Use*, ed. H. J. Dumont. (Berlin: Springer Science+Business Media, 2009), 291.

37. Thomas C. Johnson, and John O. Malala, "Lake Turkana and Its Link to the Nile," In *The Nile: Origin, Environments, Limnology and Human Use*, ed. H. J. Dumont. (Berlin: Springer Science+Business Media, 2009), 288.

38. International Rivers, "Ethiopia's Gibe III Dam: Sowing Hunger and Conflict," *International Rivers*, last modified January 2011, accessed on August 20, 2023. https://archive .internationalrivers .org /sites /default /files /attached -files /gibe3factsheet2011 .pdf.

39. Elliot Fratkin, "Ethiopia's Pastoralist Policies: Development, Displacement and Resettlement," *Nomadic Peoples* 18, no. 1 (2014): 105, accessed on August 20, 2023, http://www.jstor.org/stable/43124163.

40. Elliot Fratkin, "Ethiopia's Pastoralist Policies," 108, accessed on August 20, 2023, http://www.jstor.org/stable/43124163.

41. Carr, Claudia J. *River Basin Development and Human Rights in Eastern Africa — A Policy Crossroads* (Cham: Springer, 2017), 81.

42. Carr, Claudia J. *River Basin Development and Human Rights in Eastern Africa.*

43. Vidal, John. "Ethiopia Dam will turn Lake Turkana into 'Endless Battlefield', Locals Warn," *The Guardian*, January 13, 2015, accessed on August 20, 2023. https://www.theguardian.com/global-development/2015/jan/13/ethiopia-gibe-iii-dam-kenya.

44. Ibid, Carr, 2017

45. The Oakland Institute, "Understanding Land Investment Deals in Africa: Half a Million Lives Threatened by Land Development for Sugar Plantations in Ethiopia's Lower Omo Valley," The Oakland Institute, last modified September 10, 2011, accessed on August 20, 2023. https://www .oaklandinstitute .org /sites /oaklandinstitute .org /files /Land Deal _Brief Ethiopia Omo Valley .pdf.

46. Vidal, John. "Ethiopia Dam will turn Lake Turkana into 'Endless Battlefield', Locals Warn." https://www.theguardian.com/global-development/2015/jan/13/ethiopia-gibe-iii-dam-kenya.

47. Braden R. Allenby, "Environmental Security: Concept and Implementation," *International Political Science Review / Revue Internationale de Science Politique* 21, no. 1 (2000): 5–21.

REFERENCES

Abdelhady, Dalia Karin Aggestam, Dan-Erik Andersson, Olof Beckman, Ronny Berndtsson, Karin Broberg Palmgren, Kaveh Madani, Umut Ozkirimli, Kenneth M. Persson and Petter. "The Nile and the Grand Ethiopian Renaissance Dam: Is There a Meeting Point between Nationalism and Hydrosolidarity?." *Journal of Contemporary Water Research & Education* 155, no. 1 (July 2015): 73–82, accessed on August 20, 2023. https://doi.org/10.1111/j.1936-704X.2015.03197.x

Abd-Elhamid, H., Ismail Abd-Elaty, and Mohsen Sherif. "Evaluation of Potential Impact of Grand Ethiopian Renaissance Dam on Seawater Intrusion in the Nile Delta Aquifer." *International Journal of Environmental Science and Technology*

16, no. 5 (May 2019): 2321–2332, accessed on August 20, 2023. https://doi.org/10.1007/s13762-018-1851-3.

Abtew, Wossenu, and Shimelis Behailu Dessu. *The Grand Ethiopian Renaissance Dam on the Blue Nile*. Cham: Springer, 2019.

Allan, John Anthony. "Nile Basin Asymmetries: A Closed Fresh Water Resource, Soil Water Potential, the Political Economy and Nile Transboundary Hydropolitics." In *The Nile: Origin, Environments, Limnology and Human Use*, edited by Henri J. Dumont, 749–770. Berlin: Springer, 2009.

Allenby, Braden R. "Environmental Security: Concept and Implementation." *International Political Science Review/Revue Internationale de Science Politique* 21, no. 1 (January 2000): 5–21, accessed on August 20, 2023. http://www.jstor.org/stable/1601426.

BBC. "East Africa Seeks More Nile Water from Egypt." *BBC*, May 14, 2010, accessed on August 20, 2023. http://news.bbc.co.uk/2/hi/africa/8682387.stm.

Burke, Jason. "Rise and Fall of Ethiopia's TPLF – from Rebels to Rulers and Back." *The Guardian*, November 25, 2020, accessed on August 20, 2023. https://www.theguardian.com/world/2020/nov/25/rise-and-fall-of-ethiopias-tplf-tigray-peoples-liberation-front.

Cage, Feilding, Samuel Granados, and Michael Ovaska. "Power Struggle." *Reuters*, July 29, 2021, accessed on August 20, 2023. https://graphics.reuters.com/ETHIOPIA-DAM/movanmkbmpa/index.html.

Carr, Claudia J. *River Basin Development and Human Rights in Eastern Africa — A Policy Crossroads*. Cham: Springer, 2017.

Chellaney, Brahma. *Water, Peace, and War: Confronting the Global Water Crisis*. Lanham: Rowman & Littlefield, 2013.

DeLancey, Virginia. "The Economy." In *Sudan: A Country Study*, edited by Berry LaVerle, 5th edition, 144–220. Washington, D.C.: Federal Research Division, Library of Congress, 2015.

Ezzat, Dina. "Egyptian Diplomats Left Red-faced after Nile Dam Meeting Broadcast Live." *Ahramonline*, June 4, 2013, accessed on August 15, 2023. https://english.ahram.org.eg/NewsContent/1/64/73144/Egypt/Politics-/Egyptian-diplomats-left-redfaced-after-Nile-dam-me.aspx.

Faye, Cheikh. "Organization for the Development of the Senegal River Basin (omvs) and Integrated Water Resources Management (iwrm): What Benefits and Difficulties of the omvs for iwrm in Senegal?." *Folia Geographica* 61, no. 1 (2019): 17–35.

Elliot Fratkin, "Ethiopia's Pastoralist Policies: Development, Displacement and Resettlement." *Nomadic Peoples* 18, no. 1 (2014): 94–114, accessed on August 20, 2023. http://www.jstor.org/stable/43124163.

Gouda, Dalia M. "The Grand Ethiopian Renaissance Dam, Agriculture, and the Rural Poor in Egypt." In *Grand Ethiopian Renaissance Dam Versus Aswan High Dam: A View from Egypt*, edited by Abdelazim M. Negm and Sommer Abdel-Fattah, 197–230. Cham: Springer 2019.

Gebregziabher, Tefera Negash. "Ideology and Power in TPLF's Ethiopia: A historic Reversal in the Making?." *African Affairs* 118, no. 472 (July 2019): 463–484, accessed on August 20, 2023. https://doi.org/10.1093/afraf/adz005.

Hamada, Youssef M. *The Grand Ethiopian Renaissance Dam, its Impact on Egyptian Agriculture and the Potential for Alleviating Water Scarcity*. Cham: Springer, 2017.

International Rivers. "Ethiopia's Gibe III Dam: Sowing Hunger and Conflict." Last modified January 2011. Accessed on August 20, 2023. https://archive.internationalrivers.org/sites/default/files/attached-files/gibe3factsheet2011.pdf.

Johnson, Thomas C., and John O. Malala. "Lake Turkana and Its Link to the Nile." In *The Nile: Origin, Environments, Limnology and Human Use*, edited by. H. J. Dumont, 287–304. Berlin: Springer Science+Business Media, 2009.

Keith, Bruce, David Ford, and Radley Horton. "Considerations in Managing the Fill Rate of the Grand Ethiopian Renaissance Dam Reservoir using a System Dynamics Approach." *The Journal of Defense Modeling and Simulation: Applications, Methodology, Technology* 14, no. 1 (2016): 33–43, accessed on August 20, 2023. https://doi.org/10.1177/1548512916680780.

Liersch, Stefan, Hage Koch, and Fred Fokko Hattermann. "Management Scenarios of the Grand Ethiopian Renaissance Dam and Their Impacts under Recent and Future Climates." *Water* 9, no. 10 (September 2017): 728, accessed on August 20, 2023. https://doi.org/10.3390/w9100728.

Maupin, Agathe. *Energy Dialogues in Africa: Is the Grand Ethiopian Renaissance Dam Transforming Ethiopia's Regional Role?*. Johannesburg: South African Institute of International Affairs, 2016.

Michaelson, Ruth. "'It'll Cause a Water War': Divisions Run Deep as Filling of Nile Dam Nears." *The Guardian*, April 23, 2020. Accessed on August 15, 2023. https://www.theguardian.com/global-development/2020/apr/23/itll-cause-a-water-war-divisions-run-deep-as-filling-of-nile-dam-nears.

Paravicini, Giulia. "Ethiopia Completes Third Phase of Filling Giant Nile Dam." *Reuters*, August 12, 2022. Accessed on August 20, 2023. https://www.reuters.com/world/africa/ethiopia-completes-third-phase-filling-giant-nile-dam-2022-08-12/.

Reuters. "Sudan says Ethiopian Dam made no Impact on Floods this Year." *Reuters*, August 25, 2021. Accessed on August 20, 2023. https://www.reuters.com/world/africa/sudan-says-ethiopian-dam-made-no-impact-floods-this-year-2021-08-25/.

Sallam, Osama M. "Vision for Future Management of Groundwater in the Nile Delta of Egypt After Construction of the Ethiopian Dams." *Hydrology: Current Research* 9, no. 3 (2018): 302, accessed on August 20, 2023. https://doi.org/10.4172/2157-7587.1000302.

Shiva, Vandana. *Water Wars: Privatization, Pollution, and Profit*. Berkeley: North Atlantic Books, 2016.

Stevis-Gridneff, Martina. "Nobel Peace Prize Awarded to Abiy Ahmed, Ethiopian Prime Minister." *New York Times*, October 11, 2019. Accessed on August 20, 2023. https://www.nytimes.com/2019/10/11/world/africa/nobel-peace-prize.html.

Schmeier, Susanne. *Governing International: Watercourses River Basin Organizations and the Sustainable Governance of Internationally Shared Rivers and Lakes*. London: Routledge, 2015.

The Oakland Institute. "Understanding Land Investment Deals in Africa: Half a Million Lives Threatened by Land Development for Sugar Plantations in Ethiopia's

Lower Omo Valley." Last modified September 10, 2011. Accessed on August 20, 2023. https://www.oaklandinstitute.org/sites/oaklandinstitute.org/files/Land_Deal_Brief_Ethiopia_Omo_Valley.pdf.

Waal, Alex. "Ethiopia Civil War: Why Fighting has Resumed in Tigray and Amhara." *BBC*, September 1, 2022. Accessed on August 20, 2023. https://www.bbc.com/news/world-africa-62717070.

Wheeler, Kevin G. et al. "Understanding and Managing New Risks on the Nile with the Grand Ethiopian Renaissance Dam." *Nature Communications* 11, (October 2020): 5222, accessed on August 20, 2023. https://doi.org/10.1038/s41467-020-19089-x.

Wubneh, Mulatu. "Ethnic Identity Politics and the Restructuring of Administrative Units in Ethiopia." *International Journal of Ethiopian Studies* 11, no. 1 & 2 (2017): 105–138, accessed on August 20, 2023. https://www.jstor.org/stable/26586251.

Vidal, John. "Ethiopia Dam will Turn Lake Turkana into 'Endless Battlefield', Locals Warn." *The Guardian*, January 13, 2015. Accessed on August 20, 2023. https://www.theguardian.com/global-development/2015/jan/13/ethiopia-gibe-iii-dam-kenya.

Vijverberg, Jacobus, Ferdinand A. Sibbing, and Eshete Dejen. "Lake Tana: Source of the Blue Nile." In *The Nile: Origin, Environments, Limnology and Human Use*, edited by Henri J. Dumont, 163–192. Berlin: Springer, 2009.

Yohannes, Okbazghi. *Water Resources and Inter-Riparian Relations in the Nile Basin: The Search for an Integrative Discourse*. Albany: State University of New York Press, 2008.

Chapter 8

Revisiting the 2006 Toxic Waste Dump in Côte d'Ivoire

Balla M. Keita

In the night of August 19, 2006, a local waste management company, *Compagnie Tommy* (Tommy), dumped toxic wastes in at least eighteen sites in Abidjan, the capital city of Côte D'Ivoire, West Africa. The next day, August 20, the entire city was enveloped in a foul smell that hung thick in the air. A powerful and suffocating smell spread throughout the city, residence to some five million people. The catastrophe was felt deeply because of the close proximity of the discharge site to residential neighborhoods. Immediately, people in Abidjan began to complain about abdominal pain and skin and eye irritation. Hospitals were flooded with many patients suffering from the same symptoms. The situation worsened as time went by. More than 100,000 victims had besieged the hospitals,[1] and it all turned out to be the consequences of the waste dumped in their residences, places of employment, and schools. Caught by surprise, Ivoirian authorities called an emergency intergovernmental meeting to understand the seriousness of the crisis. Officially, the death count was put at fifteen persons,[2] although the actual casualty number may have been greater. Economically, the consequences were very serious and severe. Schools, markets, and businesses could not operate efficiently because of the foul odor of the waste as many people kept at home or places they felt safe.

Immediate clearing and cleaning efforts were delayed, causing further stress on the population because the Ministry of Health had not ascertained the exact nature and composition of the toxic waste dumped and could not, therefore, conduct a study to address the health impact of the crisis and assess the short- and long-term consequences of the foul smell and bodily irritations. The inability of Ivoirian authorities to deal effectively with this crisis immediately turned the incident into a political scandal. After many inquiries, the Ivoirian government discovered that the waste dumped in eighteen sites belonged to a Singapore-incorporated multinational commodity trading company, the second largest oil trader in the world, by the name of Trafigura Group Pte. Ltd (Trafigura).[3] Moreover, the waste was created onboard a ship by the name of Probo Koala, a ship registered in Panama and chartered by Trafigura.[4] As the investigation went on, Ivoirian authorities found that the waste came from Europe.[5]

There are several international laws that strictly forbid western and industrial nations from dumping toxic waste in poor countries and into the ocean. Encyclopedia Brittanica defines toxic waste as chemical waste material capable of causing death or injury to life.[6] Waste is considered toxic if it is poisonous, radioactive, explosive, carcinogenic (causing cancer), mutagenic (causing damage to chromosomes), teratogenic (causing birth defects), or bioaccumulative (i.e., increasing in concentration at the higher ends of food chains).[7] To understand the waste dump incident, it is important to look into how the multinational oil giant Trafigura processed the toxic waste and how it arrived in the Port of Abidjan.

ORIGIN OF THE TOXIC WASTE

A company by the name of PMI Trading, LTD (PMI) sold a big amount of raw petroleum called coker nafta. PMI was a company that collaborated closely with the state of Mexico. One particularity of coker nafta was its ability to be easily mixed with other fuels; however, in the coker nafta, a substance called mercaptan sulfur was found in high quantity. Coker nafta is an unwanted product that the oil and gas industry does not like to use. It is difficult to dispose of and it does not smell good, and therefore it is inexpensive. This is why the Singaporean-based oil and shipping company Trafigura was able to acquire the coker nafta at a great bargain. To sell the product acquired from PMI, Trafigura was compelled to refine it. A complex process called caustic washing was a necessary step to sell the fuel as refined. Caustic washing required mixing the coker nafta with caustic soda. Another option was available. The process was called Merox, but the cost seemed very high to Trafigura executives. They opted for the caustic washing process instead. The problem with caustic washing is that it produces a highly toxic waste. For this reason, Western countries, including the United States, do not allow caustic washing because environmental agencies are reluctant to accept the disposal of the toxic caustic waste, even after a proper cleaning process.[8]

Executives of Trafigura were aware of the challenges of disposing the toxic waste; however, the potential for large profit made them underestimate the process of refining and disposal. Trafigura found only two companies that were willing to do the caustic washing process.[9] Executives at Trafigura focused on profit and did not consider the important step of first finding a company willing to do the caustic washing, and second, another one to dispose of the waste. One particularity of the coker nafta was its high sulfuric density, making it unattractive in North America and Europe due to its negative impact on human health. High sulfuric petroleum is not sold at the pump in European countries and in North America. Diesel fuel found at the gas pump in North America and Europe have sulfur content of ten parts per million[10] (ppm), while the coker nafta has much higher sulfur content of between 1,000 and 3,000 ppm.[11] Unfortunately, due to the cheap cost, African countries readily request this type of fuel, which leaves a heavy scent throughout the city.

When Trafigura acquired the coker nafta from PMI in early January 2006, it sent it to the United Arab Emirates at Fujarah for caustic washing.[12] Later on, Trafigura continued the caustic washing process in North Africa. Tunisia was the North African country selected using a company by the name of TANKMED, which was willing to do caustic washing at the Port of La Skhira.[13] While processing the cleaning of the coker nafta, an incident happened in the middle of March 2006. A gas leak led to a foul odor at the facility, causing breathing difficulties to the workers that were sent to the hospital for treatment of the damages caused by the fumes.[14] Tunisian authorities were concerned about this incident, leading to an investigation of TANKMED's facilities. According to the Tunisia International Center for Environmental Studies, the foul odor was caused by the petrol washed with caustic soda. It determined that TANKMED's facilities were not specialized enough to handle these types of processes. Subsequently, local authorities decided to halt the caustic washing process at the Port of La Skhira.[15] Trafigura officials, confronted with this new challenge, decided on a new approach, consisting of conducting caustic washing at sea. The Probo Koala ship was chosen to undertake this operation. After the caustic washing onboard the ship, Trafigura still had to find a way to dispose of the waste. In addition, Trafigura needed to acquire the caustic soda and selected a U.S. company by the name of Univar to obtain said caustic soda.[16] The U.S. company warned Trafigura on the dangers of the caustic soda, which would necessitate a Dangerous Goods Certificate.[17]

Trafigura could not secure the caustic soda with Univar due in part to all the risks involved in the process. Moreover, Amnesty International and Greenpeace required that Trafigura disclose more details regarding the transport of caustic soda in U.S. coastal waters.[18] Therefore, Trafigura turned to a company by the name of WRT, a Dutch company. It allowed Trafigura to begin the process of caustic washing on the Probo Koala. By April 2006, Trafigura successfully accomplished the first caustic washing, which was attempted to be unloaded in Tunisia as washed gasoline. Due to the washing incident at the Port of La Skhira, the waste from the caustic wash was not unloaded in Tunisia. Later on, Trafigura was able to obtain caustic soda, delivered directly ship to ship (STS Operations)[19] on board the Probo Koala. An estimated 200 tons of caustic soda was delivered on the Probo Koala in STS operations by the middle of 2006. The Tunisian company refused to accept the "washed" gasoline because of the strong odor emanating from it. As the caustic wash proceeded, the waste on board the ship amounted to over 500 m^3, which was kept in Probo Koala's slop tanks. It became a challenge for Trafigura to dispose of the waste.[20]

THE TRANSPORT FROM EUROPE TO AFRICA

Trafigura reached out to four different countries in Europe to dispose of the waste in the slop tanks: Malta, Italy, Gibraltar, and France.[21] But they all refused: the waste was hazardous, and treating the chemical content presented a lot of danger. Later, Trafigura decided to reach out to Amsterdam Port Services (APS), known for its ability to process these types of waste. The authorities were involved in accordance with the International Convention for the Prevention of Pollution from Ships (MARPOL Convention 1973/78).[22] During the process, Trafigura used the term "slops" to describe the chemical waste. Slops was a routine procedure of cleaning undertaken for most cargo residues, but Trafigura's waste was due to caustic washing, not just slops. APS started to unload the waste on the barge, assuming it was normal waste, but the foul smell alerted local authorities. They detected hydrogen sulfide gas in the air,[23] which was considered hazardous and can have effect on health, causing breathing difficulties, nausea, vomiting, and irritation of nose, throat, and eye;[24] in high concentration, it could lead to coma and death.[25] Trafigura's toxic waste was ultimately dumped in Côte d'Ivoire. Trafigura's plan was to mix the coker nafta with gasoline and sell it to the West African market for a profit of $7 million per shipment. But before we get into how it ended up in Côte d'Ivoire, it is important to go over the political climate and the tensions within this prosperous West African nation.

UNSTABLE POLITICAL CLIMATE IN CÔTE D'IVOIRE

Côte d'Ivoire, formerly known as the Ivory Coast, is located in West Africa, between Ghana and Liberia, 520.96 miles north of the Equator. It has a population of 26.38 million (in 2020). Côte d'Ivoire is a country with beach resorts, rainforests, and a French-colonial legacy. It exports 40 percent of the world cocoa; thus, it is the largest exporter of cocoa. In addition, the country exports coffee and possesses mineral and petroleum resources. Socially, Côte d'Ivoire is made up of more than sixty ethnic groups. Each group has its own view on the distribution of state resources. In addition to this ethnic diversity, there are religious differences among the various ethnic groups. The southern groups are mostly Christians, while the northern ethnic groups are mostly Muslims. A small minority professes African religious practices. This mixture of different ethnic groups and religious practices on a relatively small territory carries the seeds of discord.

Environmental security is a crucial component for nations, especially African nations who are vulnerable in a fast evolving international environment dominated by powerful industrial countries. In September 2002, political differences based on ethnic, regional, tribal, and religious origins led to a civil war that lasted for ten years. Ivoirian authorities and the civil society are still struggling to bring harmony in the country. The waste dumping in the middle of a civil war exacerbated the conflict, pitting communities against the central government led by former President Laurent Gbagbo. As soon as the toxic waste was dumped, local residents began to complain about health issues, which led to fifteen deaths and several hospitalizations at first count, while 26,000 people were treated for breathing problems, skin irritation, and headaches.

The caustic washing produced highly toxic waste that was highly odorous. Moreover, Trafigura knew that very few facilities could handle this type of toxic waste and made no plans to safely dispose of the toxic waste. They first attempted to dispose of this waste in Amsterdam, where the disposal company asked for $620 million and Trafigura rejected this offer for high cost.[26] Later, Trafigura tried to dispose of the toxic waste in Lagos, Nigeria, for just $7,000 and failed again.[27] Battling with an internal political crisis and civil war, Côte d'Ivoire seemed the best country to unload the toxic waste for $18,000.[28] It is important to note that in 2006, Trafigura made a net profit of $511 million.[29] On August 19, 2006, the toxic waste was finally dumped in Abidjan illegally with the help of a local waste company. Over 100,000 people sought medical attention a few days later.[30]

When the Dutch police began an investigation on the toxic waste, Trafigura asked the Ivoirian waste company to create a false invoice of $100,000 for the disposal of the toxic waste.[31] The consequences of the waste dumped in Abidjan required extensive cleanup and decontamination of the environment and of all the sites where the toxic waste was dumped. This incident highlights a very important concern, namely, why did the Probo Koala dump toxic waste in a country not equipped with facilities to handle this type of waste? Sixteen years later, the population and government in Abidjan are still unaware of the long-term health impact of the dumping. Amnesty International and Greenpeace's 2012 report showed that the victims have suffered a range of serious health issues, including respiratory problems, abdominal pains, and digestive problems due to the effects of the chemicals dumped by Trafigura. Moreover, the company refused to disclose the content of the toxic waste, which hampered the cleanup efforts and medical response to the disaster.

ONGOING CIVIL WAR IN CÔTE D'IVOIRE

In September 2002, a rebellion erupted in the Northern part of Côte d'Ivoire. The central government was unable to stop the advance of the northern rebels. France and the UN intervened by positioning themselves between the two armed forces to prevent a civil war. The sources of the conflict stemmed from unequal distribution of state resources and old ethnic tensions that have undermined the country's stability. Since Côte d'Ivoire's ascension to independence in 1960, the Baoule ethnic group had maintained itself in the presidential office and accused of monopolizing state resources for more than thirty years. President Felix Houphouet-Boigny of the Baoule ethnic group was the first president of the country. He was able to maintain a relative peace from 1960 until his death in office on December 7, 1993.

While in office, President Houphouet-Boigny positioned his successor Henry Konan Bedie, a Baoule as well, as the Speaker of the National Assembly. A provision in the Constitution of Côte d'Ivoire stipulated that, in case of vacancy or death of the President, the Speaker of the National Assembly would assume the office in the interim immediately and organize the elections.[32] He/she could also present himself/herself as a candidate if he/she chose to.[33] However, the political leaders' unbridled personal ambitions led the country to the brink of a civil war.

The new Constitution crafted by President Henry Konan Bedie, who ascended the Presidency after the death of President Houphouet-Boigny, prevented any citizen born in Côte d'Ivoire whose parents were foreigners to run in the presidential election. Many observers saw this as a tailor-made Constitution against his opponent, Alassane Ouattara in the presidential race, because the latter's parents were born in Burkina Faso and had migrated to Côte d'Ivoire. Amendments were made successively to the Constitution by President Bedie, General Robert Guei, and President Laurent Gbagbo to keep Alassane Ouattara from running in the presidential race. In 2002, the manipulation of the Ivoirian Constitution for political objectives resulted in the northern rebellion as a way to demand social justice for the northern ethnic groups and impose Alassane Ouattara as a candidate for the presidency.[34]

The 2000 presidential election between General Guei and opposition leader Laurent Gbagbo was settled in violence. General Guei refused to accept the victory of longtime opposition leader Laurent Gbagbo. Instead, General Guei clung on to power and used the army to maintain himself in power. Laurent Gbagbo did not yield to the intimidation of General Guei and called his partisans to the street to claim their victory. From this point on, Côte d'Ivoire entered a horrifying phase of its history. The outcome was devastating for the young nation; General Guei was unable to keep himself in power from his opponent. Consequently, Gbagbo was proclaimed the winner of the presidential election by the Constitutional Council. One of the first gestures he made as a president was an official report on the electoral violence. According to this report, the number of deaths was estimated at 303 persons. In addition, 65 people were missing, and 1,546 people were wounded.[35]

The major problem with the 2000 election was the fact that the main contenders, namely, Alassane Outtara (of the political party *Rassemblement des républicains* (RDR)) and Henri Konan Bedie (of the very first Ivoirian political party *Parti démocratique de Côte d'Ivoire* (PDCI)), were excluded from this election. Bedie was still in exile in Paris, while Ouattara was struggling to clarify his eligibility to run for the presidential office. Therefore, Ouattara's partisans requested the reorganization of the elections since an important electoral section of voters was systematically excluded. Reporter R. J. Lique of Afrique-Express reported that President Laurent Gbagbo bragged that he was elected with 59.36 percent of votes, which was a misleading representation of the situation. In total, 5,475,143 were registered to vote in the elections. Gbagbo only obtained 1,065,597 votes, which was barely 20 percent of registered voters.

With a flawed election, President Laurent Gbagbo should have taken advantage of his presidential mandate to bring reconciliation among Ivoirians by practicing open politics and associating brothers that were enemies in the past in the republic. Instead, he vowed to definitely exclude Ouattara from the Ivoirian political life and significantly reduce his electoral potential.[36] Gbagbo's partisans were mobilized to form a resistance against Ouattara's supporters. Meanwhile, Gbagbo remained confident since his French socialist allies gave him a green light.[37]

The polarization of ethnic groups in Côte d'Ivoire introduced dissension within the population. The northerners were targeted by government forces during regular checkpoints; their administrative documents that were acquired legally were destroyed by law enforcement agents who saw all northerners and Muslims as foreigners. Laurent Gbagbo's presidential term in office increased the security dilemma. The rise of the Krou ethnic group to power (Gbagbo's ethnic group) allowed them to secure the instrument of violence to their advantage, thereby leaving northerners in a position of weakness and vulnerability. As a reaction to the rising injustice, northerners saw it fit to organize militarily to counter the Krou ethnic group's threatening position. The consequences of this reaction led to the security dilemma within the country. Thus, the regime of Laurent Gbagbo became increasingly perceived as an oppressor to northerners and the Muslim populations. They rallied en masse behind Alassane Ouattara because they saw themselves in the discrimination and injustices he suffered at the hands of state officials. From this point forward, it was clear that the social fracture was consumed, and the only solution to the social injustices and tensions gradually shifted into a succession of coups and finally to an armed rebellion.[38]

CÔTE D'IVOIRE AT WAR

On September 19, 2002, an organized detachment of soldiers coming from Burkina Faso launched a simultaneous attack in the capital city of Abidjan in the south, in Bouake in the center of the country, and in the northern town of Korhogo. The Loyalists' forces reacted with fierce strength, making the rebel retreat; the rebels failed to capture Abidjan, but succeeded in controlling the center city of Bouake and the northern city of Korhogo. The city of Bouake held 40 percent of the nation's military equipment. While this uprising could have easily been anticipated by most observers of the Ivoirian crisis, no one could predict the strength, courage, and determination of the rebel groups that were organized under the name of Patriotic Movement of Côte d'Ivoire (MPCI). Bouake was the second largest city, the base of important military equipment, and the home of the Central Bank. Later, the Central Bank would be robbed billions of CFA francs by the rebels who were short on cash.[39]

At the time of this insurgency, President Gbagbo was on an official trip to Italy, scheduled to meet the Pope. When the news of this attack reached him, he ended his trip and returned to Côte d'Ivoire at once. While some were trying to dissuade him from taking such an action, he argued that those who elected him were being killed and he could not stand to be away from his helpless citizens. Gbagbo's regime accused Burkina Faso's President Blaise Compaore of staging the rebellion and giving military training to the rebel soldiers in Ziniare, Compaore's native region. While these accusations were inferred at first, Moise Lida Kouassi, Gbagbo's Defense Minister, publicly accused Burkina Faso of providing a safe haven to the Ivoirian military dissidents.[40] Many observers pointed out that President Blaise Compaore of Burkina Faso was displeased with the treatment of his countrymen in Côte d'Ivoire with the complicity of officials and law enforcement agents. Once the rebellion exploded, the country was divided into two at the center, with the rebels controlling the north 60 percent of Côte d'Ivoire, and the government forces controlling the south 40 percent. However, the assassination of Army Chief of Staff Robert Guei did not sit well with soldiers that were loyal to him.

The rebels were determined to let Gbagbo's regime know that they were not amateur fighters; they were determined in their attacks against the loyalist's forces and were mentally and psychologically prepared for any outcome. The rebels justified the rebellion by the injustices and insecurity endured by the northerners, and they were prepared to secede from the south if a peaceful outcome could not be reached. The rebels demanded the resignation of President Laurent Gbagbo before any negotiation could proceed. The division of Côte d'Ivoire was now official, with the rebel forces, now called *Forces Nouvelles* (News Forces), controlling the north and the governmental forces controlling the south. Contrary to the rebellions in Liberia and Sierra Leone, the Ivoirian rebels were highly organized and made every effort to mobilize and involve the population in their struggle.

Despite the elaborate reorganization of the system of taxation, the rebels and the government would fail to reach a peaceful agreement. The political climate would decline gradually, introducing a serious rift within the political class, the army, and the population. Regrettably, the ethnic, regional, and religious division in Côte d'Ivoire was accomplished; the northern and Muslim population would never trust other ethnic groups and vice versa. For the rebel groups, their new and unexpected international recognition was proof that their rebellion was acknowledged and they now represented a political force. Another advantage was the growing sympathy they were shown by those unhappy with President Gbagbo's agenda for the country. Moreover, the often ambivalent position of France in the conflict would make many wonder about the real intentions of Paris, including the rebels and the government. Accordingly, reaching a peaceful solution between the two factions became a long and arduous process.

FAILURE OF PEACE AGREEMENTS

The failure of the rebellion to depose President Gbagbo plunged the country into chaos. Divided into two, Côte d'Ivoire developed two parallel economies and administrations. Neither side was able to settle the conflict militarily. The rise of tension and stalemate between the opposing parties compelled France to intervene to mediate the crisis. Thus, a meeting was organized in the area of Linas-Marcousis, France, on January 26, 2003, in which the rebels and the Ivoirian authorities met face to face for the first time to sign a peace agreement. The accord signed between President Gbagbo and the rebel forces became known as the Klebers accords. The objectives of the Linas-Marcousis meeting were to solve the differences between the rebels and the government. Therefore, the agreement signed stipulated the maintenance of Laurent Gbagbo as President and invited the rebels to join in a new government.

In May 2003, the army signed a ceasefire with the rebels, but the peace process did not last long. Many other meetings would follow on the African continent, in Lome, Accra, and Pretoria to harmonize the Linas-Marcoussis Accords. Yet, fierce resistance and unwillingness to compromise would be felt on both sides. Prime Minister Seydou Diarra was charged to form the government. The tense relations between the rebels and Gbagbo would be compromised by an attack on the national TV station in which nineteen people were killed. Seydou Diarra resigned and was succeeded by Konan Banny, the chairman of the West African Central Bank. The deterioration of the political climate made the UN deploy the first peacekeeping contingent in March 2004 (UNOCI).[41]

The hostilities between the rebel forces and the government forces escalated. In November 2004, Gbagbo's army launched an air attack on the rebel position in Bouake. This operation became known as *Dignité*. In this attack by the loyalist forces, a French base was bombarded accidentally, killing nine French soldiers. French President Jacques Chirac retaliated by incapacitating the Ivoirian air forces, destroying their military aircraft. In the economic capital of Abidjan, the French troops advanced near President Gbagbo's residence and were met with thousands of Gbagbo's supporters around the prestigious Hôtel Ivoire. Some of the enraged mob threw themselves on the ground to prevent the French mechanized battalion from advancing while others gathered around Gbagbo's residence. They accused France of attempting to overthrow President Gbagbo. Nervous of this unexpected outcome, the French troops opened fire on the crowd, killing many unarmed civilians. The Internal Displacement Monitoring Centre (IDMC) made a profile of the internal displacement in Côte d'Ivoire. The IDMC report pointed out that a diplomat at the hotel said French troops had fired shots in the air to disperse the protesters. Ivoirian state television broadcast images of injured people taken to hospitals. Over the weekend, it also showed bodies of protesters who, it said, had been killed by French forces. However, it was unclear how many casualties there were.[42] This turn of events antagonized France's position in the Ivoirian conflict and discredited its intentions.

NEW HOPES IN THE OUAGADOUGOU PEACE AGREEMENT

After the failure of previous agreements to bring peace, President Gbagbo and the rebel forces found themselves in a political stalemate. The presidential election of 2005 was postponed three times because of the political manipulations, which prevented them from seeing the big picture. Nevertheless, President Gbagbo, fearing to jeopardize his strategic position, decided to engage in a dialogue with the armed rebellion leader Soro Guillaume directly, through the mediation of President Blaise Compaore of Burkina Faso. The tense relations between Burkina Faso and Côte d'Ivoire compelled President Gbagbo to seek the intervention of South African President Thabo Mbeki. This new attempt was a good step in the right direction. As a result, on March 4, 2007, the Ouagadougou Agreement was signed between President Gbagbo and rebel leader Soro Guillaume with the anointment of President Blaise Compaore. Previously, Compaore was accused by Gbagbo's camp of providing a safe haven to the rebels. Indeed, from the very beginning, the Ouagadougou Peace Agreement (OPA) brought a relief to both sides of the conflict, the population and most observers. Soro Guillaume was designated as the new Prime Minister of President Gbagbo and they agreed to work in harmony to unify the country. OPA was expected to be the road map that would lead Côte d'Ivoire out of the crisis, through a democratic election that would finally settle the conflict.

All political actors agreed for the first time on rescheduling the 2005 presidential election in November 2010. Despite all the political promises, the rebel forces did not respect the deadline of disarmament as proposed in the recommendations of OPA. President Gbagbo complained at first but later ignored the rebels' failure to disarm completely as agreed upon in the OPA. Disappointed by the rebels' failure to disarm before the presidential elections as underlined in the recommendations of OPA, President Gbagbo realized that his failure to ensure the implementation of the disarmament put him in a critical political position. His back against the wall, he was compelled to organize the presidential election. This error of judgment would prove fatal in the postelection crisis that will lead to the death of more than 3,000 people.[43] It is in this unstable political climate in Côte d'Ivoire that the toxic waste was dumped by Trafigura in 2006, despite international laws and agreements which prevent such incidents.

INTERNATIONAL LAW AND TOXIC WASTE MANAGEMENT

Various international laws aim to protect humans and the environment from the illegal movement of dangerous waste. The two main international laws that apply to the illegal dumping of toxic waste by Trafigura in Côte d'Ivoire are the Basel Convention and the MARPOL Convention. But first, let us review a few general principles of international law.

International Human Rights Law

International human rights law, along with other goals, also aims to protect humans from the harmful effects of toxic waste, which affects food, water, health, and work. In 1995, the Commission on Human Rights adopted its first resolution concerning the adverse effects of exposure to hazardous substances and toxic waste.[44] The UN Human Rights Special Procedures, in seeking to resolve the human rights issue created by the exposure of people to harmful substances without their prior informed consent, developed a mandate and appointed a Special Rapporteur on toxic and human rights. The Special Rapporteurs are appointed every three years and their mandate can be extended for another three years.[45]

The Vienna Declaration and Program of Action: Adopted in 1993 by the World Conference on Human Rights, the Vienna Declaration and Program of Action recognizes that illicit dumping of toxic and dangerous substances and waste potentially constitutes a serious threat to human rights to life and health of everyone. Consequently, the World Conference on Human Rights calls on all states to adopt and vigorously implement existing conventions relating to the dumping of toxic and dangerous products and waste and to cooperate in the prevention of illicit dumping.[46]

The UN Conference on The Human Environment: The Sixth Principle of the United Nations Conference on the Human Environment (Stockholm Declaration), adopted in 1972, states that the discharge of toxic substances or of other substances and the release of heat, in such quantities or concentrations as to exceed the capacity of the environment to render them harmless, must be halted in order to ensure that serious or irreversible damage is not inflicted upon ecosystems.[47] It adds that the just struggle of the peoples of ill countries against pollution should be supported.[48]

INTERNATIONAL LAWS REGARDING MANAGEMENT OF TOXIC WASTE

The Basel Convention: The Basel Convention on the Control of Transboundary Movements of Hazardous Wastes and their Disposal, mentioned above, was adopted on March 22, 1989, under the UNEP as a multilateral agreement, in Basel, Switzerland.[49] The Basel Convention encourages environmentally sound management of exported and imported waste, especially in developing countries. The Basel Convention was adopted to address deposits of toxic wastes from developed nations dumped in Africa and other developing countries. The Convention became effective in 1992.[50]

The Convention requires that the exporting country notify the receiving country and any transit countries of the proposed shipment.[51] Should an accident occur during the shipment of the waste, the Basel Convention requires that the liable parties notify the potentially affected countries of the accident (articles 8 and 9).[52] To facilitate waste management in an environmentally sound manner, there are some restrictions imposed by the Basel Convention. For instance, exports of waste can only occur under the following circumstances: if the exporting country does not have adequate disposal or recycling capacity; if the exporting country does not have disposal and recycling facilities that can manage the waste in an environmentally sound manner; or if the wastes are required as a raw material for recycling or recovery industries in the importing country.[53]

According to the Basel Convention, a transboundary movement of hazardous or other wastes is deemed illegal if it is done without notification to the state to which the waste is to be taken, requesting consent of that state; on the basis of consent obtained through falsification, misrepresentation, or fraud; if it does not conform in a material way with the documentation; or if it results in deliberate improper disposal (such as dumping) (articles 2 and 9).[54]

Because the toxic waste dumped in Abidjan was hazardous, the Basel Convention applied. Also, all the countries involved were parties to the Basel Convention: Trafigura was headquartered in Switzerland, the Probo Koala was from Panama, and Côte d'Ivoire was the location of the dumping.[55] Nevertheless, Trafigura argued that the Basel Convention did not quite apply to the situation in Côte d'Ivoire and Trafigura tried to reject most of its liability in the case. Despite the arguments put forth, Trafigura was still at fault under the MARPOL Convention.

The MARPOL Convention: It is correct to state that the Basel Convention does not explicitly apply to waste created aboard ships. However, the International Convention for the Prevention of Pollution from Ships (MARPOL) does apply to waste created aboard ships, and particularly to the case of Trafigura's dumping of toxic waste in Côte d'Ivoire. Adopted in 1973 (Convention), modified in 1978 (Protocol), and amended in 1997 (Protocol – Annex VI), MARPOL entered into force on October 2, 1983 (Annexes I and II).[56] According to the International Maritime Organization, MARPOL is designed to prevent pollution of the marine environment by ships from operational or accidental causes, including pollution by oil discharge and other harmful substances.[57]

Annex III contains general requirements for the issuance of detailed standards on packing, marking, labeling, documentation, stowage, quantity limitations, exceptions, and notifications. Annex III also defines *harmful substances* as those which are identified as marine pollutants in the International Maritime Dangerous Goods (IMDG) Code.[58] MARPOL requires that harmful substances be discharged to a reception facility.[59] The Convention requires that ship wastes be managed and disposed of in an environmentally sound manner to prevent pollution to the sea and harm to human health. It provides that slops and wastes generated at sea are brought ashore rather than being disposed of at sea and that member states have an obligation both to provide adequate reception facilities and to ensure that materials discharged ashore are subsequently handled correctly.[60]

As earlier mentioned, Trafigura argued that the wastes in question were ship waste because they were generated on a ship.[61] Furthermore, Trafigura asserted that Côte d'Ivoire was at fault for failing to provide appropriate port facilities for disposing of the waste. A Dutch Court disagreed, noting that the slops could not be regarded as being produced by the normal operation of a ship.[62] Moreover, the Court concluded that Trafigura acted willfully and maliciously by allowing the disposal to continue, fully knowing that the wastes in question were chemical slops with extreme acidity and a very high chemical oxygen demand (COD) content. The Court accentuated the price difference between the quote Trafigura received in Amsterdam as aforementioned ($620 million), based on the actual composition of the waste, and the price Trafigura eventually paid to dispose of its waste in Abidjan ($18,000).[63]

Furthermore, per Article 11(d) of MARPOL, the International Maritime Organization lists waste-receiving facilities and their capacities; Côte d'Ivoire is not listed as having an adequate facility to handle oily waste, a fact that Trafigura was aware of, and as the Dutch Court noted, Trafigura already knew in April 2006 that the discharge of slops for processing was, at the very least, not simple.[64]

According to Trafigura, as the result of litigation, the company paid $198 million to the Government of the Ivory Coast for restoration and for the victims (it was reported that victims never got a hold of that money as it remained in the hands of the corrupt members of the Ivoirian Government);[65] Trafigura settled by paying 1 million Euros in the Netherlands; and Trafigura similarly paid 30 million Pounds to settle claims by 29,614 claimants represented by Leigh Day & Co law firm in the United Kingdom. It was also reported that part of the money paid to Leigh Day & Co for distribution to the claimants never reached its intended beneficiaries.[66]

Similarly, Italy paid a Nigeria town $100 a month to store toxic waste.[67] Two Italian firms stored 18,000 drums of hazardous waste in a small fishing village of Koko, Nigeria.[68] The Nigerian government, just like the Ivoirian Government, was shocked at this news and ordered the Italian ship docked in Lagos, Nigeria, to be seized.[69] According to a British environmental group, 28 percent of the content of the toxic waste had polychlorinated biphenyl (PCB) and a highly toxic compound called dioxin. The Nigerian government was embarrassed just like the Ivoirian government, when the story made international headlines in 1988.[70] Nigeria demanded that Italy removed the waste and arrested fourteen Nigerians involved in smuggling. This shows how wealthier and powerful nations preyed on the vulnerabilities of less developed regions.[71]

LESSONS OF THE WASTE DUMP INCIDENT AND THE WAY FORWARD

It was in a difficult political environment in Côte d'Ivoire that Trafigura's toxic waste was dumped. Conflicts in developing countries are conducive to malpractice and disregard for the environment. While Trafigura negotiated with a local Ivoirian company to dispose of the waste, it knew that Tommy did not have the expertise nor the know-how to treat the waste from the slops. Preoccupied with the political conflict and the failure to organize the presidential election in 2005, Ivoirian authorities had little concern about the welfare of their population.

This akes a case for political instability in Africa as a major threat to its environmental security and emphasizes the need to build more lasting and independent institutions and regulatory agencies and expand their capacities for environmental monitoring and law enforcement using technologies, to provide safeguards against those who would like to take advantage of the situation. The fact that African countries have very weak institutions to monitor and enforce environmental as well as other laws made it possible for Trafigura and the local company to sneak in and dump the toxic waste under the toxic political cloud that permeated the country at the time.

What other environmental implications, other than the dumping of the waste, may be associated with the Côte D'Ivoire civil war that can speak to the problem of environmental security in the country generally? Destruction of environmental resources, land access, production problems, and so forth.

African nations need to change their laws on petroleum imports in their respective states. The sulfur content in the petroleum they import must be no more than ten parts per million, just like in Europe and North America. It will prevent powerful corporations from attempting to bring cheap fuel into their countries. In most African countries, the suffocating fuel smell engulfs their cities because of the high sulfur content they allow in their countries. It is not only damaging to the environment, but it also presents serious health risks to the population.

CONCLUSION

The 2006 waste dump incident cannot be solely blamed on Trafigura; Ivoirian authorities must share a part of the responsibility because they readily accept any fuel into the country for profit. Petroleum products with a sulfur content of ten parts per million should be banned in all African countries. This is the way forward. Legislations need to be enacted to address this issue immediately. Major oil and gas corporations prefer to deal with clean fuel because of the pressure on international environmental organizations and strict laws in Europe and North America. If Côte d'Ivoire had laws that required the country to only accept sulfur content in its petroleum at ten parts per million, the waste dump in 2006 would never had happened because Trafigura would have avoided Côte d'Ivoire due to its strict laws. African nations make themselves an easy prey to major oil and gas corporations because of their willingness to accept any type of petroleum product into their country. They are concerned with getting the fuel at a cheap price, regardless of the health and environment hazard to their countries. This must change.

NOTES

1. Amnesty International and Greenpeace Netherlands. The Toxic Truth: About a Company Called Trafigura, a Ship Called the Probo Koala, and the Dumping of Toxic Waste in Côte d'Ivoire © Amnesty International 2012, p. 1.
2. Ibid, p. 3.
3. Ibid.
4. Ibid, p. 1.
5. Ibid, p. 3.
6. Aldag, Ramon J. "Toxic Waste." *Encyclopedia Britannica*, Apr. 13, 2020, https://www.britannica.com/science/toxic-waste (accessed on November 10, 2022).
7. Ibid.
8. Ibid, *Amnesty International and Greenpeace Netherlands*, 2012.
9. Ibid, p. 25.
10. United States Environmental Protection Agency: Diesel Fuel Standards and Rulemakings, https://www.epa.gov/diesel-fuel-standards/diesel-fuel-standards-and-rulemakings (accessed on November 29, 2022).
11. Free Patents Online: Production of benzene, toluene, and xylene (BTX) from FCC naphtha, United States Patent 5685972, https://www.freepatentsonline.com/5685972.html (accessed on November 29, 2022).
12. Ibid, *Amnesty International and Greenpeace Netherlands*, 2012, p. 27.
13. Ibid, *Amnesty International and Greenpeace Netherlands*, 2012, p. 28.
14. Ibid, *Amnesty International and Greenpeace Netherlands*, 2012, p. 29.
15. Ibid, *Amnesty International and Greenpeace Netherlands*, 2012, p. 30.
16. Ibid.
17. Ibid, *Amnesty International and Greenpeace Netherlands*, 2012, p. 31
18. Ibid, *Amnesty International and Greenpeace Netherlands*, 2012, p. 29.
19. Ibid, *Amnesty International and Greenpeace Netherlands*, 2012, p. 32.
20. Ibid, *Amnesty International and Greenpeace Netherlands*, 2012, p. 33.
21. Ibid, *Amnesty International and Greenpeace Netherlands*, 2012, p. 35.
22. Ibid.
23. Ibid, *Amnesty International and Greenpeace Netherlands*, 2012, p. 37.
24. Ibid.
25. Ibid.

26. Cardesa-Salzmann, A. (CEDAT, Universitat Rovira i Virgili) 2012. The Trafigura Case, EJOLT Factsheet No. 45, p. 8.
27. Ibid.
28. Ibid.
29. Ibid.
30. Ibid.
31. Amnesty International Netherlands. Trafigura: A Toxic Journey © Amnesty International, 2016.
32. Henry Konan Bédié, Wikipedia: available at Henri Konan Bédié - Wikipedia (accessed on November 29, 2022).
33. Ibid.
34. Lique, R.J., "Côte d'Ivoire: Les Causes d'une Rebellion Previsible." *Afrique–Express.* http://www.afrique-express.com/afrique/Côtedivoire/rebellio-mcpi.html (accessed on November 29, 2022).
35. Ibid.
36. Ibid. p. 1.
37. Ibid. p. 1.
38. Keita, Balla M, *A Political History of Côte D'Ivoire from 1936 to 2011: The Politics of Ethnicity, Region, and Religion* (Edwin Mellen Press: NY 2013).
39. Ibid.
40. Ibid.
41. Ibid.
42. Internal Displacement Monitoring Centre (iDMC), "Côte d'Ivoire: A Profile of the Internal Displacement Situation," (April 10, 2008): 50–51.
43. Ibid.
44. United Nations Human Rights Office of the High Commissioner, About the Mandate, Special Rapporteur on Toxics and Human Rights, https://www.ohchr.org/en/special-procedures/sr-toxics-and-human-rights/about-mandate (accessed on November 10, 2022).
45. Ibid.
46. United Nations Human Rights Office of the High Commissioner, International Standards, Special Rapporteur on Toxics and Human Rights, https://www.ohchr.org/en/special-procedures/sr-toxics-and-human-rights/international-standards (accessed on November 10, 2022).
47. Ibid.
48. Ibid.
49. United Nations Environment Programme: Basel Convention: Controlling transboundary movements of hazardous wastes and their disposal, available at Basel Convention > The Convention > Overview (accessed on November 10, 2022).
50. Ibid.
51. Ibid.
52. Ibid.
53. Ibid.
54. Ibid.
55. Ibid.

56. International Convention for the Prevention of Pollution from Ships, November 2, 1973, 12 I.L.M. 1319, amended by Protocol of 1978 Relating to the International Convention for Prevention of Pollution from Ships, Feb. 17, 1978, 1340 U.N.T.S. 61 [hereinafter MARPOL Convention] International Convention for the Prevention of Pollution from Ships (MARPOL) (imo.org).
57. Ibid.
58. Ibid.
59. International Convention for the Prevention of Pollution from Ships, November 2, 1973, 12 I.L.M. 1319, amended by Protocol of 1978 Relating to the International Convention for Prevention of Pollution from Ships, Feb. 17, 1978, 1340 U.N.T.S. 61 [hereinafter MARPOL Convention] International Convention for the Prevention of Pollution from Ships (MARPOL) (imo.org).
60. Ibid.
61. Bratspies, Rebecca, Corrupt at its Core: How Law Failed the Victims of Waste Dumping in Côte d'Ivoire, *Columbia Journal of Environmental Law*, Vol. 43:2, 417-473 (2018), 457.
62. Bratspies, Ibid., p. 458.
63. Ibid.
64. Ibid.
65. The Probo Koala incident, Trafigura news, The Probo Koala incident | Trafigura, https://www.trafigura.com/news-and-insights/responses/the-probo-koala-incident/#:~:text=Trafigura%20notably%20paid%20USD198%20million,law%20firm%20in%20the%20UK (accessed on November 12, 2022).
66. Ibid.
67. Buck, Stephanie, In the 1980s, Italy paid a Nigerian town $100 a month to store toxic waste – an it's happening again (May 26, 2017), https://timeline.com/koko-nigeria-italy-toxic-waste-159a6487b5aa (accessed on February 27, 2023).
68. Ibid.
69. Ibid.
70. Ibid.
71. Ibid.

REFERENCES

Akindès, Francis. "The Roots of the Military-Political Crises in Côte d'Ivoire." *Research Report no. 128*. Uppsala: The Nordic Africa Institute, 2004.

Aldag, Ramon J. "Toxic Waste." *Encyclopedia Britannica*, April 13, 2020, https://www.britannica.com/science/toxic-waste, accessed on November 10, 2022.

Amnesty International and Greenpeace Netherlands. *The Toxic Truth: About a Company Called Trafigura, a Ship Called the Probo Koala, and the Dumping of Toxic Waste in Côte d'Ivoire © Amnesty International,* 2012.

Amnesty International Netherlands. *Trafigura: A Toxic Journey © Amnesty International,* 2016.

ARDA. *Country profile: Côte d'Ivoire*, 2006, http://www.thearda.com/international-Data/countries/Country_61_3.asp, accessed on November 20, 2022.

Bratspies, Rebecca. "Corrupt at its Core: How Law Failed the Victims of Waste Dumping in Côte d'Ivoire, Columbia." *Journal of Environmental Law*, 43(2), 2018: 417–473, 457.

Buck, Stephanie. In the 1980s, Italy Paid a Nigerian Town $100 a Month to Store Toxic Waste – An it's Happening again. May 26, 2017, https://timeline.com/koko-nigeria-italy-toxic-waste-159a6487b5aa, accessed on February 27, 2023.

Business & Human Rights Resource Centre. Dutch Probo Koala Toxic Waste Cases Finally Settled out of Court. November 16, 2012, Dutch Probo Koala toxic waste cases finally settled out of court - Business & Human Rights Resource Centre (business-humanrights.org), accessed on November 30, 2022.

Cardesa-Salzmann, A. (CEDAT, Universitat Rovira i Virgili). The Trafigura Case, EJOLT Factsheet No. 45, 8, 2012.

France 24: Trafigura Found Guilty in Ivory Coast Toxic-waste Case, March 23, 2010, Trafigura Found Guilty in Ivory Coast toxic-Waste Case (france24.com), accessed on November 10, 2022.

Free Patents Online: Production of Benzene, Toluene, and Xylene (BTX) from FCC Naphtha, United States Patent 5685685972, https://www.freepatentsonline.com/5972.html, accessed on November 29, 2022.

Henry Konan Bédié, Wikipedia: available at Henri Konan Bédié - Wikipedia (accessed on November 29, 2022).

ICG (International Crisis Group). "Côte d'Ivoire: Can the Ouagadougou Agreement Bring Peace?" Africa Report. Dakar/Brussels: ICG, June 27, 2007.

ICG (International Crisis Group). "Côte d'Ivoire: Is War the only Option?" Africa Report. No. 171. Dakar/Brussels, March 3, 2011, https://www.crisisgroup.org/africa/west-africa/c%C3%B4te-divoire/cote-d-ivoire-war-only-option, accessed on November 29, 2022.

Internal Displacement Monitoring Centre (iDMC), "Côte d'Ivoire: A Profile of the Internal Displacement Situation." April 10, 2008: 50–51.

International Convention for the Prevention of Pollution from Ships, November 2, 1973, 12 I.L.M. 1319, amended by Protocol of 1978 Relating to the International Convention for Prevention of Pollution from Ships, February 17, 1978, 1340 U.N.T.S. 61 [hereinafter MARPOL Convention] International Convention for the Prevention of Pollution from Ships (MARPOL) (imo.org), accessed on November 28, 2022.

International Federation For Human Rights, Probo Koala Case: Five Years after this Tragedy, No True Accountability for the Perpetrators and no Effective reparation for the Victims, Press Release, August 19, 2011, Probo Koala Case: Five Years after this Tragedy, No True Accountability for the Perpetrators and No Effective Reparation for the Victims (fidh.org), accessed on November 9, 2022.

Karan, C. "The Probo Koala Toxic Ship Incident & Consequences, Case Studies." September 28, 2012, The Probo Koala Toxic Ship Incident & Consequences (marineinsight.com), accessed on November 9, 2022.

Lique, R.J. "Côte d'Ivoire: Les Causes d'une Rebellion Previsible." *Afrique–Express*, http://www.afrique-express.com/afrique/Côtedivoire/rebellio-mcpi.html, accessed on October 3, 2022.

Reuters. "Amsterdam Faces Full Charges in Toxic Dumping Case, Africa Crisis." March 10, 2010, Amsterdam faces full charges in toxic dumping case | Reuters, accessed on November 5, 2022.

Skogseth, Geir. "Report: Côte d'Ivoire: Ethnicity, Ivoirité and Conflict." *LANDINFO*, November 2, 2006.

The Probo Koala Incident, Trafigura News, The Probo Koala Incident | Trafigura, https://www.trafigura.com/news-and-insights/responses/the-probo-koala-incident/#:~:text=Trafigura%20notably%20paid%20USD198%20million,law%20firm%20in%20the%20UK, accessed on November 12, 2022.

UN Environment Programme, Press Release. UN Environment Releases Independent Audit of Sites Affected by Toxic Waste Dump in Côte d'Ivoire, January 30, 2018, UN Environment releases independent audit of sites affected by toxic waste dump in Côte d'Ivoire (unep.org), accessed on November 27, 2022.

United Nations Environment Programme: Basel Convention: Controlling Transboundary Movements of Hazardous Wastes and their Disposal, Basel Convention > The Convention > Overview, accessed on November 10, 2022.

United Nations Human Rights Office of the High Commissioner, About the Mandate, Special Rapporteur on Toxics and Human Rights, https://www.ohchr.org/en/special-procedures/sr-toxics-and-human-rights/about-mandate, accessed on November 10, 2022.

United Nations Human Rights Office of the High Commissioner, International Standards, Special Rapporteur on Toxics and Human Rights, https://www.ohchr.org/en/special-procedures/sr-toxics-and-human-rights/international-standards, accessed on November 10, 2022.

United Nations Human Rights Office of the High Commissioner, Toxic Waste: UN Expert Releases Report on "Probo Koala" Incident, Press Release, September 16, 2009, Toxic waste: UN Expert releases report on "Probo Koala" incident | OHCHR, accessed on November 10, 2022.

United States Environmental Protection Agency: Diesel Fuel Standards and Rulemakings, https://www.epa.gov/diesel-fuel-standards/diesel-fuel-standards-and-rulemakings, accessed on November 29, 2022.

VOA News, Africa, Ivory Coast Toxic Waste Victims Still Await Payments, November 13, 2015, Ivory Coast Toxic Waste Victims Still Await Payments (voanews.com), accessed on November 30, 2022.

Chapter 9

Implementation and Enforcement of Sustainability Provisions in Regional Trade Agreements in Africa

Moses Onyango Ogutu

The objective of most regional economic communities (RECs) in Africa is to promote integration in their respective regions.[1] The RECs work closely with pan-African level organizations such as the African Union (AU) to achieve continental objectives such as Agenda 2063, Africa's blueprint, and master plan for socioeconomic and political transformation.[2] The potential of regional trade to contribute to sustained growth, poverty reduction, and inclusive development has been evidenced in several regions including Asia and Latin America.[3] However, in Africa, expected results have been slow, often due to a combination of weaknesses in both the public and private sectors.[4] Sustainable development goals (SDGs) adopted in 2015 represent a triad of economic, political, and social policies, which complement each other to achieve growth, trade, and equality. Well-designed and complementary policies, both domestic and foreign, regulatory frameworks in infrastructure, trade facilitation, and social protection when effectively enforced will play a significant role in achieving sustainable development objectives such as environmental protection through trade.[5] Furthermore, ascension to bilateral, regional, or multilateral trade regimes has proven to streamline institutions leading to domestic implementation of foreign commitments to comply and gain from the benefits of such commitments.[6] Studies have also shown that countries at the frontline of sustainability tend to be more open to trade,[7] recognizing the significant role regional trade can have in achieving the SDGs, contemporary trade agreements, especially those promoted by high-income countries, incorporate sustainable development provisions addressing a range of sustainability issues such as environmental protection, respect and promotion of labor standards, food security, human rights, and public health. However, those promoted by middle- and low-income countries either lack comprehensive commitments or in some cases completely omit sustainability provisions.[8]

Nevertheless, international trade does not automatically lead to inclusive economic development as such contribution depends significantly on the interface between private and public policies that direct and complement trade.[9] At the Third International Conference on Financing for Development in Addis Ababa, Ethiopia, in July 2015, delegates observed that "with appropriate supporting policies, infrastructure and an educated workforce, trade can help promote productive employment and decent work, women's empowerment and food security, as well as a reduction in inequality, and contribute to achieving the sustainable development goals."[10] Given the role supporting policies and chapters on sustainable development play in translating RTAs into tangible results for the marginalized, and ensuring equitable growth and sustainable development, it is imperative that African trade agreements adopt and codify measures that promote sustainability and inclusivity in their design frameworks. This is in spite of concerns of developmental economists such as Ha-Joon Chang, who argue that by insisting on the adoption of "best standards" on issues such as labor standards and environmental sustainability in trade agreements, developing countries will simply not be able to trade.[11] It is also recognized that despite Africa's emissions of carbon being among the lowest in the world,[12] African GHG emissions, mainly stemming from agricultural activities, are among the fastest growing.[13]

In recent times, African states have been presented with the dilemma of how to balance between industrialization and environmental sustainability. For example, choosing between massive industrialization policies that would require utilization of brown energy such as coal to generate jobs and implementing green policies such as the adoption of renewable energies. Still, the adoption of measures to ensure environmental security and sustainability in RTAs is particularly relevant as trade policy is increasingly being called upon to respond to the growing climate crisis whose social and economic impacts are particularly devastating for low- and middle-income countries such as those in Africa.[14] This is especially true in the case of the AfCFTA whose successful implementation has been widely touted to present positive developmental consequences for the citizens of its member countries.[15] In the case of the AfCFTA, studies have found that without measures to ensure equitable growth, benefits arising from trade under the AfCFTA will not be distributed equally among participating countries.[16]

This chapter achieves four objectives. First, it provides a review of the literature on the prevalence and coverage of sustainability and inclusivity provisions in preferential trade agreements. Second, it analyzes the implementation and enforcement measures of such provisions. Third, the chapter analyzes the coverage, implementation, and enforcement of sustainability provisions, including environmental protection measures in the design frameworks of key regional trade agreements in Africa—the East African Community (EAC), the Common Market for Eastern and Southern Africa (COMESA), the Southern African Development Community (SADC), and the AfCFTA. Finally, the chapter concludes by providing guidelines on how trade agreements can promote and implement environmental security. In this case, an argument is made to include a chapter on comprehensive and consolidated Protocol on Trade and Sustainable Development in the design framework of the AfCFTA. Such inclusion will ensure that member states also commit to the promotion of inclusivity and sustainable development, in addition to free trade.

SUSTAINABILITY AND INCLUSIVITY PROVISIONS IN THE DESIGN FRAMEWORK OF RTAS

FTAs are increasingly going beyond mere mentioning of "sustainability" or "inclusivity," to incorporate detailed sustainability provisions in their main body text, through chapters, specific paragraphs, or related side agreements.[17] Inclusion of sustainability elements in RTAs varies and depends on the specific issues the parties intend to address such as environmental and marine conservation or labor and human rights protection. In the design frameworks of RTAs, elements to address provisions relating to SDGs are incorporated through:

(i) *Preambular references* such as commitments to promote social well-being, environmental protection, or sustainable development in general (often taking the format of GATT Article XX).
(ii) *References to Multilateral Environmental Agreements (MEAs)*, such as the Montreal Protocol for the protection of the ozone layer, or the Washington Convention on International Trade in Endangered Species of Wild Fauna and Flora. These references often include commitments by the contracting parties to ratify or implement those agreements (at times, the MEAs are made an integral part of the RTAs).
(iii) Objective Commitments *to Cooperate on Sustainability Issues* such as environmental or labor matters. These often take the form of MoUs. Examples of MoUs include the MoU on Labor and Social Security Cooperation in the China–Chile FTA, MoU on Labor Cooperation in the Hong Kong–China–Chile FTA, and the MoU on Environmental Cooperation in China–Chile agreement.
(iv) In the absence of side commitments to cooperation such as MoUs, some RTAs include specific *Side Agreements*. For example, the North

American Agreement on Environmental Cooperation (NAAEC) set up by the parties to the North American Free Trade Agreement (NAFTA).

(v) More recent RTAs promoted by the EU include a dedicated *Chapter on Environment, Labour or Sustainable Development* covering all or specific sustainability issues where the parties specify their commitments. Examples include the EU–Korea FTA; the FTA between Korea and Colombia; Hong Kong–China–Chile FTA; and the agreement between Vietnam and the Eurasian Economic Union (EEC). In Africa, examples, include the economic partnership agreements (EPAs), such as that with the EAC. Issues covered in the chapters range from fisheries, protection of ozone layer, conservation of biodiversity, scientific and information technology, forestry, aquaculture, timber, and so forth.

(vi) Nearly all RTAs include a general exception clause modeled on GATT Art. XX specifying that actions by contracting parties "necessary to protect human, animal or plant life or health" are not inconsistent with the trade-related obligations of the treaty.

IMPLEMENTATION AND ENFORCEMENT MECHANISMS

The mere mention or inclusion of sustainability provisions in FTAs does not guarantee their contribution to sustainable development. Such contributions result from implementation and enforcement of relevant policies. A review of the treaty texts of RTAs has shown that implementation and enforcement mechanisms take various forms, as discussed below.[18]

i) *Obligations to uphold environmental or labor Laws.* These obligations take the form of clauses that prevent the contracting parties from trying to increase trade and investment flows by weakening domestic environmental laws and regulations; for example, Korea–Colombia FTA; China–Chile FTA; EU FTAs; and most recent FTAs.

ii) *Dispute settlement mechanisms.* Such mechanisms allow parties to request for consultations to solve matters arising out of the implementation of sustainable development obligations; for example, when a contracting party tries to increase trade by weakening their domestic environmental laws or regulations.

iii) *Procedural guarantees.* These include provisions that require parties to ensure that persons with recognized interests under their laws have appropriate access to judicial, quasi-judicial, or administrative proceedings for the enforcement of the parties' environmental and labor and provide for appropriate administrative and procedural protections in accordance with its law (examples include United States–Australia RTA and Colombia–Korea agreement).

iv) Incorporation of the *right to regulate environmental matters.* In this case, the contracting parties prefer to preserve their right to go beyond the existing level of environmental protection by introducing new, more far-reaching environmental regulations.

v) Advocating for *transparency and public participation.* In this format, RTAs require contracting parties to provide public access to relevant sustainability issues such as environmental policies and labor laws and measures to include public participation in the policymaking process.

vi) *Institutional mechanisms for implementation.* Examples include the Joint Committee on Trade and Sustainable Development tasked with implementing sustainable development chapters in EU RTAs; the Commission for Environmental Cooperation (CEC) set up under NAFTA's NAAEC; and the Council on Sustainable Development in the Colombia–Korea FTA. In the Hong Kong–China–Chile FTA, the "Parties agree to establish a collaborative framework" and designate contact points to facilitate communication for the purposes of the chapter on environment.

Nevertheless, most FTAs further include specific environmental exceptions in certain chapters. For example, in the ASEAN–Australia–New Zealand FTA, "each Party retains all authority to interpret its laws, regulations, and administrative provisions pursuant to the general exceptions" (Chapter 6 Article 2).[19]

Having reviewed the prevalence, coverage, implementation, and enforcement of sustainability provisions in the design frameworks of RTAs in general, the next section analyzes the coverage of such concepts in African RTAs.

COVERAGE OF SUSTAINABILITY AND INCLUSIVITY PROVISIONS IN AFRICAN RTAS

Despite the link between trade and sustainable development, many Africa-initiated RTAs do not incorporate sustainable development and inclusivity provisions. Similar to most agreements between high- and low-income regions,[20] African RTAs mostly incorporate environmental safeguard provisions. Sustainability provisions are, however, present in FTAs between African countries (and RECS) and developed country partners such as the EU or the United States. The inclusion of such provisions is often at the insistence of the developed country trade partner.[21] For example, the economic partnership agreement between the EU and SADC contains a sustainable development chapter covering social and environmental matters (Chapter II).[22] Comparably, the economic partnership agreement between the EU and the EAC contains extensive coverage of sustainable development in all aspects of cooperation (fisheries, agriculture, environment and water, human rights and labor laws, etc.).[23] Meanwhile, in the African Growth and Opportunity Act (AGOA) between the United States and African partner countries, eligibility is based on the beneficiary country's respect for inclusion and enforcement of international labor rights and enacting policies that help reduce poverty such as environmental protection.[24] Moreover, access to both the EU and U.S. Generalized System of Preferences (GSP) mechanisms is based on respect for labor and human rights and other social sustainability and inclusivity elements. Non-compliance with such policies is often enforced through dispute settlement mechanisms or sanctions.[25]

Environmental provisions are the most prevalent in the body texts of African RTAs. While African countries do cooperate on sustainable development issues, such cooperation occurs through fragmented and overlapping policies that often lack enforcement mechanisms. Studies have shown that inclusion of sustainable development provisions in FTAs enhances implementation and enforcement since the parties to the FTA commit to comply with the provisions of the agreement either directly or within an agreed period.[26]

As aforementioned, studies have shown that most FTAs that contain sustainability provisions are those occurring between developed countries or promoted by them.[27] The next section reviews coverage of sustainability provisions in RTAs promoted by African countries. The review further analyzes the drivers behind these RTAs, their enforcement mechanism, and challenges facing the implementation of the adopted measures. The specific RTAs covered include the COMESA, the EAC, and the SADC.

COVERAGE OF SUSTAINABILITY PROVISIONS IN THE COMMON MARKET FOR EASTERN AND SOUTHERN AFRICA

Before the emergence of the AfCFTA, the COMESA was the largest subregional economic grouping in Africa in terms of population and geographical coverage. The major objective of the COMESA is to attain sustainable growth and development of its twenty-one member states by promoting a balanced and harmonious development of its production and marketing structures. The COMESA applies a development approach to regional integration, starting with market integration, industrialization, and infrastructure development. The vision of the COMESA is "to have a fully integrated internationally competitive regional economic community with high standards of living for its entire people, ready to merge into the African Economic Community" and its mission is "to achieve sustainable economic and social progress in all Member States through increased cooperation and integration in all fields of economic activity."[28] The COMESA member states also aim to promote cooperation in all fields of economic activity through the adoption of joint policies and programs to raise the standards of living of its peoples; promote peace, security, and stability; and among others, contribute toward the establishment, progress, and realization of continental objectives such as the African Economic Community, the process that birthed the AfCFTA.[29]

The objectives of COMESA reflect its priorities to promote sustainable economic development, and its protocols and policies demonstrate interlinkages between social and economic development and environmental protection provisions.[30] From its objectives, and various policy documents, the COMESA has identified key strategic and priority areas that address sustainable development including removing barriers to trade, building productive capacity for global competitiveness; addressing supply-side constraints related to infrastructure, peace, and security; and addressing cross-cutting issues such as gender and social affairs, climate change, statistical development, knowledge-based society and human capital, cooperation and partnerships, and aid for trade.

The COMESA treaty makes several references to sustainable development, though the words "inclusive" or "inclusion" are missing. Sustainability and inclusivity issues are addressed through various protocols, policies, regulations, and programs focused on agriculture, environment and natural resources, and gender and social affairs. Examples include: the COMESA Gender Policy 2002, which aims to facilitate the mainstreaming of gender perspectives into all policies, structures, systems, programs and activities to make them gender-responsive and contribute to the effective achievement of sustainable socioeconomic development in the region while ensuring compliance with regional and international gender commitments; the Regional Strategy for Mainstreaming Gender in Agriculture and Climate Change; the Programme on Climate Change Adaptation and Mitigation in the Eastern and Southern Africa Region, jointly initiated with the EAC and SADC; and regulatory policies and guidelines in transport, ICT, and energy.

Nonetheless, the COMESA lacks a comprehensive sustainable development strategy (incorporating all aspects of sustainable development), especially in the body text of its agreement. However, the body text covers environmental aspects of sustainable development such as cooperation in the management of natural resources to "efficient management and sustainable utilisation of natural resources" by taking "necessary measures to conserve natural resources."[31]

COVERAGE OF SUSTAINABILITY PROVISIONS IN THE EAST AFRICAN COMMUNITY

The EAC aims to achieve a market and political integration for the attainment of sustainable growth and development. The treaty establishing the EAC stipulates various principles to enhance harmonization and integration of its partner states. The first among these principles is:

> the attainment of sustainable growth and development . . . by the promotion of a more balanced and harmonious development; the strengthening and consolidation of cooperation in agreed fields that would lead to equitable economic development within the partner States and which would, in turn, raise the standard of living and improve the quality of life of their populations; the promotion of sustainable utilization of the natural resources and taking of measures that would effectively protect the natural environment of the Partner States.[32]

As with the COMESA treaty, "inclusiveness," "inclusive development," or "inclusivity" does not appear in the EAC treaty, but inclusivity-related features are observed in principles aimed at achieving sustainable development. These principles cover everything from "partnerships to achieve sustainable development," mainstreaming of gender and recognition of the role of women, to the promotion of peace and security and stability.[33] Inclusivity and accountability principles are covered in the Vision 2050, which aims among others to optimize the utilization of resources to accelerate productivity and the social well-being of its citizens.[34]

The EAC treaty text has specific chapters addressing SDGs. For example, cooperation in environment and natural resources management (Chapter 19); cooperation in health (prevention of communicable and noncommunicable diseases and the strengthening of health systems and social welfare (with employment, poverty alleviation programs, and working conditions) and cultural activities (Chapter 21); enhancing the role of women in socioeconomic development by encouraging and enhancing their participation in decision-making (Chapter 22); inclusion of private sector and civil society (Chapter 25); and cooperation in agriculture and food security (Chapter 18).[35] Various road maps such as the Peace and Security Protocol and the EAC Vision 2050 have been designed to consolidate, focus, and foster the implementation of these chapters and articles. Institutional mechanisms have also been established to foster the implementation of the objectives of the sustainable development chapters. The EAC can be said to be a pathfinder and provides a model for the incorporation of sustainability and inclusivity in the design frameworks of RTAs in Africa.

COVERAGE OF SUSTAINABILITY PROVISIONS IN THE SOUTHERN AFRICAN DEVELOPMENT COMMUNITY

In contrast to other African regional economic communities (e.g., the COMESA and EAC), "development" is a key pillar of cooperation for the SADC. Regional integration policies in the SADC are implemented through various Regional Indicative Strategic Development Plans (RISDP). The RISDP are designed to provide the SADC and its member states with consistent and comprehensive long-term economic and social policies to achieve the SADC's developmental objectives.[36] The main sustainable development priorities for the SADC as identified in its various RISDPs are poverty eradication, reducing inequalities, enhancing food security and agriculture, addressing climate change and its impacts, providing access to water, ensuring adequate energy supply, fostering investment and job creation, promoting market access and trade, and infrastructure development.

Most of the SADC's sustainable development commitments do not appear in the treaty text but as protocols, declarations, or commitments. The Protocol on Environmental Management for Sustainable Development addresses issues relating to the environment in the region by recognizing the significance of sustainable use and management of the environment in the fight against poverty and food insecurity. Others include the Protocol on Wildlife Conservation and Law Enforcement (1999); the Protocol on Fisheries (2006); the Declaration on Agriculture and Food Security (2004); the Protocol on Forestry (2002); the Protocol on Science, Technology and Innovation, which outlines the framework for cooperation to effect contribution of science and technology to social and human development; the Protocol on Gender and Development, which aims among others to provide for the empowerment of women, to eliminate discrimination, and to achieve gender equality and equity through the development and implementation of gender-responsive legislation, policies, programs, and projects; the Declaration on Productivity (1999), which is intended to craft appropriate policy frameworks that engender productivity; and the Protocol on Health (1999), which aims to coordinate regional efforts in health-related areas and issues.

Although various protocols and declarations relating to sustainable development are in place, not all of them are in force as the requisite two-thirds of SADC Member States have not ratified them. Moreover, where the protocols are in force they are not necessarily being implemented or enforced by the member states. One of the reasons for the lack of enforcement is because the SADC Tribunal is defunct, and the new Tribunal protocol is yet to be ratified by the requisite two-thirds of SADC Member States. Furthermore, even if the new tribunal protocol is ratified it does not allow for legal or natural persons to bring disputes against state parties. In its current form, the protocol only accommodates state-to-state disputes. The implication is that citizens, for example, may not directly bring a dispute against a state or a corporation violating environmental measures. However, such cases may be brought for prosecution in the countries, where they occur or where the parties have specific clauses for resolution. The SADC is working on creating a more general compliance mechanism, which will incorporate the tribunal. The approval and implementation of the protocol will go a long way in ensuring member states' compliance with regional commitments to promote sustainable development, including those relating to the environment.

Most recently, the SADC is in the process of developing its Vision 2050 and RISDP 2020–30.[37] The Vision 2050 and the related RISDP 2020–30 focus on five strategic priority areas divided into a foundation, pillars, and a cross-cutting theme. Peace, Security, and Good Governance act as the foundation while Industrial Development and Market Integration, Infrastructure Development in Support of Regional Integration, and Social and Human Capital Development are pillars I, II, and III, respectively. A final priority area addresses cross-cutting issues including gender, climate change, and disaster risk management.

The SADC Vision 2050 envisions a peaceful, inclusive middle- to high-income industrialized region, where all citizens enjoy sustainable economic well-being, justice, and freedom.[38] As evident, the terms "inclusive" and "sustainable" economic well-being are explicitly mentioned in the Vision statement.[39] Moreover, the draft RISDP 2020–30 explicitly references the Southern African SDGs and makes every effort to align its regional strategic targets with these goals. The SADC RISDP 2020–30 illustrates how the SADC is working toward the UN's Agenda 2030 and its four key pillars: Social inclusion (goals 1–7), economic growth (goals 8–11), environmental responsibility (goals 12–15), and governance (goals 16 and 17).[40] In particular, the RISDP 2020–30 Blueprint states the following in its introduction,

> Throughout these Blueprints, care is taken to align the action areas of each pillar's aspirations with the SDG's and the AU's Agenda 2063 priority areas to ensure that regional programmes and projects are aligned with the global and continental development aspirations.

Additionally, within the Blueprints, which are essentially the draft RISDP 2020–30, under the action area of "Sustainably Developing SADC's Blue and Green Economies," the document urges SADC to, "Monitor Member State compliance with the Paris Agreement" of 2015, "The Future We Want," Agenda 2063, and Agenda 2030.[41]

From the review, it is observed that the SADC, through its various protocols and declarations and recently its Vision 2050 and RISDP 2020–2030, is the Africa regional REC that most comprehensively addresses the SDGs and sustainable development more broadly. The SADC Vision 2050 and RISDP 2020–30 serve to consolidate the REC's sustainable development efforts by coordinating all its different declarations and protocol's SDG-related efforts into one aspirational (SADC Vision 2050) and one strategic document (SADC RISDP 2020–30). Nevertheless, this does not account for the founding treaty of SADC not having been amended to include a chapter on sustainable development. On the other hand, the EAC is the only African REC that has amended its treaty to incorporate a Sustainable Development Chapter, in line with contemporary FTAs. Like the SADC, the EAC also has a Vision 2050, which is designed to consolidate, focus, and foster the implementation of its chapter on sustainable development, as well as institutional mechanisms for implementation. Although the ECOWAS is not covered in this review, the ECOWAS treaty also lacks a sustainable development chapter or protocol, and the revised treaty does not make any reference to sustainable development.

The lack of a comprehensive sustainable development chapter in the treaty text should not be taken to mean the lack of cooperation on those aspects. On the contrary, as the review has shown, African RECs address sustainability and inclusivity issues through various protocols and regulations. The argument presented is for the harmonization of these seemingly fragmented policies through a comprehensive chapter (or protocol) addressing sustainable development issues, as with other regional integration issues addressed by the AfCFTA.

BUILDING A CASE FOR INCLUSION OF SUSTAINABILITY PROVISIONS IN AFRICAN FTAS, INCLUDING THE AFRICAN CONTINENTAL FREE TRADE AREA

The agreement establishing the AfCFTA came into force on May 30, 2019, and became operational in July 2019. As the largest FTA in history with a market share of more than $3 trillion, the AfCFTA presents many opportunities to Africa. These include boosting employment opportunities, increased food security through reduction to barriers on trade in agricultural products, better allocation of resources, improved competition, reduced-price differentials, and reduced vulnerability of countries to external trade shocks by reducing the trade balance deficits of African countries among others. In July 2020, the World Bank noted that the AfCFTA could boost regional income by 7 percent or $450 billion, speed up wage growth for women, and lift thirty million people out of extreme poverty by 2035, and more broadly, reduce poverty and broaden inclusion.[42]

The objectives of the AfCFTA are geared toward the promotion of sustainable development. To highlight a few examples of the general objectives, objective (a) aims to "create a single market for goods, services," facilitated by movement of persons to deepen the economic integration of the African continent following the Pan African Vision of "An integrated, prosperous and peaceful Africa" enshrined in Agenda 2063; objective (c) is to promote and attain sustainable and inclusive socio-economic development, gender equality, and structural transformation of the State Parties; and objective (g) aims to promote industrial development through diversification and regional value chain development, agricultural development, and food security (Article 3 General Objectives).[43]

Sustainability standards have received much attention in FTAs, and modern trade agreements (including the EPAs between African RECs and the EU) incorporate an exhaustive chapter integrating all SDGs with sub-articles covering issues from environmental protection, food security, peace and security, and health cooperation to fisheries management. Moreover, provisions on trade and sustainability, for instance, are often designed to promote higher sustainability standards in market access.[44] Although the AfCFTA is geared toward the promotion of sustainable and inclusive economic development in Africa, sustainability and inclusivity issues are loosely mentioned without clear structures on their implementation. The seeming lack of comprehensive coverage of sustainability and inclusivity issues in the design framework of the AfCFTA means that the agreement is much focused on market integration, but without a comprehensive framework to ensure that existing trade contributes to sustainable development. Furthermore, this means that such issues are left to be addressed by other continental initiatives such as Agenda 2063 and other continental policies that sometimes overlap and are challenging to coordinate.

Understandably, the AfCFTA is still at its infancy stages and negotiations are ongoing for various protocols to the agreement. Thus, it is important that the agreement not only incorporate a Protocol on Trade and Sustainable Development, but also devises implementation and enforcement mechanisms. The framework of the Protocol on Trade and Sustainable Development could be designed in a manner that aligns the AfCFTA with the 17 SDGs and their 169 goals and 230 targets. For instance, the protocol could encompass strategies to address food security, health (including rules on medicines and medical equipment, which have gained relevance in light of the COVID-19 pandemic), and environment and climate change. The provisions on environmental protection are particularly relevant in the context of recognized impacts of climate change, including conflicts relating to access to environmental resources such as forests and fishing activities. The protocol could also include binding rules on labor and other aspects of human rights, borrowing on designs of FTAs that have comprehensive provisions, for example, the framework of the EPA between the EU and various RECs in Africa. The protocol should further incorporate implementation and enforcement mechanisms to ensure compliance. This would allow for the consolidation of regional frameworks, protocols, declarations, and policies on sustainable development and eventually result in a regional framework for Trade and Sustainable Development under the AfCFTA.

The implementation of the AfCFTA is based on, among others, the principles of the preservation of the *acquis* and taking the REC's Free Trade Areas as building blocks. In this case, in addition to other regions, the EAC and SADC provide models that the AfCFTA can build on to comprehensively address sustainability and inclusivity issues. For instance, as the chapter has highlighted, the revised EAC treaty contains all-encompassing chapters addressing SDG-related issues in the body text. On the other hand, the SADC has developed comprehensive protocols addressing SDG issues, including its Vision 2050 and RISDP 2020–2030.

CONCLUSION

This chapter sought to provide a comprehensive review of how trade agreements cover sustainability and inclusivity elements and implementation and enforcement mechanisms, the coverage of sustainability elements in the design framework of African FTA/RTAs, and to argue for the inclusion of specific, comprehensive, and consolidated chapter or protocol (Trade and Development Protocol) in the design framework of African RTAs, including the AfCFTA. The argument is that such inclusion would harmonize the many fragmented sustainability policies and ensure a commitment to the implementation of sustainability policies in addition to free trade.

Sustainable development has become an important topic in international trade policy, and FTAs have answered the call by expanding the scope of sustainability provisions in the body text of their agreements. FTAs are increasingly covering sustainability and inclusivity elements including human rights, labor, and environmental provisions. Moreover, the agreements have moved from loose mentioning of sustainability provisions (e.g., preambular mentions only) to substantive coverage such as the inclusion of full chapters addressing sustainability issues, followed by enforcement mechanisms, procedural guarantees, and dispute settlement mechanisms.

Nevertheless, a review of the COMESA, EAC, SADC, and the AfCFTA agreements indicates that in as much as trade is touted as a means to achieving SDGs, with the exception of a few, Africa-led or initiated trade agreements are yet to include specific measures to address SDGs in the body text of the agreements. Rather, they are present in ancillary documents such as protocols, declarations, and regional strategy documents. This is different from contemporary FTAs promoted by high-income countries that often include sustainable development chapters.

Consequently, I have highlighted a clear case and structure on how African RTAs can address sustainability through trade policy. Arguably, African nations are already party to FTAs that extensively cover sustainable development provisions (such as EPAs), and the RECs (such as EAC and SADC) provide models that can be localized by the AfCFTA or future agreements. Trade policy communities can build upon models advanced by the RECs to innovatively cooperate on developing guidelines and the structure of sustainable development chapters in RTA and utilize regional mechanisms to ensure sustainability.

NOTES

1. African Union (Online). *Regional Economic Communities (RECs).* https://au.int/en/organs/recs (Accessed on March 12, 2023). See also, Africa Regional Integration Index (Online). *Regional Economic Communities.* https://www.integrate-africa.org/rankings/regional-economic-communities/ (Accessed on March 12, 2023).
2. African Union. Agenda 2063: The Africa We Want. https://au.int/sites/default/files/documents/36204-doc-agenda2063_popular_version_en.pdf; See also, Africa Regional Integration Index (Online). *Regional Economic Communities.* https://www.integrate-africa.org/rankings/regional-economic-communities/ (Accessed on March 12, 2023).
3. See, for example, Helble & Ben, 2017; United Nations, 2014.
4. African Union. *Report on the Status of Regional Integration in Africa, 2019.* https://au.int/sites/default/files/newsevents/workingdocuments/35970-wd-2.2_new_execitive_summary_-_report_on_status_of_regional_integration_-_eng.pdf (Accessed on March 12, 2024).
5. Draper, P., Khumalo, N., & Tigere, F. *Sustainability Provisions in Regional Trade Agreements: Can they be Multilateralised?* 2017. http://e15initiative.org/publications/sustainability-provisions-in-regional-trade-agreements-can-they-be-multilateralised/; see also, UNCTAD, 2016; United Nations, 2014.
6. Ibid, Draper et al., 2017; see also, Portela & Orbie, 2014; Zhou & Cuyvers, 2011.
7. Erixon, F., & Lamprecht, P. The Role of Trade Policy in Promoting Sustainable Agriculture. *European Centre for International Political Economy (ECIPE),* (2020): 1–42. https://ecipe.org/publications/sustainable-agriculture/.
8. Ibid, Draper et al., 2017.
9. United Nations, 2014.

10. United Nations, 2015, p. 23.

11. Ha-Joon Chang. Kicking Away the Ladder: The "Real" History of Free Trade. *Foreign Policy In Focus* (Silver City, NM: Interhemispheric Resource Center, December 2003). http://www.personal.ceu.hu/corliss/CDST_Course_Site/Readings_old_2012_files/Ha-Joon%20Chang%20-%20Kicking%20Away%20the%20Ladder-The%20%E2%80%9CReal%E2%80%9D%20History%20of%20Free%20Trade.pdf.

12. Sovacool, Benjamin K., Andrew Hook, Mari Martiskainen, Andrea Brock, & Bruno Turnheim. The Decarbonization Divide: Contextualizing Landscapes of Low-Carbon Exploitation and Toxicity in Africa. *Global Environmental Change* 60, (2020): 102028.

13. Ibid.

14. Mauricio, Mesquita, & Moreira - Marcelo Dolabella. Can Trade Policy Contribute to Fighting Global Warming in Latin America and the Caribbean? *Inter-American Development Bank*, October 27, 2022. https://blogs.iadb.org/integration-trade/en/can-trade-policy-contribute-to-fighting-global-warming-in-latin-america-and-the-caribbean%ef%bf%bc/.

15. See Echandi, Roberto, Maliszewska, Maryla, & Steenbergen, Victor. *Making the Most of the African Continental Free Trade Area: Leveraging Trade and Foreign Direct Investment to Boost Growth and Reduce Poverty*. Washington, DC: World Bank, 2022. https://openknowledge.worldbank.org/handle/10986/37623.

16. Abrego, L., Amado, M.A., Gursoy, T., Nicholls, G.P., Perez-Saiz, H., & de Zamaroczy, M. The African Continental Free Trade Agreement: Welfare gains estimates from a general equilibrium model. *IMF Working Paper,* 2019. https://www.elibrary.imf.org/view/journals/001/2019/124/article-A001-en.xml.

17. Berger, A., Brandi, C., Bruhn, D., & Chi, M. Towards "Greening" Trade? Tracking Environmental Provisions in the Preferential Trade Agreements of Emerging Markets. In *Deutsches Institut für Entwicklungspolitik Discussion Papers,* 2017; see also, Draper et al., 2017.

18. Baker, P. *Handbook on Negotiating Sustainable Development Provisions in Preferential Trade Agreements*. United Nations, 2018. https://artnet.unescap.org/publications/books-reports/handbook-negotiating-sustainable-development-provisions-preferential; see also, Berger et al., 2017; Draper et al., 2017.

19. Ibid, Baker, 2018.

20. Most RTAs that incorporate sustainability and inclusivity provisions are those promoted by developed countries. See, for example, discussion by Baker, 2018, and Draper et al.

21. Ibid, Baker, 2018.

22. EU and SEconomic Partnership Agreement between the European Union and its Member States, of the one part, and the SADC EPA States, of the other part. *Official Journal of the European Union*, L 250/3. https://doi.org/10.1371/journal.pone.0065542.

23. European Commission. Economic Partnership Agreement between the EU and the Eastern African Community (EAC). October 2014, 2.

24. US Government, U. S. G. Trade and Development Act of 2000. 1–56. https://agoa.info/downloads/legal/2385.html.
25. See Portela & Orbie, 2014; see also, Zhou & Cuyvers, 2011.
26. Ibid, Draper et al., 2017; Portela & Orbie, 2014; Zhou & Cuyvers, 2011.
27. Ibid, Draper et al., 2017; see also, Portela & Orbie, 2014; Zhou & Cuyvers, 2011.
28. COMESA. *COMESA Communication Policy and Strategy 2016–2020.*, 2016. https://www.comesa.int/wp-content/uploads/2019/02/COMESA-Communication-Strategy.-_final_web_email.pdf.
29. COMESA. *Treaty Establishing the Common Market for Eastern and Southern Africa (Revised).*, 2019. https://www.tralac.org/resources/by-region/comesa.html.
30. COMESA. *Role of COMESA in advancing Africa's Sustainable Vision and Mission of COMESA. 1* (2009): 1–3.
31. Ibid, COMESA, 2019.
32. EAC. *Treaty for the Establishment of the East Africa Community (As amended on 14th December, 2006 and 20th August, 2007).* (1999): 1–129. https://www.eacj.org/?page_id=33.
33. EAC. East African Community vision 2050: regional vision for socio-economic transformation and development. *The East African Community*, 142 (2015). http://repository.eac.int/handle/11671/567.
34. Ibid.
35. Ibid, EAC, 1999.
36. See, for example, The SADC Regional Indicative Strategic Development Plan (RISDP) 2020–2020203030. https://www.sadc.int/document/sadc-regional-indicative-strategic-development-plan-risdp-2-0.
37. SADC. 2020. Draft SADC Vision 2050 and RISDP 2020–30 Blueprints. Published by Imani Development International Ltd. https://imanidevelopment.com/sadc-vision-and-risdp-2020-30/ . (Accessed on July 25, 2020).
38. SADC. "Vision 2050". SADC Secretariat, 2020. Available at: https://www.sadc.int/sites/default/files/2021-08/SADC_Vision_2050.pdf.
39. Ngwawi, 2012.
40. See page 17 of the latest draft of the SADC RISDP 2020–30. Available at: https://imanidevelopment.com/wp-content/uploads/2020/03/4th-Draft_RISDP-2020-30-Blue-Prints2.pdf.
41. See SADC. 2020. Draft SADC Vision 2050 and RISDP 2020–30 Blueprints.
42. World Bank. *The African Continental Free Trade Area: Economic and Distributional Effects.*, 2020. https://www.worldbank.org/en/topic/trade/publication/the-african-continental-free-trade-area.
43. African Union. *Agreement Establishing the African Continental Free Trade Area.* (2017): 1–77. https://au.int/en/treaties/agreement-establishing-african-continental-free-trade-area.
44. Erixon, F., & Lamprecht, P. The Role of Trade Policy in Promoting Sustainable Agriculture. *European Centre for International Political Economy (ECIPE)*, (2020): 1–42. https://ecipe.org/publications/sustainable-agriculture/.

REFERENCES

Abrego, L., M.A. Amado, T. Gursoy, G.P. Nicholls, H. Perez-Saiz and M. de Zamaroczy. "The African Continental Free Trade Agreement: Welfare gains estimates from a general equilibrium model." *IMF Working Paper*. 2019. https://www.elibrary.imf.org/view/journals/001/2019/124/article-A001-en.xml.

Africa Regional Integration Index (Online). *Regional Economic Communities*. Available at: https://www.integrate-africa.org/rankings/regional-economic-communities/ (Accessed on March 12, 2023).

African Union. *Agreement Establishing the African Continental Free Trade Area*. 2017: 1–77. https://au.int/en/treaties/agreement-establishing-african-continental-free-trade-area.

African Union. *Report on the Status of Regional Integration in Africa: Vol. Eco/STC/MA*. 2019. www.ecowas.int/wp-content/uploads/2017/11/2016-Convergence-.

African Union. *Report on the Status of Regional Integration in Africa*. 2019. https://au.int/sites/default/files/newsevents/workingdocuments/35970-wd-2.2_new_executive_summary_-_report_on_status_of_regional_integration_-_eng.pdf. (Accessed on March 12, 2024).

Baker, P. *Handbook on Negotiating Sustainable Development Provisions in Preferential Trade Agreements*. United Nations, 2018. https://artnet.unescap.org/publications/books-reports/handbook-negotiating-sustainable-development-provisions-preferential.

Berger, A., Brandi, C., Bruhn, D., and Chi, M. "Towards 'greening' trade? Tracking environmental provisions in the preferential trade agreements of emerging markets." In *Deutsches Institut für Entwicklungspolitik Discussion Papers,* 2017.

COMESA. (2009). *Role of COMESA in advancing Africa's Sustainable Vision and Mission of COMESA. 1,* 2009: 1–3.

COMESA. *COMESA Communication Policy and Strategy 2016–2020.*, 2016. https://www.comesa.int/wp-content/uploads/2019/02/COMESA-Communication-Strategy.-_final_web_email.pdf.

COMESA. *Treaty Establishing the Common Market for Eastern and Southern Africa (Revised).*, 2019. https://www.tralac.org/resources/by-region/comesa.html.

Draper, P., Khumalo, N. and Tigere, F. *Sustainability Provisions in Regional Trade Agreements: Can they be Multilateralised?* 2017. http://e15initiative.org/publications/sustainability-provisions-in-regional-trade-agreements-can-they-be-multilateralised/.

EAC. *Treaty for the Establishment of the East Africa Community (As amended on 14th December, 2006 and 20th August, 2007).* (1999): 1–129. https://www.eacj.org/?page_id=33.

EAC. "East African Community vision 2050: Regional vision for socio-economic transformation and development." *The East African Community* 142, (2015). http://repository.eac.int/handle/11671/567.

Echandi, Roberto, Maliszewska, Maryla, and Steenbergen, Victor. *Making the Most of the African Continental Free Trade Area: Leveraging Trade and Foreign Direct*

Investment to Boost Growth and Reduce Poverty, Washington, DC: World Bank, 2022. https://openknowledge.worldbank.org/handle/10986/37623.

Erixon, F. and Lamprecht, P. "The role of trade policy in promoting sustainable agriculture." *European Centre for International Political Economy (ECIPE)*, (2020): 1–42. https://ecipe.org/publications/sustainable-agriculture/.

EU and SADC. "Economic partnership agreement between the European Union and its member states, of the one part, and the SADC EPA States, of the other part." *Official Journal of the European Union, L 250/3.*, 2016. https://doi.org/10.1371/journal.pone.0065542.

European Commission. "Economic partnership agreement between the EU and the Eastern African Community (EAC)." October 2015, 2.

EU and SADC. "Economic partnership agreement between the European Union and its member states, of the one part, and the SADC EPA States, of the other part." *Official Journal of the European Union, L 250/3.* https://doi.org/10.1371/journal.pone.0065542.

Ha-Joon Chang. "Kicking away the ladder: The 'real' history of free trade." *Foreign Policy in Focus* (Silver City, NM: Interhemispheric Resource Center, December 2003). http://www.personal.ceu.hu/corliss/CDST_Course_Site/Readings_old_2012_files/Ha-Joon%20Chang%20-%20Kicking%20Away%20the%20Ladder-The%20%E2%80%9CReal%E2%80%9D%20History%20of%20Free%20Trade.pdf.

Helble, M. and Ben, S. *WIN–WIN How International Trade Can Help Meet the Sustainable Development Goals* (M. Helble & S. Ben (eds.)). Asian Development Bank (ADB) Institute., 2017.

Jack, Goldstone. "The battle for earth's climate will be fought in Africa." *The Wilson Center.*, 2021 (Online). Available at: https://www.wilsoncenter.org/article/battle-earths-climate-will-be-fought-africa. Accessed on March 12, 2023.

Mauricio, Mesquita and Moreira - Marcelo Dolabella. "Can trade policy contribute to fighting global warming in Latin America and the Caribbean?" *Inter-American Development Bank*, October 27, 2022. https://blogs.iadb.org/integration-trade/en/can-trade-policy-contribute-to-fighting-global-warming-in-latin-america-and-the-caribbean%ef%bf%bc/.

Ngwawi, J. "Vision 2050 SADC ponders future." *SADC Today* 14, no. 4 (2012). https://www.sardc.net/editorial/sadctoday/documents/v14n3.pdf.

Portela, C. and Orbie, J. "Sanctions under the EU Generalised System of Preferences and foreign policy: Coherence by accident?" *Contemporary Politics* 20, no. 1 (2014): 63–76. https://doi.org/10.1080/13569775.2014.881605.

SADC. *Draft SADC Vision 2050 and RISDP 2020–30 Blueprints.* Published by Imani Development International Ltd., 2020. https://imanidevelopment.com/sadc-vision-and-risdp-2020-30/. (Accessed on July 25/07/2020).

SADC. "Vision 2050." *SADC Secretariat*, 2020. Available at: https://www.sadc.int/sites/default/files/2021-08/SADC_Vision_2050.pdf.

UNCTAD. *Trading into Sustainable Development: Trade, Market Access, and the Sustainable Development Goals,* 86., 2016. https://sustainabledevelopment.un.org/content/documents/2509ditctab2015d3_en.pdf%0Ahttps://unctad.org/en/pages/PublicationWebflyer.aspx?publicationid=1474.

United Nations. *International Trade and Inclusive Development: Building Synergies*, 200., 2014. http://www.cepal.org/en/publications/37040-international-trade-and-inclusive-development-building-synergies.

United Nations. "Addis Ababa action agenda of the third international conference on financing for development." *Conference Report* 53, no. 9 (2015): 1689–1699. https://doi.org/10.1017/CBO9781107415324.004.

US Government, U. S. G. *Trade and Development Act of 2000,* 2000: 1–56. https://agoa.info/downloads/legal/2385.html.

World Bank. *The African Continental Free Trade Area: Economic and Distributional Effects*., 2020. https://www.worldbank.org/en/topic/trade/publication/the-african-continental-free-trade-area.

Zhou, W. and Cuyvers, L. "Linking international trade and labour standards: The effectiveness of sanctions under the European union's GSP." *Journal of World Trade* 45, no. 1 (2011): 63–85. https://kluwerlawonline.com/journalarticle/Journal+of+World+Trade/45.1/TRAD2011003.

Chapter 10

Dimensions of Environmental Conflicts in the Niger Delta of Nigeria, 1956–2015

Emmanuel Ezi Obuah and Reginald Chikere Keke

The exploration and exploitation of natural resources of economic value and disagreements arising over ownership, control, and utilization have continuously fascinated scholars in an attempt to draw the interconnections between exploration and management with little consideration of environmental conflicts arising from exploitation. Natural resource conflicts often emanate from resource scarcity,[1] the paradox of resource abundance and development[2] and failure in resource governance for the common good. This work looks at the dimensions of environmental conflicts in the Niger Delta centered on exploration of oil as an invaluable natural and nonrenewable resource. The Niger Delta region was an integral part of the British colony of Nigeria when in 1914, the British colonial government enacted the Oil Mineral Ordinance (OMO), and in 1938, Shell D'Archy got the full right of exploration and license to drill oil in the Niger Delta, in Ijo, Northwest Owerri, and Bomu in Ogoni. In 1959, Shell BP signed an Oil Profit Agreement with Nigeria's Colonial Government.

The history of Niger Delta oil-related environmental conflicts can be traced to January 1956, when Royal Dutch Shell or Shell D'Archy, after several failed attempts, struck oil in commercial quantity in Itokopiri in the Oloibiri District of present-day Bayelsa State.[3] By December of the same year, a second oil well was discovered in Afam in Eleme in present-day Rivers State. The period from 1956 to 2015 has seen the historical evolution of the oil sector, leading to different dimensions of environmental conflicts. In 1956, oil was discovered in commercial quantity in Nigeria and this triggered the commencement of sinister environmental conflicts, while 2015 marks the date when the most genuine program ever in the history of Nigeria to clean up oil spills was launched by Vice President Yemi Osibajo in the Niger Delta.

Environmental conflicts and their consequences ranging from local spills to global climate change and pollution to poverty in the Niger Delta are unprecedented—the results of unfettered crude oil exploitation and exploration. Oil as nature's gift with its attendant wealth and prosperity has been a painful source and cause of conflicts particularly environmental, which has become an endemic threat to the lives and livelihoods of the people in the Niger Delta region.

In Nigeria, the decision to find, exploit, and distribute oil wealth and resources is in the big hand of powerful state and non-state actors with diverse interests, thus making oil a social resource inescapably political. The dynamics of unhealthy oil politics, rivalry, conflict, and cooperation determine the social and geographical distribution of the proceeds from oil to the Niger Delta, thus aggravating environmental risks and conflicts in the region. Therefore, there is a seeming nexus in the region between oil politics and dimensions of environmental conflicts consequent on crude oil exploration. This is because while the state shares the oil wealth with other sectors in the state, as little as 11 percent derivation is allocated to the region, which is very insignificant in resolving oil-related conflict encounters in the region inclusive of environmental conflicts such as oil pollution, land degradation, poor air quality index, potable water contaminations, loss of mangrove vegetation, environmental pollution, excessive heat and sound pollution, health challenges and even death caused by the extractive activities of the oil companies, and government's failure in its constitutional responsibility of mitigating the negative effects of oil and gas exploration in the region.

Oil exploration and refining release heavy gaseous hydrocarbons such as methane and ethane, sulfur, nitrogen, carbon dioxide, salt, and other harmful metals into the land and atmosphere in large quantities with hazardous repercussions on the environment leading to different dimensions of environmental conflicts in the region. Some of the prominent oil spills recorded in the Niger Delta include Bomu II blowout, 1970; Forcados terminal spill, Funiwa 5 oil well blowout, and Oyakana pipeline oil spill, all in 1980; Akoma pipeline spill, 1985; Oshika pipeline oil spill, 1993; and the Goi Trans Niger pipeline spillage, 2004. Oil spillages prevent oxygen from entering the water, causing the death of plant and marine life by suffocation.[4]

Leakages and fire incidents associated with oil and gas exploration and transportation are sources of environmental conflicts in the region as well. The Nigerian Liquefied Natural Gas (NLNG) pipeline crossing through Kala-Akama and Okrika leaked and caught fire and burnt for three days in 2004, killing plant and animal life and leading to the death of whole plant species and several endangered animal species. The activities of the NLNG in Finima in Bonny has resulted in the loss of giant animals like hippopotamus, elephants, white-crested monkey, and crocodiles and the forceful migration of other wildlife due to anthropogenic disturbances.[5] The proliferation of environmental conflicts has made the extraction of oil in the region literally a mixed blessing or even an outright curse.[6]

Environmental conflicts in the Niger Delta are commonly caused by resource depletion, deforestation, soil erosion and degradation, desertification, flooding, pollution, health hazards, destruction of the natural ecosystem, poverty, and militancy occasioned by crude oil exploration and related activities. This means that diverse procedures and strategies for conflict prevention and resolution must be adopted in oil-related conflicts in the Niger Delta with primary attention to strategic leadership and comprehensive management of environmental and natural resources used for the general good to forestall conflicts, through assurances and effort of governmental and nongovernmental bodies and communities at national and local levels.

Furthermore, technical resourcefulness, the capacity to overcome deteriorating environmental conditions, and curbing of human greed through corruption could be significant features in addressing environmental burdens that lead to the diverse dimensions of environmental conflicts in the region. In the Niger Delta, it is abundantly true that the maintenance of good environmental conditions, improvement in the lives and livelihood of the people through resource control, and inclusiveness in resource management may hold one of the vital keys to future peace in the region. The Niger Delta is witnessing extremely fast-changing weather conditions. The land, sea, rivers, and the atmosphere are heating up with the air and the water polluted, thus destroying marine life, eroding the soil, and leading to desertification of the forest, water scarcity, and poor air quality. These conditions contribute to environmental conflict.

In the Niger Delta, the uncontrollable appropriation of oil resources, ineffective laws, and failure in corporate social responsibility (CSR) by multinational oil companies (MOCs) to mitigate environmental conflicts makes the Niger Delta one of the most volatile environmental "wars"[7] zones in the world caused by crude oil exploitation. The region can only be saved by genuine efforts of government policies that take the people into consideration, dedicated leadership, effective CSR, and international support. Mitigating environmental conflicts will entail a holistic approach that assimilates public policy, alternative dispute resolution, conflict analysis, project evaluation, dialogue, public involvement, education, and other creative interventions into an inclusive strategy of integrated environmental and resource management of the area in the interest of the people.

To this end, to contextually place the work in a historical trajectory, the chapter discussed the history and geography of the region, the theoretical framework for environmental conflict, the environmental conflict profile of the region, the dimensions of Niger Delta environmental conflicts, and a conclusion.

HISTORY AND GEOGRAPHY OF THE NIGER DELTA

According to Hailey, the history of a place is "written clearly in its geographical features."[8] Therefore, any worthwhile study of the dimensions of environmental conflicts in the Niger Delta should begin with the geographical environment to ascertain its potential and propensities for environmental conflicts, the continuing struggle of groups to modify and adapt to their environmental conditions which in turn determines their economic, sociocultural, and political activities and consequently their history. Environmental conflicts in the Niger Delta are a continuous dialogue between its geography and history.

The Niger Delta is geographically located in the southernmost part of Nigeria, built up by deposits piled by the River Niger over sixty-five million years ago and sitting directly at the apex of the Gulf of Guinea on the Atlantic Ocean, in the Bight of Biafra[9] covering a greater part of the Nigerian coastline. Consisting of intricate wetlands hosting the largest mangrove forest in the world of about 5,400–6000 squares kilometers,[10] it is the most densely inhabited delta on earth. The Niger Delta is Africa's largest delta and the third-largest wetland on the planet, covering about 70,000 square kilometers (27,000 square kilometers), 7.5 percent of Nigerian land mass, traversed and criss-crossed by a large number of prominent rivulets, streams, canals, and creeks, viz. Rivers Nun and Pinnington, Fish Town Rivers, Rivers—Sangana, Santa Barbara, Bartholomew, Sombrero, New Calabar, Orashi, Bonny, Imo, Andoni, Opobo, Kwa Ibo, Middleton, and Forcados,[11] collectively called "Oil Rivers."[12]

The Niger Delta radiates southward in a triangular shape,[13] historically divided into three broad subregions of Western, Eastern, and Central Delta in the core states of Bayelsa, Delta, and Rivers in the South–South geopolitical zone of Nigeria.[14] It comprises the Itsekiri Kingdom, Warri, the Kingdom of Nembe (Brass), Elem Kalabari, Grand Bonny, Old Calabar, Okrika, and Opobo, referred to as "city-states."[15] Cartographically, the region is delimited from north by Aboh, west by the Benin River estuary, east by the Imo River, and south by Palm Point near Akassa and Nun estuary. It is situated between latitudes 4.01°N and 7.90°N and longitudes 4.50°E and 10.56°E and is synonymous with the riverine people of Nigeria.[16]

The intricate and complex dynamics of national oil politics made the Obasanjo administration promulgate the Niger Delta Development Commission (NDDC) Act of 2000, which among other things further defined the region as made up of nine states in three geopolitical zones in Nigeria, namely, the south–south (Akwa Ibom, Bayelsa, Cross River, Delta, Edo, and Rivers States); south–west (Ondo); and south–east (Imo and Abia) all of which are oil-producing and oil-bearing states. This new delineation, and demarcation, has been criticized in academic and policy circles as "political

Niger Delta[17] or "peripheries." However, Okeoma cited in Okere had argued that "Niger Delta is not a political . . . but a geological affair,"[18] meaning that oil exploration and production take place in a specific geographical and geological area in Nigeria called the Niger Delta.

The Niger Delta contains four ecological zones based on the vegetation: the mangrove forests; fresh and saltwater swamps; and coastal inland and lowland rain forests. The "offshore" zone of coastal waters extends into the Atlantic Ocean. This region, apart from its geo-strategic location and value, is very rich in natural endowments as it is the life wire of Nigeria's hydrocarbons, which accounts for over 9 percent of the nation's GDP, 90 percent of foreign exchange earnings, and 95 percent of the national budget.[19] Despite these resources, the area suffers from the Resource Curse Syndrome. The physical features entail a lowland rain forest with muddy sand beaches, and fresh and saltwater swamps predominantly in the Western and Central Delta with seasonal flooding from Rivers Niger, Orashi, and Nun due to torrential rainfall and the intermittent opening of the Lagdo dam to release excess water.

In terms of linguistics, the region is affiliated with the Kwa subfamily of the broad family of African languages called the Niger-Congo[20] with Igbo used as a second language in some Atlantic communities like Okrika, Kalabari, Nembe, Ikwerre, and Bonny owing to migration and interaction spanning over 500 years. The weather is humid and belongs to the semi-hot equatorial climate with temperatures ranging from 20 degrees Celsius to over 30 degrees Celsius, and the annual rainfall decreases from south (3,050 mm) to north (2,550–3030 mm) with an average rainfall of about 150–400 inches spread over 8–10 months between March and November with a very short dry season from December to February. This dry season period is often accompanied by intermittent rainfall.

In terms of population, the region has about thirty million people[21] with 250 different dialects which cut across 5,000 communities,[22] predominantly populated by the Ijo (Ijaw) ethnic group. Before the discovery of oil in 1956, the spine of the Niger Delta people was subsistence agriculture, which was the primary cause of environmental changes and degradation, fishing, and local manufacturing.[23] Today oil exploration and related activities are the secondary and predominant causes of land degradation, oil spills, and pollution that have devastated the lands leading to the internationalization of Niger Delta conflicts because of environmental conflicts caused by crude oil activities in the Niger Delta.

From the 1990s to 2002, violent agitations between the host communities and the international oil companies (IOCs), host communities and the Nigerian State, and militant groups and the Nigerian state, coupled with the internal crisis between the host communities, brought the Niger Delta conflicts to international limelight. This period saw the hanging of Ken Saro-Wiwa, an environmental activist, and the formation of militant groups like the Movement for the Emancipation of the Niger Delta, the Niger Delta Volunteer Force, Pan Niger Delta Resistance Movements, Movement for the Survival of the Ogoni People, and the suspension of Nigeria from the Commonwealth of Nations, which led the international community to look into the conflicts including environmental conflicts, which were consequences of oil politics in the region. International organizations such as the Green Peace Movement, the Rain Forest Action Group, Friends of the Earth, the UN, Amnesty International, and others started to beam their searchlights into the Niger Delta crisis to mitigate the conflicts brought about by oil activities. This internationalized the Niger Delta conflict in general and environmental conflicts in particular.

THEORETICAL FRAMEWORK

Literature abounds on theories and frameworks for analyzing how a natural resource can be an effectual trigger to violent conflicts, particularly environmental conflicts, such as grievance, greed, resource governance, abundance, and scarcity theories. Environmental conflicts are disputes instigated by an environmental paucity of a resource by man-made disruption of its normal renewal level. Usually, the overuse of the environmental resource exceeds its rejuvenation rate, which results in overstraining the ecosystem and in the destruction of the living space of humans, animals, and plants. This leads to disputes between individuals, companies, communities, and states thus creating a disequilibrium in the environment between human beings and the conditions that create human support for survival. This means a loss of quality of life in the socioeconomic and cultural milieus in the region. This is particularly true of the Niger Delta oil conflicts.

This study is hinged on qualitative historical method and the frustration–aggression theory. The qualitative historical method looks at the history of oil and the despicable neglect of the region from 1956 to 2015 while the frustration–aggression theory credited to John Dollard[24] has valuable insights in explaining the various dimensions of environmental conflicts in which unmet needs of the people over time have galvanized into collective actions of the Niger Delta people to try to seek justice and to ensure the implementation of government environmental policies. The central idea of this theory is that violence does not just happen as violent behavior stems from unfulfilled needs, historically.

The frustration–aggression theory takes its essence in the work of Dollard et al. Dollard led a research group at the Yale Institute of Human Relations and published a book that has become valuable in explaining human conflict behavior. Dollard's hypothesis opined that human frustration ultimately leads to aggression, and aggression always denotes that frustration has occurred at some earlier time. The theory suggests that individuals become aggressive when there are obstacles (perceived and real) on their path to success or attaining the desired objective in life. The theory indicates that aggressive behavior is not motivated by genuine hostility, but by frustration. The theory implies that "there is no smoke without a fire" and that "a hungry man is an angry man."

Frustration could be seen as (a) an individual's perception of a hostile environment, coupled with (b) his pessimism about the realization of goals and (c) the perception that the fate of these goals is in the hands of others. Aggression, on the other hand, is any behavior intended to harm another person who does not want to be harmed. The frustration–aggression theory is a good theory to explain why the youth of the Niger Delta took to arms against the Nigerian state and unleashed unprecedented mayhem and violence in the history of oil-related conflicts in Nigeria in the period under review. The frustration of the youths, prompted by the deprivation (by the government and MNOCs) from benefiting from the natural endowment of crude oil, had led to aggression and the ensuing orgy of violence in the region. This viewpoint is echoed by those who stated that the quest for emancipation, which lies at the heart of the Niger Delta struggle, is not yet directed toward secession or excision from the Nigerian State but is merely a protest against criminal neglect, marginalization, oppression, and environmental degradation as well as economic and sociopolitical hopelessness, and frustration in the Niger Delta.[25]

After several peaceful attempts to call the attention of the government, the IOCs, and the international community to their plight to no avail, their grievances graduated to the level of frustration that erupted into conflicts of various dimensions in the region. The London Conference of 1958, the six-day war in 1966 led by Isaac Adaka Boro against the Nigerian State, resource agitations in the 1990s that led to the death of Ken Saro-Wiwa, and the call for resource control and militancy in the 2000s are some of the violent behaviors, which are the outcome of frustration consequent on survival instincts. Generally, a state or condition where legitimate desires of people are denied, directly or indirectly, because of the structure of the society or political system or government's failure to provide legitimate avenues for channeling the grievances of local populations through legitimate devices might lead people to resort to violent strategies which would, in the end, result in conflicts. In the Niger Delta long years of oil exploration have not benefited the people, and they have felt neglected, impoverished, marginalized, underdeveloped, frustrated, and even death, which have translated to aggression in the form of kidnappings, piracy, youth unrest, pipeline vandalism, and other vices in society. This theory helps to explain the incidence of frustration with the attainment of national environmental policy goals and how that frustration might be followed by aggressive behavior or violence. Such violence may be directed at persons or institutions perceived to be directly or indirectly responsible for the frustration. This theory has the potential for understanding how the socioeconomic, environmental, and political conditions of oil-bearing communities of the Niger Delta have led to a sense of frustration, with implications for a violent reaction.

The judiciary, legislature, and executive arms of government and their necessary agencies (such as specialized development commissions) should provide occasions for legitimate channeling of complaints for groups in a democracy. The concern here is not just about access to participation in environmental policymaking but the implementation process and how such access or its denial might shape behavior toward the implementation of environmental public policy.

Frustration–aggression theory incorporates aspects of the political process theory in terms of the political background in which the unfulfilled needs of groups come to define their pattern of action. This theory demonstrates that violence in the Niger Delta is primarily a direct consequence of the failure of the government to effectively implement relevant oil-related national environmental policy frameworks for proper resource governance.[26] Although economic, political, social, or cultural factors also contribute to the choice of violence, it is because of how even these factors and conditions are connected to environmental problems in the region. The key feature that underscores and provides the *raison d'etre* for the violence is the unmitigated environmental devastation that activities of oil companies have imposed on the lives of millions of Niger Delta people. Successive Nigerian governments have furthered environmental conflict in the region through their non-implementation or enforcement of enacted national policies to mitigate the dire consequences of environmental destruction in the region.

ENVIRONMENTAL CONFLICT PROFILE OF THE NIGER DELTA

Historically, the background for environmental conflict in the region was set in 1956 with the discovery of oil in commercial quantity and the first shipment of about 4,000 barrels in February 1958 from Bonny terminal.[27] In 1955, 1961, and 1962 the Mobil Oil Corporation, Texaco, Nigeria National Oil Company, and Safrap (ELF), respectively, started operations in Nigeria. In 1977, the Federal Government established the Nigerian National Petroleum Corporation (NNPC).[28] From the 1970s oil boom, crude oil became the principal foreign exchange earner, and its exploration and politics became fierce with the neglect of agriculture. The 1979 Land Use Act vested all land in the Federal Government and deprived the people of their rights to traditional and customary ownership. This translated into poverty and hunger in the region. The 1980s saw the mismanagement of oil proceeds by government officials, and the 1990s witnessed agitations for more resource benefits and inclusion by the Niger Delta people following oil pollution and environmental degradation which starved them of their livelihoods. In the 1990s, oil wars led by militant groups such as Henry Okah's Movement for the Emancipation of the Niger Delta (MEND), Ateke Tom's Niger Delta Vigilante (NDV), and the Niger Delta Avengers (NDA), equipped with rocket launchers and machine guns with over 100,000 fighters, battled for supremacy and started hitting strategic Shell Oil Company's targets in the Forcados, Chevron Okan platforms, and Exxon Mobile's Qua Iboe terminals, which are Nigeria's largest oil installations. This led to a drop of about 800,000 barrels per day in 2016 (from 2.2 million barrels per day to 1.4 million barrels per day), which was the lowest production levels in twenty-five years[29] up to that time. Grievances concerning environmental degradation culminated in the execution of environmental activist and playwright, Ken Saro-Wiwa by the military regime of General Sani Abacha. This led to Nigeria's suspension from the Commonwealth of Nations, and Nigeria became a pariah state.

The 2000s witnessed more intense agitations, killings, and kidnapping of both local and expatriate workers by militants, formation of more militant groups by Ateke Tom and Alhaji Asari Dokubo, the bombing of oil platforms, oil bunkering, and oil theft culminating in the people killing policemen, which resulted in the military destruction and occupation of Odi in Bayelsa State. In 2007, two Indian workers in Indorama Petrol Chemical Company in Eleme were kidnapped, for which ₦40 million ransom was paid to the kidnappers. In the same year, five Russian workers were kidnapped at the Aluminium Smelting Company of Nigeria in Ikot Abasi in Akwa Ibom State, and in 2008, Mr. Melad Nasari, a Lebanese employee of Homan Engineering Company in Port Harcourt was kidnapped.[30] All these kidnappings were a form of protest which set the region up for environmental conflicts, making it the most volatile oil region in the world from 2004 to 2009 with negative implications for international oil supplies.

In 2009, President Yar'Adua initiated the amnesty program for all Niger Delta militants. This program was continued by President Jonathan. Between 1956 and 2015, oil and natural resources exploration and other ancillary activities degraded the environment of the Niger Delta leading to national and international oil conflicts and diversified the nature of environmental conflicts in the region. The Niger Delta had a large reserve of economic trees like red mangroves (*Rhizophora mangle*), mahogany (*Swietenia macrophylla*), iroko (*Chlorophora excelsa*), abura (*Fleroya ledermannii*), and cotton (*Gossypium*) trees used for fuel, building poles, saw logs, and transmission poles. Oil exploration has caused the disappearance of these trees in many areas and reducing, greatly, biodiversity, thereby making the region prone to environmental conflicts. Ekeke and Alfred are of the view that the region has one of the richest biodiversity in the world with 36 families and 250 species of fish.[31] These fishes are the main source of subsistence for the people, and their depletion is a consequence of oil pollution, which had discharged about 9–13 billion tons of oil into the Niger Delta environment in the last fifty years,[32] which makes the region predisposed to environmental conflicts of massive dimensions.

In the Niger Delta, conflicts over environmental degradation are a major source of direct environmental conflict exacerbated by a lack of potable water, floods, poverty, deforestation, soil erosion, and pollution. Natural gas flaring/venting volume stands at 8 bcm of gas.[33] The result is overheating of the environment, chronic skin diseases, health hazards, and poor air quality index, caused by the emission of GHGs, all of which makes the conflict profile of the region prominent. The worst violent conflicts in the Niger Delta have been that of environmental decay, infrastructural deficiencies, and lack of government presence in terms of projects that benefit the people. These have resulted in political and military instability, piracy, kidnapping, militancy, unemployment, poverty, health challenges, international oil insecurity, corruption, and destruction of the cultural environment, all consequent on oil exploration and exploitation. Over fifty years of oil exploration cannot afford the dredging or canalization of the rivers and hence once the River Niger overflows its banks, it would lead to flooding creating "environmentally internally displaced persons" who have been displaced from their homes and cut off from the urban areas with nowhere to turn to. This raises the conflict profile of the region through lacks and wants, and thus the rise of frustration and aggression.

From the review of literature and discussion so far, there is a linkage between the dimensions of environmental conflict— the struggle being for scarce natural resources, in this case crude oil and natural gas, kidnappings, and vandalization and destruction of oil installations by militant groups. Crude oil exploration often leads to the destabilization and interference of the natural ecosystem that negatively affects the biotic and human live support of the region, leading to different types of conflicts with very little mitigating efforts by the government. The general lack of positive and conscious cohesion between the people, government, and the MOCs to better the lots of the people increases the environmental conflict profile of the region. Consequently, there is an urgent need to resolve environmental conflicts[34] to be a win–win situation for all actors in the interest of life.

Government policies relevant to the oil industry in the Niger Delta such as Mineral Oils (Safety) Regulations, 1963; Petroleum Regulations, 1967; Oil in Navigable Waters Regulations, 1968; Oil in Navigable Waters Act No. 34, 1968; Petroleum Decree (Act), 1969; Petroleum (Drilling & Production) Regulations, 1969; Petroleum (Drilling & Production Amendment) Regulations, 1973; and Petroleum Refining Regulation, 1974, should be implemented to ensure the protection of the environment and guarantee that oil companies drill for oil in ways that protect the environment and the livelihoods of the people. Unfortunately, even these laws are defective in checking gas flares and oil spills and in controlling the disposal of hazardous wastes by oil companies. They have imposed huge burdens on both the environment and host communities, which have responded with violence resulting in conflicts of immense proportions.[35]

DIMENSIONS OF ENVIRONMENTAL CONFLICTS

Conflicts are disagreements between people, groups, and/or states over something important to them caused by divergent interests, power, poor communication, and/or misunderstanding. Conflict may lead to violence, injuries, destruction, disruption of socioeconomic, environmental, and political systems, and even to death. This study chapter limits conflict to environmental disputes and disagreements between the Niger Delta people, the government, and the MOCs whose oil exploration and exploitation activities destroy the environment with little mitigating efforts or improvement in the well-being of the people. In the region, resource exploitation, scarcity, population growth, structural inequalities, marginalization, and abject poverty are the result of conflicts of different dimensions, especially environmental. Nigeria's national environmental program is embedded in the National Policy on Environment (NPE), which states that: "Nigeria is committed to a national policy that ensures sustainable development based on proper management of the environment to meet the needs of the present and future generations."[36] This is not being met in the region due to oil exploration degrading the environment causing Nigeria's economy, between 2003 and 2008, to lose more than $100 billion to conflict and illegal oil bunkering.[37]

In the Niger Delta, conflicts over possession or access to oil and oil proceeds result in basic economic and sociopolitical conflicts. Environmental degradation is one of the dimensions of environmental conflicts consequent on crude oil activities, caused by human-made environmental changes through oil spills, digging to lay oil pipes, rock blasting, harmful drilling chemicals, loss of settlements, rise in sea levels, clearing of virgin forests leading to the death of flora and fauna, and land shortages for farming. All these negatively impact on the environment, causing environmental conflicts. This manifests in a lack of fresh water, poor soil for cultivation, loss of biodiversity, poor air quality indexes, and climate change. Environmental degradation as a dimension of environmental conflict leads to scarcity because the oil resources are in fixed quantity, unequal distribution of purchasing power and property rights, of oil resources, and the depletion of the capital stock as it is nonrenewable at least in the short run. Reyes[38] opined that environmental degradation often results in different forms of resource scarcity: physical scarcity, geopolitical scarcity, socioeconomic scarcity, and failure to adopt sustainable methods for their management. In the Niger Delta, the four types of resource scarcity mentioned above often lead to environmental conflicts, which are man-made disturbances of the normal resource regeneration rate. This has resulted in overuse and destruction of the environment due to failure of the government to adopt sustainable ways of management of this finite resource to the benefit of the Niger Delta people. This failure instigates socioeconomic scarcity and geopolitical conflicts between the host communities, the government, and the IOCs that impinge on their rights to decent living and good environmental quality.

The dimensions of oil conflicts in the Niger Delta are very complex because these conflicts have always been sandwiched between four major actors: the environment, the Niger Delta people, the oil firms, and the Nigerian state, and consequently, in the last two decades, the international community owing to oil politics. Each time there is an oil-related conflict, these four actors are immediately affected directly or indirectly by a confrontation of cooperation and agitations, making the region one of the most volatile areas of crude oil conflicts in the world between 1995 and 2009. Daniel and Ashbindu opined that there is a deep connection between environmental conditions, resources, and conflicts, especially in areas rich in natural resources on which the people depend for life and livelihood.[39]

Environmental conflicts in Nigeria stem from the fact that all oil resources within her territorial confines belong to the state,[40] with physical, legal, cultural, and national interests playing decisive roles in the ownership, production, sale, and distribution of oil wealth in the country. The south–south states of Nigeria play host to the oil reserves and feel that the oil belongs to them, and they should be the greatest beneficiaries of the proceeds and thus the route to their development and modernization. The Nigerian state does not have the expertise to harness the crude oil and decided to go into joint ventures with the IOCs, who feel that their responsibility to the people is little or absent as they are in business with the Nigerian state and not host communities.

Another dimension of environmental conflict is the health challenges of Nigerians. Environmental health-related risks have become a prime concern in Nigeria, with sundry environmental glitches of air pollution, water pollution, oil spillage, deforestation, desertification, erosion, and flooding caused mostly by anthropogenic undertakings. Data from the Institute of Health Metric and Evaluation on Global Burden of Disease (GBD) used to ascertain the causes of death and disability-adjusted life years (DALYs) in Nigeria from 2007 to 2017 showed that most of the highest-ranked causes of DALYs in Nigeria are related to environmental risk factors. Respiratory infection associated with air pollution has advanced from the fourth in 2007 to the highest-ranked cause of death in 2017. Other leading causes of death associated with environmental risk factors include chronic respiratory diseases; cardiovascular diseases; enteric infections; diarrhea diseases; and communicable, maternal, neonatal, and nutritional diseases, which have resulted in roughly 800 thousand deaths and 26 million people living with DALYs per annum in Nigeria.[41] According to the World Health Organization (WHO), the interactions between humans and the environment affect the quality of life, health disparities, and a healthy life span. Making the environment healthier can prevent about 13 million deaths yearly and avoid 13–37 percent of the world's disease burden, such as 40 percent of deaths from malaria, 41 percent of deaths from lower respiratory infections, and 94 percent of deaths from diarrhea disease.[42] This can be particularly true of the Niger Delta where crude oil exploration is threatening the survival of the people.

The host communities and people, oil companies, and the Nigerian state are incredibly set for competition, cooperation, and conflict with the environment. The Niger Delta inhabitants suffer most as their most cherished treasure, the environment, and their economic and social values are being ruined in many ways, leading to environmental conflicts which now also affect the entire world through climate change, international oil theft, and kidnapping of expatriate workers.

Before 1956, the people of the Niger Delta lived peacefully in their environment and engaged in fishing and farming as their predominant occupations. However, this peaceful communion of the people with their environment dramatically changed after 1956 when oil was discovered, and oil companies started extensive exploration and exploitation in the area. It must be noted that any activity of oil-related operations in the littoral region conflicts with the environment on which the people depend largely and there must be immediate mitigation to restore equilibrium, which has been the situation in the Niger Delta region.

Oil exploration brought pollution as one of the dimensions of environmental conflict both on-shore and off-shore causing sicknesses, loss of farmlands, loss of animal and plant species, contamination of potable water sources, death, and displacement from ancestral homes, resulting in socioeconomic and cultural shocks on the people whose livelihoods and lives are being threatened daily. The people conflict with the environment, the state, and the oil major companies, thus creating youth unrest in the region.

Land degradation and loss of vegetation are other aspects of the environmental conflict with grave consequences in the region. In addition to oil spills, clearing of lands to locate oil wells, subsurface land movements, continuous ground puncture, digging of about three feet to lay pipes over kilometers, seismic operations, dredging of rivers and swamps and seas, and geophysical blasting of rocks, which send sound waves to locate oil traps, produce narcotic effects which kill fishes, marine animals, destabilize sedimentary rocks, reduce plant photosynthetic activities, destroy animal habitats,[43] and interfere with human livelihoods and flood homes, thereby constituting an environmental conflict of gigantic proportions in the region.

Another feature of environmental conflict that has subsisted over the years in the Niger Delta since the advent of oil exploration is poor management of industrial emissions and the flaring of gas into the Niger Delta environment by the IOCs. Industrial emissions from drilling equipment, burning of fuel for heating and power, chemical reactions, and leaks release harmful chemicals like carbon dioxide, carbon monoxide, sulfuric acid, methane, ethane, benzene, and chloromethane with gaseous properties into the atmosphere, which cause greenhouse effects, climate change[44] overheat, and acid rain in the region. These leaks and emissions of gases and chemicals reverse the physical, thermal, chemical, and biological quality of underground water, making it unhealthy for plant and animal consumption within the environment; hence, diseases and death are rife in the region. This is a serious conflict reducing the life expectancy of the people. The people also illegally refine crude in a process christened "gbo fire," which causes overheating and permanent bright light due to flaring and destruction of the environment. People suffer from heat waves, inhaling dangerous gases, lack of adequate sleep, and lung and heart diseases.

Furthermore, oil spillage sabotage and illegal oil refining by the OICs, youths, and international oil thieves present another perspective of environmental conflict in the region. According to the Nigerian Minister of Environment, Mohammad Abubakar, Nigeria had between March 2015 and March 2021 recorded 4,919 oil spills, of which operational maintenance contributed 106, sabotage is 3,628, and yet to be determined is 70 (if you add these, you still have a shortfall of 1,835, except if the statistics are for something else, which needs to be clearly spelled out). These spills resulted to a total of 235,206 barrels of oil spilled into the environment and costing the Nigerian government about 4.75 trillion naira. Between 2015 and 2018, a total of 400,000 barrels of oil was spilled , making it the highest in the world during the period followed by Mexico.[45] It has been alleged that Nigerian top government officials, security operatives, and international partners are involved in oil theft in the Niger Delta region.[46] These activities discharge oil into the rivers and lakes, which destroy the marine life and the environment that the people depend entirely on for survival, with Shell's cleanup claims of remediation and reclamation seemingly being ineffective.[47] This too is another aspect of the conflict.

The sociopolitical dimension of environmental conflict in the Niger Delta can be seen from the extreme level of corruption in the Nigerian state and the oil industry, and it is driving hunger, poverty, unemployment, and political unrest in the rich oil and gas surplus region. Roughly 90 percent of national revenue is generated from oil, and according to Amnesty International, 70 percent of the population in the Niger Delta lives on less than one USD per day because of environmental degradation and corrupt leadership, among other factors.[48] In the oil sector, there are cases of firms declining to publish the financial statements of oil revenues and how the money is used.

Financial corruption has made the oil firms (IOCs) not live up to their billings of SCR to the host communities as most projects are overinflated, executed poorly, dysfunctional, or abandoned. It is expected that these firms meet up some level of SCR to their host such as providing education, good roads, potable water sources, scholarship, and other community development initiatives, which will help cushion the negative effects of oil exploration on the lives and livelihood of the people. This is not being met, resulting in conflicts. Some of the oil firms bribe community chiefs and youth leaders to intimidate the people so that CSR cannot be carried out, while others feel it is not their responsibility to address the economic and environmental imperatives in their areas of operation. This made the people lose faith and trust in the IOCs. The IOCs must connect positively with their stakeholders in the host communities to fashion ethical practices that create socioeconomic value for the people as a way of mitigating the effects of oil activities, to avoid conflicts and create an environment peaceful and conducive to all stakeholders in the sector. This will greatly reduce incidences of violence. Thus, CSR has become a rational economic venture by IOCs for the development of diverse host communities in the region.

Corruption in the Nigerian oil sector is evident in favoritism in awarding oil contracts and oil blocks to political cronies, collection of unlawful levies and dues, deceitful exchanges, misappropriation, manipulation of financial records, conspiracy in oil theft with government officials, security agencies, and international collaborators, bribery, and false oaths.[49] These end up short-changing the people from the deserved rewards of oil proceeds from the region such as employment, better living standards, availability of social amenities, reclamation of a degraded environment, and mitigating the negative effects of oil externalities[50] and climate change on the people.

Another angle of local environmental conflicts is its propensity to lead to international environmental conflicts through global warming and ozone depletion, the destruction of natural ecosystems, and biodiversity. The consequences of large-scale environmental destruction of virgin forests and mangroves caused by oil companies' activities trigger global warming and ozone depletion, along with a chain of other disturbing environmental conditions. These, in turn, aggravate existing social conditions of poverty, scarcity, prostitution, ethnic cleavages, uneven distribution of political and economic resources, and the spread of microorganisms and infectious diseases stemming from declining physical and socioeconomic health of the people. These create environments suitable for international conflicts through cross-border migration and infiltration of criminals and other miscreants. Global warming, for instance, has the potential to reduce agricultural output and cause droughts and rising sea levels, conditions that could exacerbate poverty and human and animal suffering and lead to clashes between herders and farmers as seen in the Niger Delta. As of 2006, environmental conflicts have reduced crop yields by 10 percent, farmers' incomes by 5 percent, and decreased microbial loads by 20–61 percent[51] in the Niger Delta region. The Nigerian government is far from ending these environmental challenges in the region, which directly and indirectly affects the nation.

CONCLUSION

The Nigerian government, the IOCs, and the Niger Delta people must establish the framework for the data collection on environmental conflicts which, in turn, will help researchers and policymakers identify patterns and nexuses between environmental resources and conflict, thereby facilitating conflict resolution and prevention in the interest of all stakeholders and the international community. Any meaningful development aimed at mitigating the negative effects of environmental conflicts consequent on oil exploration must incorporate the needs and yearning of the people for it to be sustainable in the long run. There must be purposeful and sincere leadership, an end to corruption in the oil sector, and development of infrastructure and social amenities that will ameliorate the environmental conflicts and other social ills in the region. Employment for the teeming youths will reduce pipeline vandalism, and oil theft, and most importantly the international community should compel the Nigerian state to implement international best practices in oil exploration and reclamation of the degraded parts of the environment in the Niger Delta. This is necessary to alleviate poverty, which is currently rampant amid plenty in the oil-rich and gas-surplus Niger Delta. Institutionalized interactive conflict management mechanisms, capacity-building, transparency, environmental justice, and people-centered projects are needed to ensure proper resource governance. Sustainable resource exploration and exploitation procedures based on international best practices will certainly cut down resource-based conflicts affecting the environment in the Niger Delta, Nigeria, and the world generally.

NOTES

1. Percival, Val, & Homer-Dixon, Thomas, "Environmental Scarcity and Violent Conflict: The Case of South Africa." *Journal of Peace Research*, 35, no. 3 (1998): 279–298.

2. Di John, Jonathan, "Mineral Resource Abundance and Violent Political Conflicts: A Critical Assessment of the Rentier State Model." *Crisis States Research Centre*, work paper series, 1, no. 20 (2002). *Crisis State Research Centre, London School of Economics and Political Science*, London.

3. Keke, Reginald, *A Diplomatic History of the Niger Delta from AD 1500–1960: A Study in Inter-State and Inter-Group Relations.* Unpublished Ph.D. Thesis University of Port Harcourt, Nigeria, 2018), 263.

4. Collins Ugochukwu, *Negative Impacts of Oil Exploration on Biodiversity in the Niger Delta Area of Nigeria.* Research Gate, 2008, 143. www.researchgate.net/publication/247897086.2008.

5. Nenibarini, Zabbey, *Impact of Extractive Industries on the biodiversity of the Niger Delta Region, Nigeria.* A paper presented at the National Workshop on Coastal and Marine Biodiversity Management, Calabar, Cross River State, 7–9 September 2004.

6. Ndimele Prince Emeka, *The Political Ecology of Oil and Gas Activities in the Nigeria Aquatic Ecosystem.* London: Elsevier, 2018, 11.

7. Le Billon, Philippe, "Economic and Resource causes of Conflicts." In *The Sage Handbook of Conflict Resolution*, ed. Jacob Bercovitch, Viktor Kremenyuk, and William Zartman, (Washington, DC: Sage Publications, 2009), 210–224.

8. Hailey, Lord, *An African Survey: A Study of Problems arising in Africa South of the Sahara.* London: Oxford University Press, 1938, 1.

9. Alagoa, Joe, *A History of the Niger Delta: An Historical Interpretation of Ijo Oral Tradition.* Port Harcourt: Onyoma Research Publications, 2005.

10. Afolabi, David, *The Nigerian Mangrove Ecosystem. Third Regional Workshop on the Gulf of Guinea Large Marine Ecosystem* (GOGLME). (Lagos. Nigeria, 1998); Nyananyo, Benard. Forest Resources in Alagoa, Joe. & Derefaka, Abi. (Eds.). *Lands and Peoples of Rivers State: Eastern Niger Delta.* Port Harcourt, Onyoma Research publication, 2002.

11. Talbot, Amery, *The Peoples of Southern Nigeria.* (Frank Cass, 1969), 239.

12. Ofonagoro, Walter, *Trade and Imperialism in Southern Nigeria, 1885–1929.* Ibadan, 1978, 23.

13. Cookey, Sylvanus, "Colonialism and Process of Underdevelopment in Nigerian." *Journal of Asian and African Studies,* 14, (1974): 11.

14. Okorobia, Atei, & Olali, Stephen, "Ethno-Nationalism and Identity Conflict in Nigerian History: The Niger Delta Situation to 2012." *Mediterranean Journal of Social Sciences*, no. 2, (2013): 431–447.

15. Dike, Kenneth, *Trade and Politics in the Niger Delta 1830–1885. An Introduction to the Economic and Political History of Nigeria*. London: Oxford, 1956, 30–31.

16. Ikime, Obaro, *Niger Delta Rivalry Itsekiri-Urhobo Relations and the European Presence.1884–1936*. London, 1969.

17. Isunonah, Adefemi, "Armed Society in the Niger Delta: Armed Forces and Society." *Journal of the Niger Delta,* (Port Harcourt: The Institute of Niger Delta Studies) no. 39, (2012): 331–338.

18. Okere, T.I., & Njoku, K. C, *The Niger Delta: From the Slave Trade to the Crude Oil*. Gabtony & Ass. Nig, 2009, 55.

19. Tekena, Tamuno, *Oil Wars in the Niger Delta, 1849–2009*. Lagos, Sterling: Horden, 2011.

20. Greenberg, Joseph, *The Languages of Africa*. (Bloomington, 1963); Williamson, Kay. (1988). *The Linguistic Pre-history of the Niger Delta*, in Joe, Alagoa, Anozie, Frederick & Nzewunwa, Nwanna. (Eds.). *The Early History of the Niger Delta*. Hamburg: Helmut Buske - verlag, 1988.

21. National Population Census Report. (NPC). Publisher: National Population Commission, 2006.

22. Anene, Joseph. (1966). *Southern Nigerian Transition 1885–1906: Theory and Practice on a Colonial Protectorate*. London: Cambridge, 1966, 5.

23. Ceptac Project. *Towards a Culture of Peace in the Niger Delta*. (Nigeria Multi Project Coy AGV, 2002).

24. Dollard, John, Leonard W. Doob, Neal E. Miller, Orval H. Mowrer, & Robert R. Sears. *Frustration and Aggression*. New Haven: Yale University Press, 1939

25. Human Rights Watch, *The Price of Oil: Corporate Responsibility and Human Rights Violations in Nigeria's Oil-Producing Communities*. New York: Human Rights Watch Publication, 1999.

26. Ezirim, Gerald, "Resource Governance and conflict in the Niger Delta Implications for the Gulf of Guinea Region." *African Journal of Political Science and International Relations* 5, no. 2 (2011): 61–71.

27. Shell Petroleum Development Corporation, *Shell in Nigeria's Briefing Note*, Lagos, 1982.

28. Nigerian National Petroleum Corporation, (NNPC), *History of the Nigerian Petroleum Industry*, Abuja, 2012.

29. Terry Hallmark, Forbes, *Oil and Violence in the Niger Delta isn't talked about much but has a Global Impac*t. Feb. 13, 2017.

30. Isaac, Albert, Nathaniel Danjibo, & Olumayowa Albert, "Back to the Past: Evolution of Kidnapping and Hostage-taking in the Niger Delta, Nigeria." *Beijing Law Review* 11, no. 1 (2020). https://doi.org/10.4236/blr.2020.111015.

31. Ekeke, Davies, & Alfred-Ockiya. "Sand Dredging Impact on the Fish Catch in Bonny Rivers Estuary, Nigeria." *Environmental Research Journal* 2, (2008): 299–305.

32. Kadapa, A.A, "Environmental Impacts of Exploration and Exploitation in the Niger Delta of Nigeria." *Global Journal of Science Frontier Research Environment and Earth Science* 12, no. 3 (2011): 19–18.

33. World Bank, *Nigeria's Gas Flaring Reduction Target by 2020*. 2017, (n.d.) Retrieved 17 October 2022. http://www.worldbank.org/en/news/feature/2017/03/10/nigerias-flaring-reduction-target-2020.

34. Chris Master, & Carol Pollio, *Resolving Environmental Conflicts*, 2nd edn. London, New York: Taylor and Francis Group, CRC Press, 2012.

35. Fidelis, Allen, *Implementation of Oil Related Environmental Policies in Nigeria*, Cambridge Scholars Publishing, 2012, 2–6.

36. Fidelis, Allen, *Implementation of Oil Related Environmental Policies in Nigeria: Government Inertia and Conflict in the Niger Delta.* London: Cambridge Scholars Publishing, 2012. ISBN (10): 1-4438-3442-4.

37. Ibid

38. Reyes, Judith, *Resources, and the Environment. Scarcity and Sustainability*, in Robert, Estell Robert (eds.), *Global Change and Challenges. Geography for the 1990s*. London, New York, 1991.

39. Daniel Schwartz, & Ashbindu Singh, *Environmental Conditions, Resources, and Conflicts: An Introductory Overview and Data Collection.* United Nations Environment Programme, (UNEP), 1999. ISBN: 92-807-1766-9.

40. PetroleumDecree(Act),1969. https://neiti.gov.ng/cms/wpcontent/uploads/2022/08/Petroleum-Act-1969.pdf.

41. Kim, Knowlton, "Globalization and Environmental Health, Encyclopaedia of Environmental Health." *Elsevier Inc.* (2011): 995–1001.

42. Prince Emeka Ndimele, *The Political Ecology of Oil and Gas Activities in the Nigerian Aquarian Ecosystem*, Academic Press UK. (2018): 55–59.

43. Ajaegbu, Francis, Ikemefuna, Obizoba, & Ekwoaba, Joseph, "Conflict and Environmental Challenges facing the oil companies in Nigeria Niger Delta Region." *International Journal of Business and Management Tomorrow*, 2, no. 3 (2012): 2–8.

44. Premium Times, July 6, 2021. *Nigeria records4919 oil spills in 6 years, 4.5 trillion barrels stolen in 4 years.* National Oil Spill Detection Agency (NOSDRA) Report. https://www.premiumtimesng.com/news/headlines/471901-nigeria-records-4919-oil-spills-in-6-years-4-5trn-barrels-stolen-in-4-years-minister.html?tztc=1.

45. Sahara Reporter News. New York, *Top Nigerian Government Officials, Security Operatives Involved in Oil Theft*, 19 October 2022.

46. Amnesty International, *CLEAN IT UP: Shell's False Claims about Oil Spills and Response in the Niger Delta*, 2015, London Amnesty International.

47. Prince Emeka Ndimele, *The Political Ecology of Oil and Gas Activities in the Nigerian Aquarian Ecosystem*, Academic Press UK, 2018, 87.

48. Eze, O. J. "Analysis of Oil Export and Corruption in Nigeria Economy." *International Journal of Economics, Commerce and Management* 3, no. 7 (2015): 112.

49. Ezirim, Gerald. "Resource Governance and Conflict in the Niger Delta Implications for the Gulf of Guinea Region." *African Journal of Political Science and International Relations* 5, no. 2 (2011): 61–71.

50. Emmanuel Obuah, & Reginald Keke, "Mitigating the Negative Externalities of Oil Drilling Activities in the Niger Delta." *Open Access Library Journal* 9, no. 5 (2022). https://doi.org/10.4236/oalib.1108629.

51. United Nations Environment Programme, (UNEP), 2006. https://www.unep.org/resources/annual-report/unep-2006-annual-report.

REFERENCES

Afolabi, David. *The Nigerian Mangrove Ecosystem. Third Regional Workshop on the Gulf of Guinea Large Marine Ecosystem (GOGLME)*. Lagos, Nigeria, 1998; Nyananyo, Ben. Forest Resources in Alagoa, Joe and Derefaka, Abi. (Eds.). *Lands and Peoples of Rivers State: Eastern Niger Delta*. Port Harcourt, Onyoma Research Publication, 2002.

Ajaegbu, Francis, Ikemefuna, Obizoba and Ekwoaba, Joseph, "Conflict and Environmental Challenges Facing the Oil Companies in Nigeria Niger Delta Region." *International Journal of Business and Management Tomorrow*, 2, no. 3, 2012, 2–8.

Alagoa, Joe. *A History of the Niger Delta: An Historical Interpretation of Ijo Oral Tradition*. Port Harcourt, Onyoma Research Publications, 2005.

Amnesty International. *CLEAN IT UP: Shell's .False Claims about Oil Spills and Response in the Niger Delta*. (London, Amnesty International, 2015.

Anene, Joseph. *Southern Nigerian Transition 1885–1906: Theory and Practice on a Colonial Protectorate*. London: Cambridge, 1966.

Ceptac Project. *Towards a Culture of Peace in the Niger Delta*. Nigeria Multi Project Coy AGV, 2002.

Chris, Master and Pollio, Carol. *Resolving Environmental Conflicts*, 2nd edn. London, New York: Taylor and Francis Group, CRC Press, 2012.

Collins, Ugochukwu, *Negative Impacts of Oil Exploration on Biodiversity in the Niger Delta Area of Nigeria*. Research Gate, 2008, 143. www.researchgate.net/publication/247897086.2008.

Cookey, Sylvanus. "Colonialism and Process of Underdevelopment in Nigerian". *Journal of Asian and African Studies,* 14, 1974: 11.

Daniel, Schwartz and Singh, Ashbindu. *Environmental Conditions, Resources, and Conflicts: An Introductory Overview and Data Collection*. United Nations Environment Programme, (UNEP), 1999. ISBN: 92-807-1766-9.

Di John, Jonathan. "Mineral Resource Abundance and Violent Political Conflicts: A Critical Assessment of the Rentier State Model." *Crisis States Research Centre*, Work Paper Series 1, no. 20, 2002. Crisis State Research Centre, London School of Economics and Political Science, London.

Dike, Kenneth. *Trade and Politics in the Niger Delta 1830–1885. An Introduction to the Economic and Political History of Nigeria*. London: Oxford, 1956, 30–31.

Dollard, John, Doob, Leonard W. , Miller, Neal E. , Mowrer, Orval H., and Sears, Robert R. *Frustration and Aggression*. New Haven: Yale University Press, 1939.

Ejituwu, Nkparom and Stanley, Okoroafor. *The House of Skulls*. Port Harcourt: M & J. Orbit. Print, 2009, 21.

Ekeke, Davies and Alfred-Ockiya. "Sand Dredging Impact on the Fish Catch in Bonny Rivers Estuary, Nigeria." *Environmental Research Journal,* 2, 2008, 299–305.

Emmanuel, Obuah. "Reginald Keke, Mitigating the Negative Externalities of Oil Drilling Activities in the Niger Delta." *Open Access Library Journal,* 5, no. 9, 2022, DOI:10.4236/oalib.1108629.

Eze, O. J. "Analysis of Oil Export and Corruption in Nigeria Economy." *International Journal of Economics, Commerce and Management,* 3, no. 7, 2015, 112.

Ezirim, Gerald. "Resource Governance and Conflict in the Niger Delta Implications for the Gulf of Guinea Region." *African Journal of Political Science and International Relations,* 5 no. 2, 2011, 61–71.

Fidelis, Allen. *Implementation of Oil Related Environmental Policies in Nigeria: Government Inertia and Conflict in the Niger Delta.* London: Cambridge Scholars Publishing, 2012, 2–6.

Greenberg, Joseph. "The Languages of Africa. (Bloomington, 1963); Williamson, Kay. (1988). The Linguistic Pre-history of the Niger Delta." In Joe, Alagoa, Anozie, Frederick and Nzewunwa, Nwanna (Eds.), *The Early History of the Niger Delta.* Hamburg: Helmut Buske - verlag, 1988.

Hailey, Lord. *An African Survey: A Study of Problems Arising in Africa South of the Sahara.* London: Oxford University Press, 1938, 1.

Hyellai, Titus, Xiaoli, Pona Duan, Ayantobo, Olusola O., Daniel Tetteh, Narh. "Environmental Health Situation in Nigeria." *Current Status and Future Needs,* 7, no. 3, 2021, e06405330. Retrieved November 22, 2022. https://www.sciencedirect.com/science/article/pii/S2 844021004357#:~:text=Nigeria%2s%20faced%20with%20e nvironmental,which%20has%20harmed%20the%20population.

Ikime, Obaro. *Niger Delta Rivalry Itsekiri-Urhobo Relations and the European Presence. 1884–1936.* London, 1969.

Isunonah, Adefemi. "Armed Society in the Niger Delta: Armed Forces and Society." *Journal of the Niger Delta.* Port Harcourt: The Institute of Niger Delta Studies, no. 39, 2012, 331–338.

Kadafa, A.A. "Environmental Impacts of Exploration and Exploitation in the Niger Delta of Nigeria." *Global Journal of Science Frontier Research Environment and Earth Science,* 12, no. 3, 2011, 19–18.

Kim, Knowlton. "Globalization and Environmental Health." *Encyclopaedia of Environmental Health, Elsevier Inc.,* 2011, 995–1001.

Le Billon, Philippe. "Economic and Resource Causes of Conflicts." In *The Sage Handbook of Conflict Resolution,* edited by Jacob Bercovitch, Viktor Kremenyuk, and William Zartman. Washington, DC: Sage Publications, 2009, 210–224.

National Population Commission Report, (NPC). Publisher: National Population Commission. 2006.

Nenibarini, Zabbey. *Impact of Extractive Industries on the biodiversity of the Niger Delta Region, Nigeria.* A paper presented at the National Workshop on Coastal and Marine Biodiversity Management, Calabar, Cross River State, September 7–9, 2004.

Ndimele Prince Emeka. *The Political Ecology of Oil and Gas Activities in the Nigeria Aquatic Ecosystem.* London: Elsevier, 2018, 11.

Nigerian National Petroleum Corporation, (NNPC). *History of the Nigerian Petroleum Industry.* Abuja, 2012.

Ofonagoro, Walter. *Trade and Imperialism in Southern Nigeria, 1885–1929.* Ibadan, 1978, 23.

Okere, Theophilus and Njoku, Kingsley. *The Niger Delta: From the Slave Trade to the Crude Oil*. Gabtony & Ass. Nig, 2009, 55.

Okorobia, Atei and Olali, Stephen. "Ethno-Nationalism and Identity Conflict in Nigerian History: The Niger Delta Situation to 2012." *Mediterranean Journal of Social Sciences,* 2, 2013, 431–447.

Percival, Val and Homer-Dixon, Thomas. "Environmental Scarcity and Violent Conflict: The Case of South Africa." *Journal of Peace Research*, 35, no. 3, 1998, 279–298.

Petroleum Decree (Act), 1969.

Premium Times, July 6, 2021. Nigeria Records 4919 Oil Spills in 6 Years, 4.5 Trillion Barrels Stolen in 4 Years. National Oil Spill Detection Agency (NOSDRA) Report. https://www.premiumtimesng.com/news/headlines/471901-nigeria-records-4919-oil-spills-in-6-years-4-5trn-barrels-stolen-in-4-years-minister.html?tztc=1.

Prince Emeka Ndimele. *The Political Ecology of Oil and Gas Activities in the Nigerian Aquarian Ecosystem*. Academic Press UK, 2018, 55–59.

Reyes, Judith. "Resources and the Environment. Scarcity and Sustainability." In *Global Change and Challenges. Geography for the 1990s*, edited by Estell Robert. London, New York, 1991.

Sahara Reporter News. *Top Nigerian Government Officials, Security Operatives Involved in Oil Theft*. New York, October 19, 2022.

Shell Petroleum Development Corporation. *Shell in Nigeria's Briefing Note*. Nigeria, Lagos, Shell, 1982.

Talbot, Amery. *The Peoples of Southern Nigeria. A sketch of their History, Ethnology and Languages with an Abstract of the 1921 Census*. Frank Cass Publishers, 1969, 239.

Tekena, Tamuno, *Oil Wars in the Niger Delta, 1849–2009*. Lagos, Sterling, Horden, 2011.

United Nations Environment Programme, (UNEP). UNEP, 2006, https://www.unep.org/resources/annual-report/unep-2006-annual-report.

World Bank. *Nigeria's Gas Flaring Reduction Target by 2020*. 2017, Retrieved October 17, 2022. http://www.worldbank.org/en/news/feature/2017/03/10/nigerias-flaring-reduction-target-2020.

Chapter 11

Herdsmen–Farmers Conflict in Nigeria's Middle Belt Region

Leonard Sitji Bombom, Racheal Yemisi Amupitan, and Marcus Ogwu Edino

Human–environment interactions are ever-changing, especially so in this era of accelerated climate changes with concomitant consequences on the environment.[1] The dimensions of these interactions are multifaceted and pose interesting yet very challenging research opportunities for understanding the inseparable interactions between human and environmental processes and social systems on the one hand and conflicts on the other hand[2] in the face of demanding environmental shifts.

In recent years, environmental changes in Nigeria have conflagrated the competition for environmental resources, largely crop and grazing lands, by farmers and herdsmen, into violence of unprecedented dimensions.[3] Fasona and Omojola reported that the largest share (35 percent) of all conflicts in Nigeria between 1991 and 2005 was related to competition over agricultural land by herdsmen and farmers.[4] Recently, the herdsmen–farmers conflict in Nigeria has surged with an intensity not previously experienced.[5] About 80 percent of these land-related conflicts in Central Nigeria are accounted for by herdsmen–farmers conflict.[6] According to Adigun, 77 percent of farmer–herder conflicts and 71 percent of its fatalities recorded in Nigeria between 2014 and 2019 occurred in the Middle Belt states of Adamawa, Benue, Kaduna, Kogi, Nasarawa, Plateau, and Taraba.[7] It is estimated that the cost of farmer–herder conflicts in Benue, Kaduna, Nasarawa, and Plateau states alone is about $14 billion annually.[8] Between 2001 and 2018, about 300,000 people were displaced in Benue, Nasarawa, Plateau, and Taraba states.[9] Gursoy had already identified these states as having the highest number of casualties from violent farmer–herder clashes as well.[10] These statistics, among others, underscore the fact that the Middle Belt is the hotspot of violent events between farmers and herders in Nigeria.[11]

Generally, government officials and many scholars have attributed the cause of the conflict or part of it to socioeconomic (human greed),[12] political,[13] ethnoreligious,[14] and sociocultural (mutual suspicion, mistrust, and animosity among groups) or a mix of these factors.[15] This chapter argues that these variables, though important in understanding the nature and pattern of the conflict, are only the proxy causes of it. They (socioeconomic, political, historical–cultural, and ethnoreligious factors) have cast a long shadow over the root causes of the conflict, which are climate and environmental changes. This chapter takes up two challenges: (i) to examine the important conceptual issues associated with the climate and environmental changes and herdsmen–farmers conflict in the Middle Belt region of Nigeria and their interconnections and (ii) to examine potential intervening policies and their implications. The chapter attempts a distant, holistic approach that examines the factors at a larger and wider scale than many commentators have engaged in. The discussion is premised on theories of climate and environmental changes and competition over environmental resources.

THE MIDDLE BELT REGION IN NIGERIA

Recently, the Middle Belt region in Nigeria has become the epicenter of herdsmen–farmers conflict in Nigeria. Understanding its geographical, environmental, and political attributes is significant to an appreciation of the causes and nature of the conflict.

The Middle Belt region is not a politically recognized entity in Nigeria. It is more a cultural region that generally coincides with Central Nigeria. In its original conception, the Middle Belt also includes the ethnic minorities in Borno, Yobe, Gombe, and Bauchi states, among others. However, discussing the issues on a state basis, the Middle Belt region covers nine states and the Federal Capital Territory (FCT) in the central region of Nigeria. These states are Plateau, Kaduna, Kogi, Benue, Kwara, Niger, Adamawa, Taraba, Nasarawa, and the FCT (see figure 11.1), which are found within the broadly classified Guinea Savanna agroecological and vegetation zones of Nigeria. The region is dominated by over 200 so-called minority ethnic groups, whose population sizes range from a few thousands to over a million. They consider themselves largely autonomous and republican, politically.

During the colonial rule, most of the area was controlled by ethnic Hausa-Fulani administrators under the indirect rule system. Most of the Middle Belt region was overseen, on behalf of the colonial power, by Hausa-Fulani administrators who were considered over-domineering, and marginalized the original dwellers of the area (indigenes). In spite of this is, large portions of the area never came under the rulership of the Hausa-Fulani before the advent

Figure 11.1 States in Central Nigeria. *Source*: Created by the author.

of colonialism. Many commentators claim that vestiges of this domination persist in some of the regions and in the conflict over land ownership and administrative, economic, and political control in much of the area.

The Middle Belt region covers an area of approximately 379,862 square kilometers and forms the belt between the drier Sahel–Sudan savanna zone in the north and the tropical forests in the southern parts of the country. This chapter argues that this relative environmental advantage (open grassland with relatively more abundant vegetation and water resources, both important resources to the farmers and herdsmen) is at the heart of the conflict between the two groups in the region.

Generally, there are two broad vegetation zones in Nigeria the savanna and forest zones (see figure 11.2). The savanna zone is an open grassland with scattered trees and shrubs, which makes it ideal for livestock rearing. Importantly, the savanna is devoid of the tsetse fly (genus *Glossina*), which

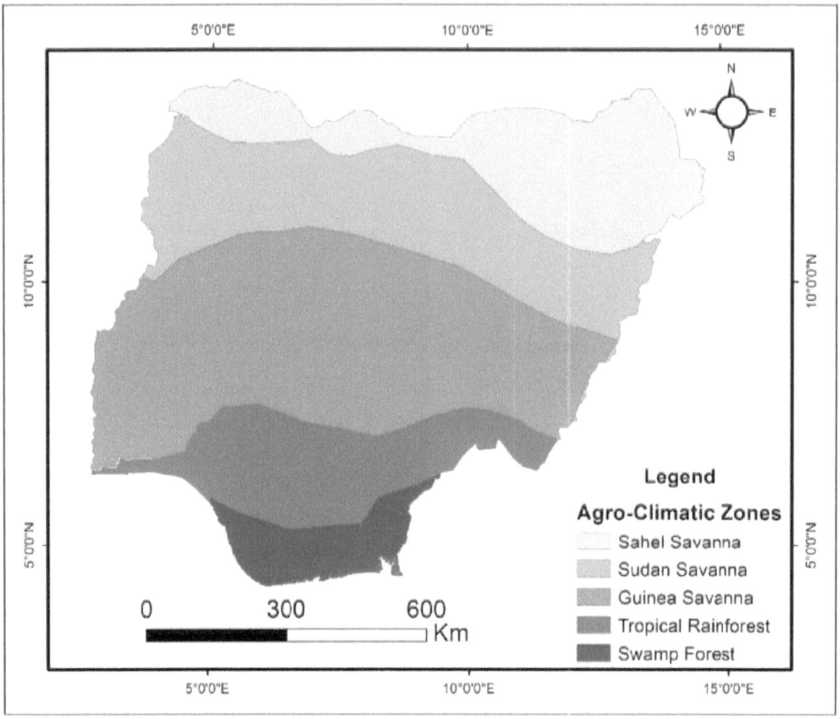

Figure 11.2 Agroecological Zones in Nigeria. *Source*: Created by the author.

causes sleeping sickness, a disease that is fatal especially to cattle and which is common in the forest zone.[16]

The savanna zone is subdivided into three classes: Sahel, Sudan, and Guinea Savanna. The Sahel is the driest and hottest part of the country and generally classified as a semidesert zone characterized by long spells of drought, overgrazing, and degradation of vegetation and soil resources. The rainy season is brief, between May and September only, and average daily temperatures exceed 30°C (86°F). The Sudan Savanna is a region of short grassland, with scattered trees and a relatively longer rainy season (May to October) than the Sahel. The Sahel–Sudan region witnesses high variability in the onset and terminations of the rainy season, receiving on average less than 1,000 millimeters of precipitation annually.[17] The Guinea Savanna zone consists of tall grasses and trees, with a rainfall regime between April/May and October, with average annual rainfall in excess of 1,000 millimeters. The favorable climatic conditions provide the necessary environmental advantages in terms of abundant vegetation and water resources that are attractive for both animal husbandry and crop cultivation. Geographically, the Guinea savanna region of Nigeria coincides with the Central Nigerian states

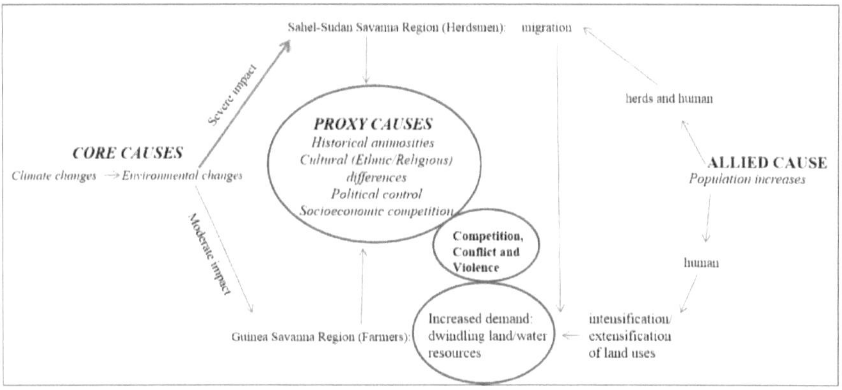

Figure 11.3 Conceptualizing Herdsmen–Farmers Conflict in Central Nigeria. *Source*: Created by the author.

(figures 11.1 and 11.2), which therefore becomes a naturally suited location for competition for environmental resources between herdsmen and farmers.

CONCEPTUAL CONSIDERATIONS IN HERDSMEN–FARMERS CONFLICT

Over the years, competition by herdsmen and farmers over environmental resources in the Middle Belt of Nigeria has become complicated and violent. Several factors have interacted to create an environment conducive for the conflict to emerge and thrive. We have categorized these factors into three causes: the core causes (climate and environmental changes), allied causes (population dynamics), and proxy causes (historical, socioeconomic, political, and cultural).

CORE CAUSES: CLIMATIC AND ENVIRONMENTAL CHANGES

Climate change is at the heart of the farmer–herder conflict in the Middle Belt of Nigeria.[18] Changes in climatic patterns have resulted in environmental changes that have led to the more southerly migration of herders from their traditional northern pastures to the Central Nigeria's region, which is predominated by farmers. The competition over agricultural land and water resources has intensified, and violent conflicts have become the result.[19]

The mounting evidence of global climate change and its attendant local environmental effects have brought the issues of environmental security to the forefront of policymaking in many countries. Studies and reports have suggested that one of the most significant consequences of climate change is the increased risk of armed conflict, especially in developing countries.[20] This risk is premised on the fact that climate change inevitably results in significant changes to the environment, which affects the availability and (re) distribution of environmental resources and confers certain advantages on some areas at the expense of others.[21]

Since the mid-1990s, the issue of resource availability has been central to the understanding of societal implications of climate change. According to Boko et al., rising temperature due to global warming (climate change) will be associated with greater rainfall variability and an increase in the frequency of severe weather events such as drought and floods.[22] They note that a likely consequence of these events is increased scarcity of renewable resources such as water and strains on the productivity and availability of arable land. The resulting environmental stress will alter the political stability of less developed states and make violence more likely.[23]

Scarcity of livelihood essentials such as water or arable land could bring about violence in several ways;[24] generally, groups could fight over a dwindling resource base when supply-induced scarcity triggers competition over renewable resources. In this case, competition over scarce environmental resources, and deteriorating living conditions, may breed frustration and a sense of deprivation. In countries like Nigeria where the foundations of livelihoods and economic activities are based on agriculture, scarcity of environmental resources such as arable and grazing land and water resources could undermine societal stability and security. As powerful groups strive to shift the resource distribution in their own favor, societal turmoil and armed conflict may emerge because increased competition could facilitate mobilization for violence.[25] Numerous case-based accounts of armed conflict in SSA related to environmental issues have been reported in the environmental security literature.[26]

Access to land and water resources has been identified as the greatest vulnerability for both herdsmen and farmers alike in (northern) Nigeria.[27] Adefolalu notes that in 1977 Sahelian vegetation (basically shrubs and dry grassland) did not exist in Niger State (in Central Nigeria), which is bounded by latitudes 9° and 11°N but now covers between 15 percent and 20 percent of the landscape of the area.[28] He, therefore, submits that the southerly march of desertification in Nigeria, a process that has significant implications for the availability of vegetation (grazing land) and water resources in the zone, is tied to a reduction in the amount of rainfall in the region. This assertion is supported by the work of Hess et al., who found that the amount of rainfall in the north-east arid zone of Nigeria has consistently reduced by 8 millimeters per year in the months of August and September between 1961 and 1990, two months that have often recorded some of the largest amounts of precipitation in the region.[29] Olaniran also reports that between 1961 and 1990, there was a reduction of 50 millimeters per year in the amount of rainfall recorded in the marginal savanna belt in the north of Nigeria.[30] This, he claims, is responsible for the more southward expansion of the drier Sahel zone and the decline in water levels of the Kainji Dam, Nigeria's foremost hydroelectric power station located in Niger state in the general Guinea Savanna belt. Olaniran further observes that the devastating droughts that besieged the country (Nigeria) between the 1930s and 1950s and between 1970s to the mid-1990s were more persistent and devastating in the northern part of Nigeria due to the already fragile nature of the ecosystem.[31] He also observes that the irregularity and variability in decreasing the amount of rainfall between 1921 and 2000 intensified over time in Northern Nigeria.

Such unreliability in rainfall patterns may be a recipe for migration to areas with better prospects of available and sustained water resources and environmental resources. Oguntunde et al. find that between 1901 and 2000, the general relationship between rainfall amount and latitude indicated a negative linear correlation, which suggests that the precipitation displays a latitudinal zonality, with the implication that rainfall decreases with increasing latitude away from the Atlantic Ocean and in line with reducing vegetal cover.[32] This means that the northern regions of Nigeria received increasingly less amount of precipitation over the years with consequent reduction in vegetal cover.

There is a consensus among researchers that there has been a negative trend in precipitation in Northern Nigeria over a long period of time. Often positive short-term trends in precipitation increase are reversed in the long term, and the variability, even in the short term, is a major cause of concern over the sustained availability of and accessibility to environmental resources in the region. As Bassett asserts the migration of Fulani herdsmen from Northern Nigeria to the lower latitudes started quite some time ago.[33] However, it is hereby argued that Bassett's thesis that the migration had resulted from relationships developed by the Fulani herdsmen and local indigenous peoples of the other regions is only partly correct. It does not explain why, in the face of a breakdown in that relationship and the current constant harassment and conflicts, the Fulani herdsmen have not retreated to the more northerly regions where they are traditionally considered to hail from. Instead, they have advanced further southward. It is safe to assert that the decreasing and unreliability of precipitation and vegetal cover in the northern Sahelian region, which the Fulani herdsmen require for their livelihoods, have increased the competition over these scarce environmental resources and encouraged a more permanent migration trend even at the discomfort of both the herdsmen and the hitherto friendly indigenous farmers of the southerly regions that the Fulani herdsmen now occupy. A return to the friendlier northern regions is not desirous because of more acute shortages in environmental resources (water and grazing land) due largely to climate changes.

As the Sahel and Sudan regions become drier, the vegetation and water resources become more negatively impacted and scarce. This has forced a larger number of herdsmen out of the region in search of better grazing land.[34] Central Nigeria becomes the natural center of attraction and destination for most of the herdsmen for several reasons. First, it is in close proximity to the Sahel–Sudan zones of origin of the herdsmen. Second, it is ideal for livestock rearing, with relatively abundant vegetation and water resources, free from livestock-killing diseases like the tsetse fly (that causes sleeping sickness) that is common in the forest zones. Like the Sahel–Sudan region, the Guinea zone has also been impacted by environmental changes. The difference is in the magnitude of the impact of climate and environmental changes, which have made the Guinea Savanna zone relatively richer in environmental resources; and third, there has been a longstanding relationship between the herdsmen and farmers in the region over decades of interaction. The initial wave of herdsmen migration into the region was welcome and both farmers and herdsmen had mutual benefits of coexistence. Increased migration by herdsmen has, however, occasioned caution, and in some locations, there was active resistance as competition over both land (for grazing for herdsmen and crop cultivation for farmers) and water resources intensify.

Recent trends indicate that herdsmen migrate earlier toward the southern and more environmentally friendly Guinea Savanna belt (Central Nigeria), which has relatively better vegetation and water resources abundance, and stay there for longer periods of time during their transhumance process than they did in the past.[35] This extended period of interaction between herdsmen and farmers beyond the traditionally expected duration suggests a response to changing environmental fortunes and has increased the stress on environmental resources, which has set the herdsmen on a collision course with the traditionally sedentary farmers of the central region.

ALLIED CAUSES: POPULATION DYNAMICS

Traditionally, the large expanse of open grazing land of the Sahel–Sudan region of Nigeria is regarded as the ancestral homeland of the Fulani herdsmen.[36] The region has been favorable for livestock cultivation for centuries. Blench suggests that as livestock and human population increased, the Fulani herdsmen had looked to other regions with more favorable conditions to settle. This was one of the motivations of the Jihad embarked upon by the group in the nineteenth century, which resulted in their conquering and dominating northern Nigeria and much of the central region of Nigeria as well.[37]

With the intervening influence of the colonialists, the Fulani herdsmen had settled into a transhumant routine of oscillating between their primary domain of Northern Nigeria and the adjoining central and some southern parts of the country to graze their livestock.

In the drier months of the year, many herdsmen leave the Sahel–Sudan north toward the central and southern parts of Nigeria where grazing land and water resources for their livestock are in relative abundance. The process would begin in January and February with the return migration to the Sahel–Sudan area usually in the months of April and May. This return migration usually coincides with the onset of the new rains and the planting season in the Guinea Savanna region. The Sahel–Sudan region would also be experiencing a new vegetation regrowth as rains set in and water becomes more abundant.[38] For a long time, the duration of proximity between herdsmen and farmers during the transhumance herding season was short and restricted to farming off-seasons at the host communities, which minimized the potentials for conflict.[39] Bassett and Turner assert that this period witnessed the establishment of good relations between host farming communities and the migrant herdsmen groups that resulted in semi-permanent and permanent settlement of herdsmen in many host communities.[40] Bassett and Turner further argue that this relationship had blossomed over time and attracted a steady flow of migrant herdsmen into the regions based on kinship with earlier settler–herdsmen who had become parts of their new communities.[41] Anecdotal evidence supports this hypothesis but also importantly presents a picture of a new and larger wave of migrating herdsmen into the Guinea Savanna area for permanent settlement, which is largely necessitated by drastic changes in climate and environmental conditions. Interestingly, however, Bassett and Turner submit that the herdsmen had failed to successfully establish long-lasting relationships with a large part of the farming communities in the central region of Nigeria, except for the Adamawa and Jos Plateaus.[42] The highlands have been known as safe havens for the cattle from tsetse fly infestations.[43] Unfortunately, these relations have dramatically collapsed. Recently, oth plateaus have had one of the most intense and sustained herdsmen–farmers conflicts in Central Nigeria. In 2002, decades of herdsmen–farmers conflict boiled over and violence ensued on the Mambilla Plateau. The conflict on the Jos Plateau and neighboring areas of Southern Kaduna and the Benue plains is still ongoing.

Population Growth and Migration

Recent evidence suggests that climate changes have resulted in drier conditions in the Sahel–Sudan region and advanced the frontiers of desertification southward into parts of the central region.[44] The most basic implication is that environmental resources (vegetation and water) have become scarcer, and the herdsmen who depend on these resources for their livelihoods have become more vulnerable. Studies have indicated that one of the most common forms of "adaptation" to such negative environmental changes is the migration of vulnerable groups to more favorable areas.[45] This has contributed substantially to the increased wave of migration of herdsmen from the Sahel–Sudan zone of the north into the Guinea Savanna zone of Central Nigeria.

Nigeria's population seems to be exploding at an alarming rate. In 1991, the country had an estimated population of 88.5 million.[46] In less than two decades, the population has almost doubled to 140 million.[47] In the Sahel–Sudan savanna region, there has been a 73 percent increase in population within the period, while in the central region the population has increased by 58.2 percent. It is also estimated that Nigeria's grazing livestock (cattle, goat, and sheep) population has increased from 53.1 million in 1990 to 137.8 million in 2021.[48] It is estimated that cattle population alone is about 20 million, the largest stock in the West African subregion.[49] The herder population itself has also witnessed a 51.9 percent increase in about half a century.[50] It is easy to assert that most of the livestock population is in the north and central regions of Nigeria. These increases provide important insight into the degree of pressure that is being exerted on environmental resources at both the source (north/Sahel–Sudan) and destination (Middle Belt/Guinea) zones of herdsmen migration.

As in most parts of SSA, most of the population (about 70 percent) in Nigeria depends on subsistent agriculture, which supports 80 percent of total food requirements in the country.[51] In Central Nigeria, most of the population is engaged in subsistent crop production. Increasing population, increasing population density, and high unemployment in the non-primary industry in the region have necessitated the expansion of cultivated land area.[52] Nyong and Fiki refer to this process as the "intensification and extensification" of land uses. Marginal lands are being cultivated and fallow periods are being completely abandoned or drastically reduced to meet some of the demand for arable crop land.[53] These measures have significantly reduced the amount of open land available for grazing.[54] Similarly, increased herdsmen and livestock population in the Sahel–Sudan require a larger expanse of grazing land and water resources for livestock.[55] These requirements are not being met as a result of depleting environmental resources due to climate change. This situation has two immediate implications: relocation of many herdsmen for permanent or temporary residency in other parts of the country and an increase in the duration of interaction between herdsmen and farmers in Central Nigeria. The potentials for conflicts therefore skyrocket.[56]

As the intensity of unfavorable environmental conditions prevailing in the Sahel and Sudan zones increases with climate change, there is an increased southerly migration of herdsmen to the Guinea Savanna region, which has already become more populated, more intensely cultivated, and has less open land area for grazing livestock than it can accommodate.

It is pertinent to acknowledge that the gradual but conspicuous migratory drift of herdsmen southward into Central Nigeria and beyond has several important implications for the conflict. First, Bassett and Turner point to climate and environmental changes in the north as important considerations in the migration process.[57] Second, this migration for a more permanent settlement in Central Nigeria by herdsmen means that the two groups will be competing for the same limited environmental resources (land and water) for the foreseeable future. The prospects of such threats to livelihoods of the farming communities in the Guinea Savanna Belt can hardly be overexaggerated. Their ancestral claim to the land and water resources of the region has been largely unquestioned since the advent of colonialism. The stout defense of these resources by their ancestors against the Fulani Jihad also adds a layer of nationalism to their desire to protect and pass on to their children the land and all its resources as it had been preserved in the face of war and bequeathed to them by their ancestors. The stakes are, therefore, high for the native people of the region. The threat, however, goes beyond the socioeconomics of the loss of their livelihoods to historical, political, and cultural factors.

Proxy causes: Historical, Socioeconomic, Political, and Cultural (Ethnoreligious) Factors

There is a complex history of relations and interactions between the Fulani herdsmen and the ethnic farming communities in the Middle Belt of Nigeria that has cast a long shadow over the root causes of the conflict between the two groups, that is, competition over environmental resources occasioned by climate and environmental changes. A potpourri of historical, political, socioeconomic, and cultural variables and dubious associations that span more than a century continues to play an important role in the perception of the competition and control over environmental resources in the region and have become the proxy causes to the conflict.

Historical–Cultural Factors

The herdsmen are mostly Muslim and belong to one of three "major" ethnic groups, the Hausa-Fulani, in Nigeria while the farmers are largely Christian or traditionalists and belong to a melange of relatively small or minority ethnic groups scattered across the Middle Belt region.[58] Before colonialism, the Fulani Jihadists had conquered and ruled over most of the north and a large portion of Central Nigeria. Blench suggested that part of the military conquest may have been informed by the necessity to open up new grazing areas for the increasing population of the Fulani and also of their herds.[59] In Central Nigeria, there is a large swath of ethnic groups (e.g., the Tiv and Idoma of the Benue Plains; the Jukun and Bachama of Taraba; the Berom, Mwaghavul, and Ngas of the Jos Plateau; the Katafs of Southern Kaduna; and numerous others), who had successfully resisted the Fulani herdsmen Jihad for long periods of time and maintained their independence and freedom before colonialism. Most of these ethnic minority groups are populated by farmers who are either Christians or Traditionalists.

With complete annexation of the region by the White colonialists, many of these ethnic groups were forced under indirect domination by the more predominant Hausa-Fulani ethnic group (whom they had resisted for decades), whose emirate system of governance the colonialists adopted for the region. Colonial history in the region is, therefore, replete with stories of condescension, domination, marginalization, and deprivation by the Hausa-Fulani over the minority groups under the tip of the colonialists who turned the other way. Consequently, there arose a deep-seated distrust and animosity between the two groups in the region.

Generally, politics and socioeconomics are tainted with the same cultural and ethnoreligious brush that has driven large wedges between the herdsmen, seen as the representative of the erstwhile dominant Hausa-Fulani Muslim group and the local, minority ethnic, and often Christian population groups in the conflict areas. The distribution of the conflict suggests some religious colorations as the more intense clashes have been in areas predominantly Christian. These include the Plateau, Benue, Taraba, Southern Kaduna, and parts of Nasarawa and Adamawa states, which are predominantly Christian populations.[60] This fact has informed the commentary that the conflict is ethnoreligious rather than socioeconomic, political, or even environmental competition.[61] It may be difficult to discount the elements of religious factors playing an important role in the conflict, but it is safe to assert that it has mostly only served to mobilize support and sustain the conflict over a long period of time and over a larger space.[62] It may be assumed that cultural (religious) affinity plays a mitigating effect on the frequency and impact of the conflict as areas of the central region whose population is largely Muslim have been impacted the least.

There is a palpable fear of the erosion of cultural and religious freedom and independence by minority groups in Central Nigeria. The establishment of Sharia laws in neighboring states did not assuage these fears, concerns, and anxieties. Instead, they have served to heighten and concretize these perceived threats to minority independence. The current militarization of the conflict in Central Nigeria conjures similar images in the central region, which has further cemented the suspicions of the minority groups that the violence is a precursor for enforced Fulani settlements in the region, reminiscent of the Jihad that had been interrupted by colonialism. The minority ethnic groups of Central Nigeria, therefore, see themselves as the natural potential victims to perceived Fulani expansionism. The final frontier, they believe, is their means of livelihoods, which is invariably tied to their lands since they are farmers and for which they are unwilling to lose control.[63]

Political and Socioeconomic Factors

Socioeconomics are tied to political patronage in an intricate manner that provides an incentive for violent competition. Political patronage and control over community resources and political/economic administrative authorities have conflagrated the conflict.[64] There is a quota system of political access engagement that many groups in the Middle Belt, Nigeria, believe is skewed toward the major ethnoreligious Hausa-Fulani Muslim group and against their interests. Employment and political appointments into federal offices are based on a quota system. Each state is accorded a certain number of positions to fill, supposedly in proportion to its population. The Hausa-Fulani group has complete control in ten states, while more than fifty other minority groups in the Middle Belt region share only a handful of states among them. Some of these states already have large Hausa-Fulani presence or influence, especially Adamawa, Nasarawa, Niger, Kwara, and Kogi. Additional presence of the Hausa-Fulani, therefore, reduces the number of political positions and access available to the minority groups in states they consider their own.

These scenarios have been enacted over the years in several states in Nigeria, some of which have snowballed into violent conflicts. This is popularly known as the indigene–settler question in Nigeria. Across the country, some ethnic nationalities regard themselves as indigenes (aborigines) of places and states and, therefore, control access to resources (political, economic, and environmental) in those states. Other citizens not of those ethnic extractions are considered settlers with only restricted access to the commonwealth. The Jos crisis in Plateau State since the 1990s is a significant reminder of this tension that exists between "indigenes" and "settlers" in almost all states of the country.[65] The current crisis in Plateau State may be traced to the conflict over a federal government appointment in 1994. A Hausa-Fulani citizen of Jos had been appointed, on the quota of Plateau State, to head a federal government welfare agency in the state. The indigenous population, fearing that the welfare program would be skewed by the appointed Chairman in favor of his minority Hausa-Fulani ethnic population in the state, protested.[66] After months of vocal protestations and threats, the Chairman was removed from office. But the seed of discord between the Hausa-Fulani group and the indigenous population, which it must be admitted had been sowed a long time ago, had germinated and has been nourished over years by violent communal conflicts between the two groups.

In February 2023, Nigerians headed to the polls for the national presidential elections. Lagos state was won by Mr. Peter Obi of the Labour Party who defeated Mr. Bola Tinubu, a former governor of the state, generally regarded as the political godfather of the state, though the latter was declared the winner of the presidential vote. Mr. Obi is Igbo, an ethnic group with a large presence in Lagos but who are recognized as "settlers" in Lagos. For the March governorship elections, however, prominent Yoruba leaders, who consider themselves the "indigenes" of Lagos, cautioned the Igbo (who in turn consider themselves indigenes of states in Southeastern Nigeria) against interfering in the political affairs of the state on the ground that they (Igbo) are only "tenants" or "settlers" and therefore non-indigenes.[67] This is the reality of the indigene–settler question in Nigeria, which also plays out in various guises in the farmer–herder conflict.[68] For example, the anti-open grazing laws in several states in Nigeria, including the Middle Belt states of Benue and Taraba, are perceived by the Fulani herder group as discriminatory and restrict their access to resources based on them being perceived as a "settler" group in those states.

It is these and similar situations that have lent credence to commentaries that even the farmer–herder conflict is based on socioeconomic opportunities or the lack of it. Consequently, scholars have identified politics,[69] human greed, economics,[70] and sociocultural (mutual suspicion, mistrust, and animosity among groups)[71] as determining factors in the farmer–herder conflicts in the region.

It is difficult to dismiss the assertions that the farmer–herder conflict is centered on socioeconomics because the competition over environmental resources revolves around livelihoods and therefore economic factors. The nature of this competition and its manifestation are important to understand the sustenance of the conflict over such a long period of time. Blench points out that many cases of deliberate and flagrant abuse of farmlands by herdsmen have been reported to law enforcement authorities to little effect.[72] With increasing influx of herdsmen into farming communities, such cases have become more frequent, and they test the already thin boundaries of patience and hospitality of the local communities.[73] Cattle rustling/theft, which has cost herders losses estimated in billions of naira,[74] has been recognized as an important factor in the conflict in the region as well.[75]

To reduce the proximity of the activities of herdsmen and farmers to minimize conflicts between them,[76] the government demarcated stock routes or grazing corridors through several communities in the Central Nigeria region. The herdsmen are expected to keep their livestock within these corridors during their transhuman migration process. However, local farming communities have cultivated areas demarcated as stock routes or grazing corridors and made incursions into grazing reserves[77] and therefore imperiled access of herdsmen to areas legally assigned to them. As local communities expand in population and sizes, land resource becomes strenuously inadequate.[78] The stock routes or grazing corridors, rich in animal manure, become easy targets.[79] It has been argued that this incursion has given the herdsmen little choice and room to maneuver their livestock through the stock routes.[80]

Very often, the land demarcated by the government as a grazing corridor is inappropriately acquired. Compensations are either inadequate or unaccounted for. The traditional landowners, recognized through ancestral inheritance, often still lay claim to the land and hardly hesitate to cultivate it when needed. The accusations and counteraccusations of mutual incursions into the preserves of each group have, therefore, escalated in recent years with little compromise on both sides. The resulting conflicts have been devastating.

The Shadow Effect

The current spate of conflicts over environmental resources is perceived as a proxy battle over social, economic, ethnic, and religious issues. Theisen, Holtermann, and Buhaug in a study that focuses on availability and changes in land and water resources report that political and economic variables, rather than environmental factors, are the strongest determinants of the risks of violent conflict.[81] This appears to be a valid point. Studies have suggested that environmental conflicts are not as common in developed economies as in developing countries simply because there are several safeguards and numerous alternative livelihoods for the population.[82] In this wise, Kahl's state exploitation hypothesis becomes a ready explanation to environmental conflicts in developing economies.[83] In simple terms, the hypothesis posits that the scarcity of environmental resources sets up an unhealthy competition among groups that provides both the incentives and opportunities to state elites to instigate violence for political, economic, and ethnoreligious control.

In an unstable country like Nigeria, the conditions are ripe for state elites at different scales and levels of governance to dig into these reservoirs of fears (for ethnic minority farming groups) and necessity (for ethnic majority herdsmen groups) to promote individual and ethnic interests against national cohesion and development interests. Environmental conflicts are probably just one of the platforms, albeit a very important one, through which these socioeconomic, political, and cultural (ethnoreligious) battles are fought.

However, it must be reiterated that the threat of competition over environmental resources is a real issue and is the ultimate underlying factor. It is the basis through which these other variables interact and act out. A change in political, economic, and ethnoreligious biases between herdsmen and farmers can only be a prelude to amicably resolve issues around resources availability, competition, and distribution occasioned by environmental changes induced by global climate change. For as long as most Nigerians depend on environmental resources of land, water, and vegetation resources for their livelihoods, the intense competition and conflict over them will continue with little restriction.

Consequently, changes in both environmental (climate change, land use, and population) variables interact with social factors (socioeconomic, ethnicity, religion, and other factors) at multiple scales to produce the conflict that has had devastating impacts on lives and livelihoods, well-being, and sustainable development in Central Nigeria. Blench laments that even though the fatality costs of the conflict between farmers and herdsmen "are reported almost daily in the newspapers, (but) no effective action has yet been taken by government to analyze or remedy the causes of these conflicts."[84] This assertion still remains true today, two decades after Blench made the submission. This chapter seeks to address this situation.

In Nigeria, political access and control have become the easiest vehicles for socioeconomic and cultural advancements for both individuals and groups. Decades of uneasy political interactions and socioeconomic and cultural associations between the herdsmen and farmer groups in the region have bred mutual distrust and animosity, which lend both groups to facile manipulation by state elites jostling for political power and control. Ethnic domination and preservation, socioeconomic advancement and strangulation, and political control and marginalization are important considerations that have overshadowed the root causes of the conflict between herdsmen and farmers in Central Nigeria, which is competition over the shrinking access to environmental resources due to drastic climate changes.

The elites on both sides of the divide exhume historical circumstances and refocus the debate from a conversation into a confrontation. Consequently, political battles for economic control of state resources become the face of access to and use of environmental resources. There is a growing sense of political awareness and socioeconomic sensitivity along with an awakening cultural identity in the central region. The minority groups are working assiduously toward reassuming a separate identity within the polity of Northern Nigeria. The issue of environmental resources control, therefore, is their stand on their identity and aspirations as co-equals to their fellow Fulani "northerners," eager to assert their "rights" to the ownership and use of their resources.

Over the years there has been a growing unease between the two groups in Northern Nigeria. While the Hausa-Fulani group wishes to continue the political alliance with the minority groups that has allowed them to dominate the country's political landscape for decades, the minority groups wish to reassert their independence and cultural identity from the alliance that they believe had been lopsided in favor of the majority Hausa-Fulani group. It is this political tussle that has been elevated to the frontlines of the environmental issues and has overtaken the more important and universal problem of climate change, environmental changes, and competition for resources.

Poorly developed and/or manipulated legal institutions, poor policy implementation and enforcement, and ineffective and compromised conflict resolution mechanisms have furthered the perception of threats to livelihoods of both groups. Damages to crop lands by herdsmen are generally uncompensated for. Encroachment by farmers into herdsmen grazing reserves or corridors is not prosecuted. The legal system has failed to settle these complaints in a timely and just manner, which leaves room for parties to resort to the use of armed force to deny or claim compensation. These factors have aggravated the situation and rendered armed conflict an inevitable outcome of even minor challenges to coexistence.

The threat and fear of the loss of livelihoods by both groups are real, coupled with limited economic alternatives to the threatened livelihoods, collaborate to inspire unhealthy competition, resist compromise and conciliation, and make the potential for violence real.[85] Farmers are afraid that the presence of herdsmen requires cutting back on land for cultivation, which is unacceptable at worst and impracticable at least in the face of increasing population and unemployment outside agriculture. Their livelihoods are inextricably tied to the lands. Herdsmen are generally aware that previous grazing regions have become overcrowded with the population of people and herds, and environmental conditions have become too precarious to stay. New grazing lands are required and neither the forests of the south nor the drying conditions of the north are conducive for their livelihoods. Their vulnerability requires them to take risks that otherwise would appear unnecessary. They are more willing, under the prevailing climate and environmental conditions, to risk conflicts in regions with more favorable environmental resources than face the loss of their livestock and livelihoods. This explains the significance that the proxy causes have assumed and the shadow that they have cast over the core causes of the conflict.

POLICY OPTIONS AND IMPLICATIONS

Several policy options have been proposed by researchers and policymakers at various levels to help resolve the conflict in Central Nigeria and other parts of West Africa. In this section, the paper considers some of these policy options and their implications.

One of the popular measures, among the farming communities in Central Nigeria, is the evacuation of herdsmen from the region. This will immediately de-escalate the tension between the warring groups, reduce the pressure on environmental resources, and take away the competition. Unfortunately, this is hardly a viable option. As had been indicated in this chapter, some herdsmen have settled in Central Nigeria over a long period of time. They have become semi-sedentary and call their current destination home. There are more serious sociopolitical consequences for asking them to relocate. Most of them can hardly claim anywhere else as home other than Central Nigeria. It is, therefore, inconceivable to evacuate all herdsmen from Central Nigeria. It is expedient to even note here that this group of herdsmen does not contribute much to the escalating conflict in the region; rather, it is the increasing wave of new immigrant herdsmen population into the region in recent years, who have had no bond with host communities that have flamed the embers of conflict. Secondly, the need to accommodate the herdsmen in the face of the current environmental challenges is in the national economic interest. The economics of meat and protein shortages and import imbalance that may accrue from importing meat to feed a large population the size of Nigeria demands serious considerations in finding lasting solutions that will benefit both farmers and herdsmen. Consequently, the federal, state, and local authorities need to be proactive in devising suitable mitigating policies and programs to effectively deal with the situation.

Recently, there have been suggestions on sedentarization of the Fulani herdsmen on a permanent basis. Proponents of sedentarizing herdsmen argue that the nomadic lifestyle is unsustainable in the modern era, and especially so with increasing population, expansion, and proliferation of settlements and increasing effects of climate and environmental changes. Consequently, there should be a deliberate policy to settle herdsmen in grazing reserves, which is regarded as their ancestral homeland.

Bassett argues that the sedentarization policy is not necessarily to the best interest of the herdsmen as research indicates that the traditional nomadic transhumance yields better livestock benefits. Bassett, therefore, suggests that measures be emplaced that would allow for effective demarcation, use, and confinement of herds within grazing corridors to reap these important benefits.[86] This presupposes that the capacity exists to ensure that herdsmen are confined within these corridors and that farmers can be kept away from these fertile lands.

One of the viable options that has been suggested is the demarcation of grazing reserves in different parts of the country. This has, understandably, been met with anger in many farming communities;[87] and many commentators have declared it "dead on arrival."[88] This option has a couple of immediate implications. First, the attraction of the grazing reserves lies in its capacity to keep the herdsmen and farmers from competing over the same set of environmental resources. Herdsmen will be expected to graze their livestock only within the confines of the reserves while farmers will be prohibited from cultivating the reserves. The key is to keep the herdsmen and farmers from co-inhabiting the same land area.[89] This is an attractive measure but requires both parties to buy into and cooperate with the government and each other to ensure success. This may appear to be a difficult proposition to contemplate in the current conflict situation. Creating reserves in farming dominated parts of Nigeria means taking someone else's farmland and/or reducing some community farmland in the face of increasing population and environmental changes resulting from climate change. Commentators have argued that grazing reserves will take prime land away from indigenous farmers in their (farming) communities, thereby conjuring land scarcity, competition, and conflict among farmers.[90]

Bassett's account of the failure of the government's grazing reserve policies in Cote D'Ivoire is instructional. The mutual suspicion and distrust between farming and herdsmen groups left little chance for success of the program. Poor government intervention and policy enforcement were the death knells on the program.[91] Close to home, however, is the fact that grazing reserves are not new in Nigeria, but they have been poorly managed.[92] Hundreds of grazing reserves were demarcated in Nigeria between 1960 and 1985, with only a handful of them currently under any semblance of survival. The infrastructure has decayed, most herdsmen have left the reserves, and in many areas, farmers have encroached on the land without any consequences.[93] IRIN reported that there are already 425 grazing reserves demarcated by the government all across the country, with farmers having already encroached on 270 of them, and only a third of the reserves are actually in use.[94] Babalobi and Akinwunmi reported that herdsmen in the Bobi grazing reserves in Niger State have adopted crop farming to supplement income from livestock herding and wished to expand their farm sizes, which apparently may not be easy to do.[95] The growing herdsmen and herds population impose restrictions on the amount of land available for cultivation within the reserves.

The physical limits of these reserves impose restrictions on land available for expansion. This is an important factor to consider in the current push for grazing reserves. First, the reserves have physical boundaries that restrict expansion in the face of increased herds and herdsmen population within them.

Second, environmental resources within the grazing reserves are finite. With increasing impacts of climate change and the desertification process, it is easy to contemplate the exhaustion of the resources within a short period of time. This is even more conceivable if the current level of environmental awareness and education of the people is considered. Overgrazing and poor soils and environmental and agriculture management practices are commonplace and have made significant contributions to the current environmental resource crisis in Nigeria.

Third, the economic burden of demarcating and maintaining grazing reserves for herdsmen has become the subject of intense national debate. The use of federal money for the direct benefits of only a small proportion of the population is considered unacceptable by a large swath of commentators. For example, Ohanwe suggests that such reserves be carved out in the home states of herdsmen,[96] which incidentally are the most affected by climate and environmental changes. At the 2014 National Conference of Nationalities in Nigeria, almost half of the delegates proposed and supported that states where herdsmen are indigenous should use their resources to create and maintain grazing reserves for them rather than the use of the federation funds. This proposal generated heated debate and tension and was largely unresolved at the end of the conference. For many commentators, this is an attractive option that should not be waved away lightly. It has two major advantages, at the micro-spatial level. The bureaucratic bottlenecks of land tenure system requirements and demarcations will be vastly reduced. Also, the cultural (ethnoreligious) differences and residue of political marginalization being experienced by other ethnic populations toward the Hausa-Fulani (the herdsmen ethnic group) will not pose much of the problems. Of course, farmers and herdsmen of the Hausa-Fulani ethnic stock have a running violent conflict between them, but the common ancestral traditional institutions are easier to employ to address the issues to the satisfaction of both parties. However, there are many drawbacks to this proposal as well, the chief among them being the fact that the ancestral homelands of the herdsmen are the most afflicted by climate change-induced shortages of environmental resources that are the main cause of the current conflict. The merits and demerits of this proposal are outside the immediate concern of this chapter, which restricts itself to broad current policy recommendations.

Fourth is the issue of land tenure system in Nigeria. Across the country, land tenure systems vary, sometimes, with pronounced differences even within a given region. Demarcating grazing reserves across the country, therefore, is a risky and tenuous endeavor. Importantly, the grazing reserves, therefore, carry an expiration date, often with no clearly defined terms on the next phase. Looking into the crystal ball, this would be a dangerous recipe soon as the mining rights conundrum on the Jos Plateau has demonstrated. Lands leased to mining companies in the early parts of the twentieth century on the Jos Plateau were meant to have been returned to their indigenous owners at the expiration of the leases. Current claims and counterclaims to ownership of these land areas have been one of the major factors in the nearly two decades of bloody conflicts on the Jos Plateau. It is, therefore, imperative that careful consideration be given to these issues before decisions are reached. A few policy considerations, which of course are open to debate, are suggested. These suggestions assume clearly defined goals and objectives for the grazing reserves with at least a significantly reasonable amount of information on the target population (herds and herdsmen) that needs to be accommodated in the reserves currently and the estimates for the near future.

First, there is the need for an inclusive process that provides platforms for dialogue and discussions between herdsmen and farmers in communities where grazing reserves are to be demarcated. This is probably the most important phase. It is obvious that the federal government proposed, planned, and established the reserves in the past without adequate involvement of the state and local governments and most importantly the herdsmen and their traditional institutions. This is mostly responsible for the failure of the reserves, which ended up being occupied by already settled Fulani people for the purposes of cultivation for crop production.

Issues of land tenure, compensations, resources development and use, rules, regulations, their enforcement policies, monitoring and evaluation of grazing reserves, their goals and objectives, and firm and concrete plans for extensions of leased lands and/or restoration to original owners at the expiration of leases need to be exhaustively discussed and decisions should be arrived at by both parties. The idea is to displace any mistrust between the groups and between either or both groups and the government. An act of good faith is paramount in ensuring the success of grazing reserves. Planning improvements in the reserves with the herdsmen being proposed for the grazing reserves will also help to ensure that they remain within the confines of the reserves. The local government authorities, traditional institutions, and grassroots participation should be the watchword at this phase. It should be the foundation upon which structures are built.

Second, a deliberate and intensive environmental protection, conservation, and resources sustainability program should be undertaken as a prerequisite for herdsmen to be admitted and accommodated in the reserves. The initial program should be followed by regular updating of knowledge and best practices in the reserves. It is safe to assert that poor environmental resources management practices and inadequate knowledge of environmental vulnerability and adaptation strategies have contributed significantly to the current environmental crisis. This is important at both ends of the geographic spectrum—the origin and destination areas. There is still a large population of herdsmen who inhabit the Sahel and Sudan regions. Education on the causes and adaptations to environmental changes is as important as it would be for herdsmen in the grazing reserves. Unless these are definitively addressed, the expiration dates of the grazing reserves will be drastically fast-forwarded.

Third, there should be a realization that the grazing reserve option is only a temporary measure. It is an incremental plan toward a complete sedentarization and integration of the herdsmen into the general economy of the country. Generally, livestock cultivation practices in developed countries began with nomadic practices and transitioned into several other forms of sedentarized practices before the current practices were developed. The goal is, therefore, to arrive at a stabilized system, where livestock (non-poultry) cultivation is incorporated into the mainstream economy. Individual ranching or even group ranching, especially family-oriented groups, should be the target. Without adequate education on ranching, interested herdsmen should be provided well-defined benchmarks of accomplishments to earn government subsidies and insurances for loans to purchase land and equipment and develop individual or group ranches, away from the grazing reserves. The grazing reserves, therefore, should be regarded as both a test run and a bonding phase for farmers and herdsmen to better understand and accept each other as partners. This will be a major step in their eventual integration in their communities of choice for ranching purposes.

The integration of herdsmen into the sedentarized and mainstream economy requires leaps of educational, technological, and economic advancement. Skilled labor, employment opportunities outside the agricultural sector, and diversification of the economy are important ingredients for a successful integration. Presently, the current rates of dependence on agriculture for livelihood cannot be sustained for long with the rates of population growth in Nigeria. There is an urgent need to ensure that the next generation and future generations are gainfully and productively employed in other sectors that will grow the economy, drastically reduce reliance on primary environmental resources, and therefore minimize competition and conflict on land resources. The desired goal is not to forcefully keep farmers and herdsmen apart by arbitrary geographic demarcations but to make it unnecessary for both groups to regard each other as rivals for environmental resources. There is a natural symbiosis that is not being tapped into by both groups as a result of the many proxy causes already identified. An adequate expansion of the economic space and opportunities for all parties will open up avenues for collaboration and mutual benefits.

CONCLUSION

This paper submits that climate change and the resultant environmental change are at the core of the conflict between farmers and nomadic herdsmen in Central Nigeria. It argues that what it terms proxy causes have taken the center stage of the discourse and, therefore, efforts at resolving the conflict avoid the major cause (core factor). Therefore, to address this problem effectively, this paper proposes a major policy thrust that centers around adaptation to environmental changes in the Sahel and provision of reserves that would minimize the movement of the herdsmen and reduce the length of duration of stay in their destinations in the Guinea Savannah (Central Nigeria).

The history of the dismal performance of government institutions to enforce policy is an unfortunate and notorious reminder of the potentials for failure rather than any possibility of success for the proposed grazing reserves, especially considering that these strategies have been tried and found wanting. This is largely because the federal government does not seem to consider involving the local authorities, traditional institutions, and the farmers and herdsmen themselves in coming out with an acceptable and sustainable solution to the conflict. Most of the decisions are taken on their behalf and solutions are proffered. The government, therefore, needs to embark on a systematic and wide-ranging consultation among herdsmen and farmers, seek compromise, conciliation, and coexistence between the groups. Educating both groups is necessary to create awareness, rebuild trust, and promote mutual cooperation.

Importantly, there should be a deliberate effort to advance the frontier of environmental education and awareness that is focused on addressing the core causes of the conflict. Sustainable harnessing and use of environmental resources, especially in the face of dwindling environmental resource bases due to climate and environmental changes, is important to secure a better productive and peaceful future. One of the exacerbating factors in the Sahel–Sudan zone is the indiscriminate and unsustainable use of environmental resources. Water resources are not being conserved, and vegetation is overgrazed. In the farming communities, marginal lands have come under greater cultivation, and unproductive agricultural practices are being employed with little attention paid to re-fertilization of the lands. This has necessitated increased area cultivation at the expense of other uses to which land could be put. These are important issues that need to be addressed in a systematic way to maximize the use of available environmental resources to mitigate the need for migration of herdsmen in the Sahel–Sudan region and "extensification" of land resources in farming communities in the Guinea Savanna region. Elaborate, systematic, and effective adaptation strategies; practices; and policies are required to reduce the vulnerability of both herdsmen and farmers to sweeping climate and environmental changes and population dynamics. The threats to the loss of livelihoods of both herding and farming groups should be removed by the provision of alternative livelihoods. These should be deliberate programs embedded in a broader policy statement specifically tailored to reduce the intensity of competition over environmental resources and minimize the potential for conflict.

It is, therefore, important for the government to provide the necessary environment, through favorable policies to grow and diversify the economy, provide employment opportunities outside of agriculture, and generate alternative livelihoods to the primary industries. This is probably the surest way to mitigate the impact of climate and environmental changes, defuse tension between competing farming and herding communities, remove the fear and risks of loss of livelihoods, reduce competition over the basic needs, and minimize opportunities for conflicts.

NOTES

1. IPCC. "Summary for policymakers, Geneva: intergovernmental panel on climate change." In S. Solomon, et al. (Eds.), *Climate change 2007: The physical science basis. Contribution of Working Group I to the Fourth Assessment Report of the Intergovernmental Panel on Climate Change*, Cambridge University Press: Cambridge, UK, 2007. http://ipcc-wg1.ucar.edu/wg1/Report/AR4WG1_Pub_SPM-v2.pdf.
2. LeBillon, P. "The political ecology of war: natural resources and armed conflicts." *Political Geography* 20, (2001): 561–584.
3. Ukpong, C. "Land as a source of conflict between farmers and herders in Oyo State, Nigeria." *Journal of Aquatic Pollution and Toxicology* 5, no. 5 (2021): 24–33.
4. Fasona, M. J. and Omojola, A. S. "Climate change, human security and communal clashes in Nigeria." International Workshop in Human Security and Climate Change, Holmen Fjord Hotel, Oslo Oct. 21–23: 3–13. 2005.
5. Brottem, L. "The growing complexity of farmer-herder conflict in West and Central Africa." *Africa Security Brief* No. 39, July, African Center for Strategic Studies, Washington DC. 2021.
6. Abbas, I. M. "'No retreat no surrender': conflict for survival between Fulani pastoralists and farmers in northern Nigeria." *European Scientific Journal* 8, no. 1 (2011): 1857–7881.
7. Adigun, O. "The trends and dynamics of Nigeria's Farmer-herder conflicts (2014–2019)." IFRA-Nigeria Working Papers Series, 2022.
8. Nwosu, C. "Between Fulani Herdsman and farmers: National security under Buhari." *The Republic,* 2017. https://republic.com.ng/aprilmay-2017/fulani-herdsmen-farmers/.
9. Nigeria Economic Summit Group. "Resolving the farmer-herder conflict." 2022. https://forum.futureafrica.com/resolving-the-herdsmen-farmers-conflicts-in-nigeria/.
10. Gursoy, G. "Farmers-Herders conflict in Nigeria: An analysis of the root causes and the effects of the conflict." Thesis presented for MA in Peace and Conflict Studies, Phillips University, Marburg, Germany, 2019.
11. Ibid, Brottem, 2021.

12. Abugu, N. A., Oliatan, O. M., Awaisu, A. H. and Bello, Y. A. "Determinants factors of resource conflict between farmers and herdsmen in Benue State, Nigeria." *Global Journal of Social Sciences* 21, (2022): 1–10.

13. Tade, O. and Yikwabs, P. "Conflict triggers between farming and pastoral communities in Nasarawa State, Nigeria." *Journal of Aggression, Conflict and Peace Research* 12, no. 3 (2020): 101–114. https://doi.org/10.1108/JACPR-10-2019-0448.

14. Krause, J. "A deadly cycle: ethno-religious conflicts in Jos, Plateau State, Nigeria." *A Working paper*, Geneva Declaration, 2011. http://www.genevadeclaration.org/fileadmin/docs/regional-publications/GD-WP-Jos-deadly-cycle.pdf.

15. *Ibid, B*rottem, 2021.

16. Bassett, T. J. and Turner, M. D. "Sudden shift or migratory drift? Fulbe herd movements to the Sudano-Guinean region of West Africa." *Human Ecology* 35, (2007): 33–49.

17. Nielsen, J. O. and Reenberg, A. "Temporality and the problem with singling out climate as a current driver of change in a small West African village." *Journal of Arid Environments* 74, no. 4 (2010): 464–474.

18. Nyong, A. "Combating climate change for national security and development in Nigeria." 33rd and 34th Convocation Lecture Delivered at the University of Jos, Nigeria on 12th May, 2023.

19. Ibid, Nigeria Economic Summit Group, 2022.

20. The Economist. "Cloudy with a chance of war." August 6, 2013. http://www.economist.com/blogs/babbage/2013/08/climate-and-conflict (accessed on June 21).

21. Ibid, Nyong, 2023.

22. Boko, M., Niang, I., Nyong, A., Vogel, C., Githenko, A., Medany, M., et al. "Africa, climate change 2007: impacts, adaptation and vulnerability." In M. Parry, O. Canziani, J. Palutikof, P. Van der Linden, and C. Hanson (Eds.), *Assessment report of the Intergovernmental Panel on Climate Change*, Cambridge University Press: Cambridge, 2007.

23. Marchiori, L. Maystadt, J-F. and Schumacher, I. "The impact of weather anomalies on migration in sub-Saharan Africa." *Journal of Environmental Economics and Management* 63, (2012): 355–374.

24. Homer-Dixon, T. "On the threshold: environmental changes as causes of acute conflict." *International Security* 16, no. 2 (991): 76–116.

25. Miguel, E., Satyanath, S., and Sergenti, E. "Economic shocks and civil conflict: an instrumental variables approach." *The Journal of Political Economy* 112, (2004): 725–753.

26. Hsiang, S. M., Meng, K. C., and Cane, M. A. "Civil conflicts are associated with the global climate." *Nature* 476, (2011): 438–441.

27. Nyong, A. "Climate-related conflicts in West Africa." *Report from Africa Population, Health, Environment, and Conflict*, ECSP REPORT, Issue 12, 2006.

28. Adefolalu, D.O. "Desertification studies (with emphasis on Nigeria)." In Vaughan, R.A. (Ed.), *Microwave Remote Sensing for Oceanographic and Marine Weather Forecast Models,* (1990): 273–323.

29. Hess, T. M., Stephens, W., and Maryah, U. M. "Rainfall trends in north east arid zone of Nigeria." *Agricultural and Forestry Meteorology* 74, no. 1–2 (1995): 87–97.

30. Olaniran, O.J. "Rainfall anomalies: The contemporary understanding." 55th Inaugural Lecture Series, University of Ilorin, Ilorin, Nigeria. 2002.

31. Ibid.

32. Oguntunde, P. G., Abiodun, B. J. and Lischeid, G. "Rainfall trends in Nigeria, 1901–2000." *Journal of Hydrology* 411, (2011): 207–218.

33. Ibid, Bassett, 2009.

34. Sambo, U. and Sule, B. "Farmers versus herders conflict and its implications on nation-building in Nigeria: A contribution to the discourse." *United International Journal for Research & Technology* 2, no. 1 (2020): 43–48.

35. Ibid, Bassett and Turner, 2007.

36. Ibid, Bassett, 2009.

37. Blench, R. "The expansion and adaptation of Fulbe pastoralism to subhumid and humid conditions in Nigeria." *Cahiers d'Etudes Africaines* 34, no. 133–135 (1994): 197–212.

38. Iro, I. "The Fulani herding system." *African Development Foundation,* Washington, DC., 1994.

39. Adisa, R. S. "Patterns of conflict and socio-psychological coping strategies among natural resource user-groups in tourism communities of the Nigerian savannah." *The Journal of Tourism and Peace Research* 1, no. 3 (2011): 1–5.

40. Ibid, Bassett and Turner, 2007.

41. Ibid.

42. Ibid.

43. Bassett, T. J. "Mobile pastoralism on the brink of land privatization in Northern Cote D'Ivoire." *Geoforum* 40, (2009): 756–766.

44. Oguntunde, P. G., Abiodun, B. J., and Lischeid, G. "Rainfall trends in Nigeria, 1901–2000." *Journal of Hydrology* 411, (2011): 207–218.

45. Black, R., Adger, W. N., Arnell, N. W., Dercon, S., Geddes, A., and Thomas, D. S. G. "The effect of environmental change on human migration." *Global Environment Change* 215, (2011): 53–511.

46. National Population Commission (1992) Census data, 1991. https://nationalpopulation.gov.ng/census-enumeration.

47. National Population Commission (2006) Census data, 2006. https://nationalpopulation.gov.ng/census-enumeration.

48. FAO. Nigeria Agriculture at a Glance. Food and Agricultural Organization of the United Nations. FAO Regional Office for Africa, 2021. https://www.fao.org/nigeria/fao-in-nigeria/nigeria-at-a-glance/en/.

49. Ajala, O. "New drivers of conflict in Nigeria: an analysis of the clashes between farmers and pastoralists." *Third World Quarterly* 41, no. 12 (2020): 2048–2066.

50. Worldometers. "Nigeriapopulation live200,386,222." 2019. https://www.worldometers.info/world-population/nigeria-population/.

51. Fayinka, F. A. "Food security in Nigeria: challenges under democratic dispensation." 9th ARMTI Annual Lecture, March 24. 2004.

52. Odunze, A. C., Tarawali, G., and Tanimu, J. "Effect of in-situ grazing and stocking density of goats on soil management: a southern guinea savanna of Nigeria experience." *Nigerian Journal of Soil Science* 4, (2003): 9–15.

53. Nyong, A. and Fiki, C. "Droughts-related conflicts, management and resolution in the West African Sahel." *Human Security and Climate Change International Workshop*, Oslo, Norway: GECHS, CICERO and PR20. 2005.

54. Blench, R. "Conflict between Pastoralists and cultivators in Nigeria." *Review Paper* Prepared for DFID, Nigeria, 2011. http://www.rogerblench.info/Development/Nigeria/Pastoralism/Fadama%20II%20paper.pdf.

55. Bassett, T. J. and Turner, M. D. "Sudden shift or migratory drift? Fulbe herd movements to the Sudano-Guinean region of West Africa." *Human Ecology* 35, (2007): 33–49.

56. Adekunle, O. A. and Adisa, S. R. "An empirical phenomenological psychological study of farmer-herdsmen conflicts in north-central Nigeria." *Journal of Alternative Perspectives in the Social Sciences* 2, no. 1 (2010): 1–27.

57. Ibid, Bassett and Turner, 2007.

58. Usman, A. "Ethno-religious and political dimensions of Farmers-Herdsmen Conflict in Nigeria." *Dutse International Journal of Social and Economics Research* 2, no. 2 (2019): 1–21.

59. Ibid, Blench, 1994.

60. Edwards, G. G. "Effect of farmers-herders conflict on food security in Plateau state." *International Journal of Science and Applied Research* 5, no. 1 (2022): 52–65.

61. Ibid, Adigun, 2022.

62. Ibid, Krause, 2011.

63. Shilgba, Leonard Karshima. A Response to Fulani Aggression in TIV Land. Sahara Reporters, March 13, 2014. https://saharareporters.com/2014/03/13/response-fulani-aggression-tiv-land-leonard-karshima-shilgba.

64. George, J., Adelaja, A., Vaughan, O., and Awokuse, T. "Explaining transhumance related violence: Fulani ethnic militia in rural Nigeria." *Journal of Rural Studies* 89, (2022): 275–286.

65. Ibid, Krause, 2011.

66. Higazi, A. "The Jos crisis: A recurrent Nigerian tragedy." Discussion Paper, No. 2., Jan. 2011. https://library.fes.de/pdf-files/bueros/nigeria/07812.pdf.

67. Vanguard Newspaper, Onanuga cautions Igbo against "interfering" in Lagos politics, 19 March 2023. https://www.vanguardngr.com/2023/03/onanuga-cautions-igbo-against-interfering-in-lagospolitics/#:~:text=Onanuga%20issued%20the%20warning%20on,%2C%20Imo%2C%20any%20Nigerian%20state.

68. Kwaghga, J. "Herdsmen/Farmers crisis: A threat to democratic governance in Nigeria." *Research on Humanities and Social Sciences* 8, no. 11 (2018): 100–108.

69. Ibid, Sambo and Sule, 2020.

70. Ibid, Abugu, 2022.

71. Ibid, Brotteim, 2021.

72. Ibid, Blench, 1994.

73. Oladele, O.T and Oladele, O. I. "Effect of pastoralist-farmers conflict on access to resources in savanna area of Oyo State, Nigeria." *Life Science Journal* 8, no. 2 (2011): 616–621.

74. Kenneth, I. O. and Ananti, M. "Marriage of inconvenience between farmers and herders in Nigeria: Can elephant and hippo tango?" *Arts and Humanities Open Access Journal* 2, no. 6 (2018): 358–372.

75. Ibid, Adigun, 2022.

76. Adisa, R. S. "Patterns of conflict and socio-psychological coping strategies among natural resource user-groups in tourism communities of the Nigerian savannah." *The Journal of Tourism and Peace Research* 1, no. 3 (2012): 1–15.

77. Abbas, I. M. "'No retreat no surrender': conflict for survival between Fulani pastoralists and farmers in northern Nigeria." *European Scientific Journal* 8, no. 1 (2011): 1857–7881.

78. Genyi, G. A. "Ethnic and religious identities shaping contestation for land-based resources: The Tiv-farmers and pastoralists conflicts in Central Nigeria until 2014." *Journal of Living Together* 4–5, no. 1 (2017): 136–151.

79. Ibid, Brotteim, 2021.

80. Ibid.

81. Theisen, O. M., Holtermann, H. and Buhaug, H. "Climate wars? Assessing the claim that drought breeds conflict." *International Security* 36, (2011): 79–106.

82. Raleigh, C. and Urdal, H. "Climate change, environmental degradation and armed conflict." *Political Geography* 26, no. 6 (2007): 674–694.

83. Kahl, C. H. "*States, scarcity, civil strife in the developing world.*" Princeton University Press: Princeton, NJ. 2006.

84. Ibid, Blench, 1994.

85. Bello, A. U. "Herdsmen and farmers conflicts in north-eastern Nigeria: causes, repercussions and resolutions." *Academic Journal of Interdisciplinary Studies* 2, no. 5 (2013): 129–139.

86. Ibid, Bassett, 2009.

87. Ibid.

88. Ibid, Shilgba, 2014.

89. Ohanwe, A. C. "Grazing reserves: A deadly recipe for disaster." NigeriaWorld, 2014. https://nigeriaworld.com/articles/2014/may/091.html.

90. IRIN. "Nigeria: Curbing farmer-nomad clashes and protecting livestock routes." *Integrated Regional Information Networks* (IRIN), 2009. http://www.preventionweb.net/english/professional/news/v.phd =11359.

91. Bassett, T. J. "The political ecology of peasant-herder conflicts in Northern Ivory Coast." *Annals of Association of American Geographers* 78, no. 3 (1998): 453–472.

92. Ibid, Ukpong, 2021.

93. Ibid, Iro, 1994.

94. Ibid, IRIN, 2009.

95. Babalobi, O. O and Akinwumi, J. A. "Mixed farming in a grazing reserve in Northern Nigeria." *Nigerian Journal of Animal Science* 3, no. 2 (2000): 125–130.

96. Ibid, Ohanwe, 2014.

REFERENCES

Abbas, I. M. "'No retreat no surrender': Conflict for survival between Fulani pastoralists and farmers in northern Nigeria." *European Scientific Journal* 8, no.1 (2011): 1857–7881.

Adefolalu, D. O. Desertification Studies. In Vaughan, R. A. (Eds.), *Microwave Remote Sensing for Oceanographic and Marine Weather-Forecast Models. NATO ASI Series*, Springer: Dordrecht 298, (1990). https://doi.org/10.1007/978-94-009-0509-2_14

Adekunle, O. A. and Adisa, S. R. "An empirical phenomenological psychological study of farmer-herdsmen conflicts in north-central Nigeria." *Journal of Alternative Perspectives in the Social Sciences* 2, no.1 (2010):1–27.

Adigun, O. "The trends and dynamics of Nigeria's Farmer-herder conflicts (2014–2019)." IFRA-Nigeria Working Papers Series, 2022. https://theses.hal.science/IFRA-NIGERIA/hal-03762007v1.

Adisa, R. S. and Adekunle, O. A. "Farmer-herdsmen conflicts: A factor analysis of socio-economic conflict variables among arable crop farmers in north central Nigeria." *Journal of Human Ecology* 30, no.1 (2010): 1–9.

Adisa, R. S. "Patterns of conflict and socio-psychological coping strategies among natural resource user-groups in tourism communities of the Nigerian savannah." *The Journal of Tourism and Peace Research* 1, no. 3 (2011): 1–15.

Adisa, S. A. *Land use conflict between farmers and herdsmen – implications for agricultural and rural development in Nigeria rural development – contemporary issues and practices*, Intech Publishers, 2012. www.intechopen.com.

Ajala, O. "New drivers of conflict in Nigeria: An analysis of the clashes between farmers and pastoralists." *Third World Quarterly* 41, no. 12 (2020): 2048–2066.

Aremu, O., Bello, E., and Akanbi, B. F. "Trend analysis and change detection point detection of rainfall across agroecological zones for sustainable development in Nigeria." Environmental Risk Assessment and Remediation, 1 no. 2, (2017): 36–46.

Ati, O. F., Iguisi, E. O. and Afolayan, J. O. "Are we experiencing drier conditions in the sudano-sahelian zone of Nigeria?" *Journal of Applied Sciences Research* 3, no.12 (2007): 1746–1751.

Babalobi, O. O. and Akinwumi, J. A. "Mixed farming in a grazing reserve in Northern Nigeria." *Nigerian Journal of Animal Science* 3, no. 2 (2000): 125–130.

Bassett, T. J. "Mobile pastoralism on the brink of land privatization in Northern Cote D'Ivoire." *Geoforum* 40, (2009): 756–766.

Bassett, T. J. and Turner, M. D. "Sudden shift or migratory drift? Fulbe herd movements to the Sudano-Guinean region of West Africa." *Human Ecology* 35, (2007): 33–49.

Bello, A. U. "Herdsmen and farmers conflicts in north-eastern Nigeria: Causes, repercussions and resolutions." *Academic Journal of Interdisciplinary Studies* 2, no. 5 (2013): 129–139.

Black, Richard, W. Neil Adger, Nigel W. Arnell, Stefan Dercon, Andrew Geddes, and David Thomas. "The effect of environmental change on human migration." *Global environmental change* 21 (2011): S3–S11.

Blench, R. "The expansion and adaptation of Fulbe pastoralism to subhumid and humid conditions in Nigeria." *Cahiers d'Etudes Africaines* 34, no. 133–135 (1994): 197–212.
Blench, R. "Conflict between Pastoralists and cultivators in Nigeria." Review Paper Prepared for DFID, Nigeria, 2011. http://www.rogerblench.info/Development/Nigeria/Pastoralism/Fadama%20II%20paper.pdf.
Boko, M., Niang, I., Nyong, A., Vogel, C., Githenko, A., Medany, M., et al. "Africa, climate change 2007: Impacts, adaptation, and vulnerability." In M. Parry, O. Canziani, J. Palutikof, P. Van der Linden & C. Hanson (Eds.), *Assessment report of the Intergovernmental Panel on Climate Change*, Cambridge University Press: Cambridge, 2007.
Brottem, L. "The growing complexity of farmer-herder conflict in West and Central Africa." *Africa Security Brief* No. 39, July, African Center for Strategic Studies: Washington DC., 2021.
Edwards, G. G. "Effect of farmers-herders conflict on food security in Plateau state." *International Journal of Science and Applied Research* 5, no. 1 (2022): 52–65.
Fayinka, F. A. "Food security in Nigeria: Challenges under democratic dispensation." 9th ARMTI Annual Lecture, March 24., 2004.
Food and Agricultural Organization (FAO). Nigeria agriculture at a glance, 2021. https://www.fao.org/nigeria/fao-in-nigeria/nigeria-at-a-glance/en/.
Genyi, G. A. "Ethnic and religious identities shaping contestation for land-based resources: The Tiv-farmers and pastoralists conflicts in Central Nigeria until 2014." *Journal of Living Together* 4–5, no. 1 (2017): 136–151.
George, J., Adelaja, A., Vaughan, O. and Awokuse, T. "Explaining transhumance related violence: Fulani ethnic militia in rural Nigeria." *Journal of Rural Studies* 89, (2022): 275–286.
Gursoy, G. "Farmers-Herders conflict in Nigeria: An analysis of the root causes and the effects of the conflict." Thesis presented for MA in Peace and Conflict Studies, Phillips University: Marburg, 2019.
Hess, T. M., Stephens, W. and Maryah, U. M. "Rainfall trends in northeast arid zone of Nigeria 1961–1990." *Agricultural and Forestry Meteorology* 74, no.1–2 (1995): 87–97.
Homer-Dixon, T. "On the threshold: Environmental changes as causes of acute conflict." *International Security* 16, no. 2 (1991): 76–116.
Hsiang, S. M., Meng, K. C. and Cane, M. A. "Civil conflicts are associated with the global climate." *Nature* 476, (2011): 438–441.
IPCC. "Summary for policymakers, Geneva: Intergovernmental panel on climate change." In S. Solomon, et al. (Eds.), *Climate Change 2007: The Physical Science Basis. Contribution of Working Group I to the Fourth Assessment Report of the Intergovernmental Panel on Climate Change*, Cambridge University Press: Cambridge, UK, 2007. http://ipcc-wg1.ucar.edu/wg1/Report/AR4WG1_Pub_SPM-v2.pdf
Iro, I. *The Fulani Herding System*, African Development Foundation: Washington, DC., 1994.
Kahl, C. H. *States, Scarcity, Civil Strife in the Developing World*, Princeton University Press: Princeton, NJ, 2006.

Kenneth, I. O. and Ananti, M. "Marriage of inconvenience between farmers and herders in Nigeria: Can elephant and hippo tango?" *Arts and Humanities Open Access Journal* 2, no. 6 (2018): 358–372.

Krause, J. "A deadly cycle: Ethno-religious conflicts in Jos, Plateau State, Nigeria." *A Working paper*, Geneva Declaration, 2011. http://www.genevadeclaration.org/fileadmin/docs/regional-publications/GD-WP-Jos-deadly-cycle.pdf.

Kwaghga, J. "Herdsmen/Farmers crisis: A threat to democratic governance in Nigeria." *Research on Humanities and Social Sciences* 8, no. 11 (2018):100–108.

LeBillon, P. "The political ecology of war: Natural resources and armed conflicts." *Political Geography* 20, (2001): 561–584.

Marchiori, L. Maystadt, J-F. and Schumacher, I. "The impact of weather anomalies on migration in sub-Saharan Africa." *Journal of Environmental Economics and Management* 63, (2012): 355–374.

Marietu, T. S. and Olarewaju, I. O. "Resource conflict among farmers and Fulani herdsmen: Implications for resource sustainability." *African Journal of Political Science and International Relations* 3, no. 9 (2009): 360–364.

Miguel, E., Satyanath, S., and Sergenti, E. "Economic shocks and civil conflict: An instrumental variables approach." *The Journal of Political Economy* 112, (2004): 725–753.

Nielsen, J. O. and Reenberg, A. "Temporality and the problem with singling out climate as a current driver of change in a small West African village." *Journal of Arid Environments* 74, no. 4 (2010): 464–474.

Nigeria Economic Summit Group. "Resolving the farmer-herder conflict." 2022. https://forum.futureafrica.com/resolving-the-herdsmen-farmers-conflicts-in-nigeria/.

Nwosu, C. "Between Fulani Herdsman and farmers: National security under Buhari." *The Republic,* 2017. https://republic.com.ng/aprilmay-2017/fulani-herdsmen-farmers/.

Nyong, A. "Climate-related conflicts in West Africa." *Report from Africa Population, Health, Environment, and Conflict*, ECSP REPORT, Issue 12, 2006.

Nyong, A. and Fiki, C. "Droughts-related conflicts, management and resolution in the West African Sahel." *Human Security and Climate Change International Workshop*, Oslo, Norway: GECHS, CICERO and PR20. 2005.

Nyong, A. "Combating climate change for national security and development in Nigeria." 33rd and 34th Convocation Lecture Delivered at the University of Jos, Nigeria on 12th May 2023.

Odunze, A. C., Tarawali, G. and Tanimu, J. "Effect of in-situ grazing and stocking density of goats on soil management: A southern guinea savanna of Nigeria experience." *Nigerian Journal of Soil Science* 4, (2003): 9–15.

Oguntunde, P. G., Abiodun, B. J. and Lischeid, G. "Rainfall trends in Nigeria, 1901–2000." *Journal of Hydrology* 411, (2011): 207–218.

Ohanwe, A. C. "Grazing reserves: A deadly recipe for disaster." *NigeriaWorld*, 2014. https://nigeriaworld.com/articles/2014/may/091.html.

Oladele, O. T and Oladele, O. I. "Effect of pastoralist-farmers conflict on access to resources in savanna area of Oyo State, Nigeria." *Life Science Journal* 8, no. 2 (2011): 616–621.

Olaniran, O. J. "Rainfall anomalies: The contemporary understanding." 55th Inaugural Lecture Series, University of Ilorin: Ilorin, Nigeria, 2002.

Raleigh, C. and Urdal, H. "Climate change, environmental degradation and armed conflict." *Political Geography* 26, no. 6 (2007): 674–694.

Sambo, U. and Sule, B. "Farmers versus herders conflict and its implications on nation-building in Nigeria: A contribution to the discourse." *United International Journal for Research & Technology* 2, no. 1 (2020): 43–48.

Shehu A. "The causes and consequences of Fulani pastoralist-farmers conflict in Nigeria." *International Journal of Innovation and Research in Educational Sciences* 5, no. 3 (2018): 357–361.

Shilgba, L. K. "A response to Fulani aggression in Tiv land." https://saharareporters.com/2014/03/13/response-fulani-aggression-tiv-land-leonard-karshima-shilgba.

Tade, O. and Yikwabs, P. "Conflict triggers between farming and pastoral communities in Nasarawa State, Nigeria." *Journal of Aggression, Conflict and Peace Research* 12, no. 3 (2020): 101–114. https://doi.org/10.1108/JACPR-10-2019-0448.

The Economist. "Cloudy with a chance of war." August 6, 2013. http://www.economist.com/blogs/babbage/2013/08/climate-and-conflict.

Theisen, O. M., Holtermann, H. and Buhaug, H. "Climate wars? Assessing the claim that drought breeds conflict." *International Security* 36, (2011): 79–106.

IRIN. "Nigeria: Curbing farmer-nomad clashes and protecting livestock routes." *Integrated Regional Information Networks* (IRIN), 2009. http://www.preventionweb.net/english/professional/news/v.phd =11359.

Ukpong, C. "Land as a source of conflict between farmers and herders in Oyo State, Nigeria." *Journal of Aquatic Pollution and Toxicology* 5, no. 5 (2021): 24–33.

Usman, A. "Ethno-religious and political dimensions of Farmers-Herdsmen Conflict in Nigeria." *Dutse International Journal of Social and Economics Research* 2, no. 2 (2019): 1–21.

Worldometers. "Nigeria population live 200,386,222." 2019. https://www.worldometers.info/world-population/nigeria-population/.

Chapter 12

Forests

The Environmental Theater for Internal Security Operations in Nigeria

Oluchukwu Ignatus Onianwa

Globally, there is a dramatic change in warfare driven by the upsurge in armed non-state actors determined to disentangle the security architecture of nations. Predicting the nature, location, and duration of this new warfare is a very difficult task for the armed forces.[1] Thus, the complex nature of environments that armed groups operate in has prompted a dynamic turnaround in the military response tactics to internal security operations.

Nature has ensured that some parts of the planet, Earth, are forest-endowed and others are forest deficient.[2] Forests are complex ecosystems that house numerous living things from the simplest to the most intricate of life-forms.[3] These forests provide a variety of forest goods[4] and support to a larger degree a range of regulating services like biological disease control, carbon sequestration, natural disaster mitigation, waste assimilation, and erosion regulation.[5] The forests also contribute to economic growth via ecotourism.[6] Rural livelihoods often depend on productive forests that support employment and income earning and poverty reduction.[7]

Forest areas are generally distant from human domains.[8] That makes it serve as a safe haven for criminal activities[9] such as kidnappings, cattle rustling, and unknown gunmen.[10] This chapter investigates forests as an environmental theater of internal security operations in Nigeria. It identifies different forests occupied by armed insurgents and counterinsurgency operations of security forces against them. It interrogates the environmental impact of armed groups versus military battles to gain control of the forests in Nigeria. To this end, both empirical (primary) and secondary sources (books, journal articles, newspapers, and internet sources) are employed for data collection.

Internal security operation is a topical national and international discourse especially in Nigeria. However, enough has not been written on the forests as an environmental theater of warfare. The invasion of the forests by the armed groups has created unimaginable environmental insecurity and unproductive human activities and rendered the economic value of the forests incapacitated. Insurgent armed groups' access to forests is occasioned by its poor management. Thus, studying the engagement, in the forest, between state armed forces and non-state insurgents and its implications is significant to scholarship.

I argued that the employment of forests as theaters of war is a new internal conflict strategy that is highly predominant in Africa, particularly in Nigeria. Forests are strong bases that non-state actors found habitable mainly in developing their military structure ranging from training of its members, arms buildup, and planning attacks and ensuring the security of its members.

Despite the heavy bombardment and ground military offensives unleashed against them, armed non-state groups have remained undeterred in their confrontation against the security forces in the forests. Technology is needed to effectively execute forest-enabled warfare mainly to assist in identifying the locations where armed insurgents reside and effectively annihilate them. Consistent review of military operational strategies hinges on the experiences of offensives against the insurgents in the forests and is to form part of the decision-making process of the security forces intended to enhance the chances of winning the forest-prone conflict. Forests need to be turned from the theater of warfare into a center of development through the building of infrastructure that would benefit the people and scare the insurgents away.

The first section of the chapter discusses forests in Nigeria and their locations and use by the insurgents. The second section examines security operations in the forests by the security forces. The third section looks at the regional security networks (RSNs) with a case study of *Amotekun* in the Southwestern Nigeria. The last section interrogates the failure of intelligence gatherings in forest-based security operations.

FORESTS AS THEATER OF CONFLICT IN NIGERIA

In SSA, many countries are surrounded with thick forests often described by scholars as ungoverned spaces. Adequate management of the forests should be of topmost priority for national governments. Consequently, these forests have turned into a den of illicit operations involving armed non-state actors and their affiliates. The forest is the axis through which they set their targets and attack communities and rob people of their belongings along the highways beside or close to the forests. This phenomenon has manifested in the fashion of village, farmlands, and market raids.[11]

The increasing rates of the use of forests as bases of illicit operations by armed groups has made it necessary to put into scholarly perspectives these forests where they are located and the different activities being carried out in the forests at different points in time. Idu and Gwagwa Forests in Abuja and Kabakawa Forest in Kaduna are bases for criminals who regularly mount roadblocks to attack travelers and rob people living in the areas that border the forests.[12] According to Ibrahim Goni, the Conservator General, National Park Service, "about 1,129 square meters of forest reserves and national parks in the country serve as hideouts for terrorists and bandits."[13]

Nine forests that stretch the entire northern part of Nigeria into some neighboring West African countries are being governed by bandits, thus worsening the insecurity challenges in the region. The nine forests are Sambisa, Alagarno, Kamuku, Kuduru, Kuyambana, Burwaye, Ajja, Dajin Rugu, and Sububu. Bandits, terrorists, and kidnappers have taken over these forests from where they launch attacks and subsequently keep kidnapped victims hostage. Bandits come from Kuduru Forest, fully armed, and transact business activities in Girya, Goburawa, Karshi, Garke, and Uguwan Idi villages, and many others unchallenged.[14]

Alagarno Forest, on the other hand, covers parts of Borno and Yobe States and is also a hiding place for insurgents. Evidence shows that former Boko Haram leader late Abubakar Shekau had over the years shuttled between Sambisa and Alagarno Forests, launching attacks on communities in Yobe and Borno States. In Kano, the Falgore Forest is located about 150 kilometers from the city and cuts across three local governments of Tudun Wada, Doguwa, and Sumaila. Dajin Rugu is a thick forest that stretches from Birnin Gwari in Kaduna State, through parts of Katsina State, and ends in Zamfara State. However, Dajin Rugu has turned into a camp for Fulani cattle rustlers, thieves, and armed robbers.[15]

Cattle rustling is prevalent in the countryside forests of Northern Nigeria. The cattle rustlers plot their criminal enterprise in highly syndicated operational camps in the rural forests of the region. Their activities have been most pronounced in the States of Zamfara, Katsina, Kano, Plateau, Benue, Nasarawa, Bauchi, and Taraba. In northwestern Nigeria, most of the recorded cases of cattle rustling take place in the Kamuku and Kuyanbana Forests.[16] Many of these attacks are launched from areas where there is an absence of state capacity or political will to exercise control.[17]

Herdsmen and farmers' conflict is essentially a forest-based violence. The perpetrators of this violence plan their attacks from their camps in the jungle from where they proceed to hit their target communities. This pattern of rural violence has been pervasive in the north–central states such as Benue, Nasarawa, and Plateau States as veritable critical flash-beds.[18]

Kidnapping is another form of crime that is well executed in the forests. Kidnappers take their victims to the forests where they feel safe to initiate torture and negotiations for their release. Some of these forests beside Akwanga–Nasarawa Eggon, Benin–Ore, and Zaria–Funtua highways have been notorious for such attacks.[19] These forests, along with those found in states like Oyo, Ogun, Ekiti, Osun, and Ondo, are mostly used as dens for these kidnappers.[20]

Nigeria forestlands have equally played host to several other criminal occurrences including highway robbery, cult activities, narcotic cropping, drug trafficking, human trafficking, ritual killing, and arms proliferation.[21] They are often perceived as fertile grounds within which terrorist organizations incubate and thrive. Therefore, containing these spaces falls within the strategic frontier of security priorities.[22]

In the southeast, multiple non-state actors operating in the zone use the forests as places of abode for the planning and execution of their plans. Of particular interest is the series of forests in the states such as Abia, Anambra, Ebonyi, Enugu, and Imo. While the Umunneochi Forest, which borders parts of Imo, Enugu, and Anambra states, is the most notorious in Abia State, the Enugu–Ugwuogo and Nike–Opi–Nsukka Forests are its equivalent in Enugu State. Also located in Ebonyi State is the Agbebor/Agba forest in Onicha and Ishielu.[23] The consistent use of these forests by armed non-state actors has caused severe disruptions of the socioeconomic conditions of the people and prompted the Nigerian Government to admit that kidnappers and bandits have taken over forests nationwide. The Federal Minister of Environment, Balarabe Abbas, disclosed this information at the ministry's top management retreat in Kaduna on January 11, 2024. The minister said, "It is quite unfortunate that most of our forests have been taken by bandits, kidnappers and other terrorists that have been unlashing mayhem on innocent Nigerians."[24] To this end, the Nigerian security forces, especially the army, air force, police, and indigenous security agencies, have intensified efforts to clear the forests of these armed non-state actors by launching massive security operations, both land and air operations, across the Federation.

FORESTS AND SECURITY OPERATIONS

It is a fact that forests are the locations where non-state actors occupy to strategize and execute heinous activities, and such operations are viewed as threats to national security and need to be eliminated. There are various categories of armed security forces engaged in stamping out armed insurgent groups in the forests. First, there are government military personnel such as armed and air forces (Army, Air Force, and even Naval Personnel)—the Nigerian Police and Nigerian Civil Defense and Security Corps, among others. Second, there are government-recognized and sponsored vigilantes and regional/community security networks operating at different levels of government (regional level, such as Amotekun; state level such as state-sponsored vigilantes; and local government levels such as hunters' associations) with different degrees of arms operating jointly as Joint Task Force (JTF) or independently but supported by one another when the need arises.

Internal security operations in different parts of the country are not new. However, such operations became intense in the democratic dispensation when different militant groups in places like the Niger Delta sprang up and launched a vigorous campaign against the Nigerian State due to perceived marginalization and unequal share of oil resources derived in that region.[25]

Nwagboso argued that

> the trajectory of the crisis in the Niger Delta is understandable from the point of view of the inability of the government to provide basic amenities and infrastructural development for the people. This sorry situation is compounded by the economic dislocation occasioned by oil exploitation and exploration activities in the region.[26]

This situation created militant groups, mostly the Movement for the Emancipation of the Niger Delta (MEND) and Niger Delta Avengers (NDA). Mgbonyenbi and Emeni argued that "indeed, the people of the region have come to consider it reasonable to form armed groups or join militant groups as a means of meeting their justice and social needs."[27]

A new phase of the Niger Delta conflict opened following the government response through the military operational team known as JTF comprising personnel from the Nigerian Army, Air Force, Navy, and Police, which has been deployed in the Niger Delta region since the early 2000s. The JTF was tasked with tackling the militant threat in the region and protecting its oil from theft.[28] The conflict was fought in the creeks and mangrove forests, making it one of the most complex military operations carried out by the Nigerian Army since the end of the civil war (1967–1970).

As the Niger Delta crisis was about to come to an end, a new dimension to internal insurgency and terrorism developed in Northern Nigeria: the rise of Boko Haram terrorism.[29] When Boko Haram emerged, the first action taken by the armed group was to find a base to operate from. Boko Haram's search for an appropriate site coincided with the planned counterinsurgency operation by the Federal Government of Nigeria. Given the rapid urbanization and population growth, the group could not build a base in the cities amid the inevitable firepower of the Nigerian military. Subsequently, the Sambisa Forest became their haven.

The Sambisa Forest, located in Borno State in northeast Nigeria, has become synonymous with Boko Haram terrorism. The journey of Boko Haram into the forest started when they were dislodged from the city of Maiduguri by military forces. Finding nowhere else to launch their deadly attacks, they invaded the forest. The security guards staffing the gate were killed and the remaining staff scampered away for their lives. The terrorists then set up various camps in the forest and made Camp Zairo their headquarters. It was from the various camps that deadly plans of attack were made; the most prominent being the abduction of women and children for rape, forced marriage, sex slavery, and suicide bombing missions.[30]

Throughout its stay in Sambisa Forest, Boko Haram built its power and military operation that threatened the entire West African subregion, a development that generated international concern, leading to the declaration of the group by the international community as one of the topmost terrorist organizations in the world. This troublesome development prompted the Nigerian government to embark on counterinsurgency operations against Boko Haram fighters wherever their location in the Sambisa forests.[31]

The Southeastern Nigeria is another zone with an intense security condition. The security of the zone has remained volatile ever since the upsurge of secessionist agitations spearheaded by a group known as the Indigenous Peoples of Biafra (IPOB) led by Nnamdi Kanu. IPOB's establishment of a security outfit known as Eastern Security Network (ESN) in the southeast further increased the intensity of insecurity in the zone. Ever since its establishment, the ESN has been operating in the forests in different parts of the southeastern region geared toward securing it from the armed gunmen wreaking havoc in the different states of the zone.[32]

While the establishment of ESN appears to be a good omen among the pro-secessionists, the security outfit is perceived by the Federal Government of Nigeria as a threat to national security that should be downgraded and incapacitated. As a result, the Nigerian Army in collaboration with other security forces has been involved in the armed operations against the ESN in forests of the southeast. Troops of 34 Brigade, Nigerian Army engaged the ESN in a military face-off at a forest near Amaifeke–Akkatta road in Orlu Imo State. This operation resulted in capturing a cache of arms and ammunition belonging to the dissident group. The firearms and ammunition were recovered during troops' operations to clear separatists' hideouts in the general area of Amaifeke–Akkatta road.[33]

On December 31, 2021, the Nigerian Navy's Special Boat Service (NNSBS) Tier 1 from the Nigerian Navy Base in Onitsha, Anambra State, invaded Ochan Forest known as the Anambra States hotspot for kidnappers and arrested five suspected kidnappers who masterminded the abduction of the traditional ruler of Ogwu-aniocha community in Anambra State, Igwe

Oliver Nnaji, kidnapped on November 15, 2021. The NNSBS operation was propelled by the intelligence report that the NN received from reliable informants within the state.[34]

On January 17, 2022, the Nigerian Army in joint operation conducted with other security agencies invaded the stronghold of the ESN at Lilu Forest in Ihiala Local Government Area of Anambra State, which is a border town between Imo and Anambra States. According to the Director of Army Public Relations, Brigadier General Onyema Nwachukwu,

> the forest hitherto is known to be used as an operational base of the dissidents. The operations, conducted in the early hours of Monday, January 17, 2022, led to the neutralization of some gunmen in a firefight that ensued. It was a well-coordinated dawn attack on the fortress, where plans were being perfected to violently enforce the illegitimate sit-at-home declaration by the groups.[35]

The forest within Ajeme Akpawfu village in the Akpawfu community in Nkanu East Local Government Area of Enugu State was another zone the military invaded to root out the gunmen terrorizing parts of the land. The troops engaged in a shoot-out within the forests where the gunmen were hiding.[36]

The Azia, Orsumoghu, Lilu, Mbosi, and Uli forests in Orsu and Ihiala Local Government Areas of Anambra and Imo States, respectively, were invaded by the Nigerian Army. During the operations, the troops uncovered decomposing remains of kidnapped and murdered victims dumped in a deep well by members of the insurgency groups. While conducting the operations, the unrelenting troops encountered and cleared four ambush sites along roads in Orsumoghu, Mbosi, Ihiala, and Uli, dislodging the insurgents from their criminal havens.[37]

On Monday, May 16, 2022, a similar action was taken against ESN by the Nigerian troops of 34 Brigade Obinze in a fierce encounter that ensued at Amaifeke and Ihioma in Orlu Local Government Area of Imo State. The ESN militia opened fire on sighting the patrol team, who were out to ensure safety and security of lives and property of citizens.[38]

A distinct feature of the internal security operations across Nigeria is the joint collaboration among the security forces. This synergy of combined teams of security agencies has not only strengthened the counterattack strategies but also expedited the forest security operations against the armed groups.[39]

The internal security operation spearheaded by the Nigerian Army is a country-wide exercise not restricted to a particular region or zone because the activities of non-state actors are widespread. The more armed forces confront the armed groups, the greater they move from one location to another particularly in southwest region. The 32 Artillery Brigade, Owena Cantonment,

Akure, in Ondo State invaded forests in Arifa community and Efon–Alaye–Itawure axis of Ekiti State. While the operation lasted, a decomposed body of a man suspected to be a victim of kidnapping, was discovered with his GSM phone still ringing repeatedly. The overall aim of the operation was to make the black spots in the area crime-free.[40]

On December 19, 2022, following the interception and abduction of some travelers by suspected Fulani militia near Science School, Kuru, K-Vom, in Jos South local government area of Plateau State, a team of police operatives moved to Akassa forests where they came into contact with and rescued the kidnapped victims following credible information provided on the hideout of the criminals.[41] Non-state actors, particularly kidnappers, believe that forests are the easiest and safest locations to function effectively. That is why once they have abducted their victims, they move them into the forests. What this shows is that it is high time that a new forests-based security strategy and tactics were developed by the security forces to deal with this type of security threat.

Internal security operation is all-embracing, involving the entire security institutions in the country. Normally only the police is supposed to tackle the issue of internal security and restoration of law and order while the armed forces focus on checkmating external threats against the country. But as the security challenges escalated, it became imperative to forge a joint force of all the security agencies in tackling the menace of the armed non-state actors in different parts of the country.

Consequently, a decision was taken to engage the operation of the Nigeria Air Force to air interdict and inflict heavy bombardment against the bandits who have made the forest an abode for planning and execution of their criminal activities. The Governor of Kaduna State in the northern region, Mallam Nasir El-Rufai, once called for the massive bombardment of all forests inhabited by terrorists. This would contain the security challenges bedeviling the country. According to him,

> we have been pressing for the past two years for the military from the air and the ground to bomb these forests while soldiers on the ground kill all those that escape the bomb. I do not believe that there is any innocent person in the forest . . . It is time for the Nigeria Air Force and Nigerian Army to bomb the terrorists out of existence . . . .[42]

To strengthen internal security operations, the air force designed operational strategies to counter the insurgents. One of these strategies is the direct bombardment of the enemy's camp inside the forests through targeted shooting from the air and shattering the groupings of the insurgents.

Under the Air Component of Operation Hadin Kai, one of the designed military operations against insurgency, the Nigerian Air Force (NAF) struck a target known as Somalia, which was a terrorist hospital and healthcare facility in the Sambisa Forest, the base of Boko Haram. The air interdiction carried out on September 9, 2022, followed credible intelligence on the location of the insurgents. Somalia, a location near Sambisa Forest, is the terrorist's medical facility, where wounded fighters as well as those critically ill were treated. According to Air Commodore NAF Edward Gabkwet, Director of the NAF Public Relations and Information, "for the NAF, its ability to continually dominate troubled parts of the country in collaboration with the surface forces in order to eliminate terrorist elements and restore normalcy should never be in doubt."[43]

On December 22, 2016, troops advanced for the capture of Camp Zairo, which was the main Boko Haram Terrorists (BHT) enclave suspected to be holding high-value persons in Sambisa Forest. The air component and artillery were effectively employed during the operation before the final assault by the troops.[44]

On July 12, 2021, based on credible and timely information on the massing up of bandits around Sububu Forest possibly in preparation for an attack, the Air Component of Operation Hadarin Daji dispatched its NAF platforms and engaged the bandits in a precision bombardment. "Several armed bandits, motorcycles, camps, and other equipment were destroyed."[45]

Air missions conducted by the military over locations of interest in Birnin Gwari, Giwa, Igabi, and Chikun local government areas of Kaduna State led to the killing of many bandits. These air missions have produced some results when employed and appear to bear out the confidence expressed by the Director of Public Relations and Information (DOPRI), NAF, Air Commodore Edward Gabkwet, that the NAF, along with other surface forces have the capacity and capability to dominate trouble spots and restore normalcy.[46]

The Air Component of Operation Whirl Punch in continuation of intensive air operations, on August 30, 2022, neutralized fourteen terrorists while conducting air interdiction missions against identified terrorist's hideouts at Alhaji Isiaka location and Kuduru in Igabi Local Government Area of Kaduna State. The strike was successful as the terrorists were eliminated while several others sustained injuries. Similar strikes were also undertaken at Udawa along Kaduna–Birnin Gwari Road in Kaduna State after receiving intelligence reports of terrorist activities and cow rustling in the area.[47]

The Deputy Theatre Commander, Air, of Operation Lafiya Dole, Air Vice Marshal Isiaka Oladayo Amao, said that

> in continuation of the counter insurgency, the air component . . . conducted 286 operational sorties against terrorists' targets from 25 December to date in the Sambisa Forest. All the areas bombarded were thoroughly monitored on intelligence surveillance before the attacks or shelling.[48]

Air bombardment accompanied with intelligence gathering has the potency of accurately targeting and hitting the enemy in a swift fashion at the exact location where they reside in the forests.

According to Air Commodore Edward Gabkwet, "the successful strikes and ground attacks around Bama were classical examples of the centrality of joint collaboration as a critical factor in operational success."[49] The Air Component of Operation Thunder Strike destroyed a camp and neutralized scores of bandits in airstrikes executed at the Kuduru Forest area of Kaduna State. The Coordinator, Defence Media Operation, Major General John Enenche, explained:

> The air interdiction mission at Kuduru forest was carried out sequel to credible intelligence reports indicating that the heavily armed bandits responsible for a previous attack on troops at Ngede Allah had originated from a camp within the forest. Accordingly, following confirmatory intelligence, surveillance and reconnaissance missions that led to the identification of the camp, the air component dispatched Nigerian Air Force attack aircraft to engage the location, scoring devastating hits, which destroyed the bandits' logistics structures and killed several of the criminals . . . .[50]

On December 16, 2022, the NAF 303 Medium Airlift Group (MAG) Ilorin commenced a joint operation codenamed "Exercise Apejopo Idi," meaning gathering of the Eagles, aimed at utilizing NAF ground and air assets to comb the bushes and vulnerable communities within Eiyenkorin–Peke–Okolowo axis involving a large movement of troops.[51] The deployment of air power against terrorists and insurgents has changed the tides of the ongoing operations in the entire Northern Nigeria, providing most Nigerians with some respite and confidence in the Armed Forces of Nigeria.[52]

While the regular air interdiction of the insurgents hiding in forests has yielded massive results and even curtailed the activities of the armed non-state actors, such operations have occasionally been criticized in different quarters. There are frequent media reports of attacks on civilian residences instead of the location of the insurgents. This has, on several occasions, raised the human rights concerns about the internal security operations in the Nigerian State.[53]

REGIONAL SECURITY NETWORKS AND INTERNAL SECURITY OPERATIONS: THE *AMOTEKUN* EXPERIENCE

Apart from the government security institutions, RSNs are another security layer involved in the forests-based conflict in Nigeria. There are reasons that caused the creation of RSN in Nigeria. First is the high rate of kidnappings and farmers–herdsmen conflict. The kidnapping activities and herders' menace have disrupted the peaceful coexistence of communities, depriving them of going to farms, markets, and schools. Ojo et al. lamented the reckless scores of terrors unleashed by suspected herdsmen, which resulted in unease calm across the nook and cranny of the southwest recently and the entire country.[54] Intense violence perpetrated by militant herdsmen has since expanded further toward the southwest and southeast regions, as herders search for grazing routes for their cattle.[55]

Another factor is the weakness of the state security forces in eradicating the menace of insecurity. The government's intransigence toward security reforms within the context of decentralization of the national security architecture for better policing quickened the agitation for RSN amid the armed non-state actors' siege of different forests across the country.[56]

The moment the regional security mechanism was developed, it generated striking views among stakeholders across the country. According to Nwagboso, "a state that is unable to guarantee the security and safety of its citizens often loses its legitimacy and most times attributed as a failed state."[57]

In support of these local security networks, Prof. Itse Sagay (SAN), the Chairman, Presidential Advisory Committee against Corruption (PACAC), believed that it is not too good for the whole nation to depend completely on the federal government for security, but should rely more and more on local initiative, so that those who feel the pain are those who try to take control of the security situation. Furthermore, Ikuteyijo (2008) observed that the inability of the police to stem the surging crime waves by enforcing the security of lives and properties has compelled the community members to turn from the formal way of enforcing a law to various self-help initiatives to secure themselves.[58]

The Southwestern region of Nigeria comprising Ondo, Oyo, Ekiti, Osun, and Ogun States is the first region to establish an RSN known as the Amotekun. The governors of these states met in consultation with the community stakeholders and security agencies of the region and came up with the idea of establishing a community-based security outfit aimed at complementing the efforts of the government-owned security forces in the fight against armed groups, making use of the forests in the region to cause security breaches in the zone.

Immediately, Amotekun was established and its members were dispatched into the forests to engage the insurgents across the region in a fierce battle and to recover arms which they stockpiled in the forests. The Amotekun Corps' Commander, Osun State Branch, Brig Gen Bashir Adewimbi (retd.), said that

> the banditry and terrorism we are experiencing in our region now is the spillover effect of what is happening around the country . . . We are strategizing to comb

> the forests and make sure we drive them away or arrest them. The operatives of the corps are determined to go into the forests and flush out these criminals using the forests as their hideout.[59]

On January 10, 2021, Amotekun embarked on military operations in Ibarapa and Oke Ogun areas of Oyo State. However, while the security team in Aiyete was attacked in the forest by insurgents, three of them (insurgents) were killed by men of the Amotekun Corps. According to the Commandant, Oyo State Security Network Agency, Operation Amotekun, Col. Olayinka Olayanju (Rtd.):

> What happened was that we launched clearance operations in different locations around the thick forests of the four local governments that have suffered kidnapping and wanton killings in recent times. When our men entered the forests at Aiyede, they were attacked and a gun battle ensued, which led to the death of three of the attackers... There were six teams in all comprising members of Amotekun, vigilantes, hunters, and Miyetti Allah Vigilante. All the six teams had elements of these groups.[60]

The forest along Lagos–Ibadan Expressway is no doubt a strategic flashpoint for kidnappers in the southwest. As a result, the Amotekun Corps in collaboration with the state security forces embarked on major operations to push out the armed groups from forests in the region.[61] According to the Police Public Relations Officer, Ekiti State Command, Sunday Abutu, "The issue of combing the forests has been on for a long time by the operatives of the command . . . ."[62]

Soldiers attached to the Joint Security Task Force in Oyo State codenamed Operation Burst and personnel of the Amotekun Corps invaded the forests in the Surulere Local Government Area of Oyo State for criminals.[63] The contributions of regional security outfits to the elimination of insurgents' operations in the forests have been enormous. They have been able to uncover a series of arms stockpiled in the forests by the armed groups being used to carry out armed operations and kidnapping across the regions. Through their efforts, kidnapping activities have been reduced to the barest minimum.

The forest-based conflict in Nigeria has enormous implications. Many communities in the country make use of the forest for agricultural activities. But as the conflict intensifies, farmlands are destroyed, thereby making it difficult for the communities to carry out farming activities. The forest war has contributed to the loss of wild animals that live in the forests across the country. These wild animals have migrated to other neighboring countries where they can find peace and a good habitable environment.

The physical military confrontation between the armed forces and the insurgents leads to forest deforestation and destruction of farmlands. Deforestation has increased because of the internal security operations that have the forests as their theater. Many trees in the forests have been cut down unknown to the owners. As the military unleashes attacks against armed groups, trees and bushes are affected by way of falling as well as burning.

Given the massive use of the forests as a hideout for insurgents, the idea of destroying the forests in its entirety is always muted by the authorities. This action gives room for deforestation on a massive scale such that it would be difficult for the forests to grow and return to their former normalcy. Most of the forests have roads constructed through them and security outposts created with the view to prevent the insurgents from coming back to resume their operation over there.

Counterinsurgency in the forest has led to the huge loss of biodiversity (trees with medicinal and scientific value, among others). Many forests in Nigeria are occupied by natural trees with medical and scientific content, which may be used to produce drugs and other medicinal contents. Most of these trees may be found to grow in only limited places where conditions are conducive for their growth and fruitfulness.

Different communities own forests across Nigeria where they serve as cultural heritage. During the planting season, the community heads assign portions of lands to their members for cultivation of crops. Such cultural practices have been sustained from one generation to another, creating a sense of harmony and unity among the community members. Even the proceeds coming from the forests are shared equally among different clans, signifying peace and unity. However, the communal living that is cherished in forest areas is lost as the military invades the forests in search of armed groups. Such internal security operations, which sometimes come unexpectedly, put the communities in disarray as they begin to seek means of protecting themselves from impending operations.

INTELLIGENCE GATHERING IN FOREST-BASED SECURITY OPERATIONS: A CRITIQUE

Intelligence is the product resulting from the collection, collation, evaluation, analysis, integration, and interpretation of collected information.[64] Intelligence gathering and sharing are inevitably major weapons in the fight against insecurity, especially violent crimes that are associated with terrorism and insurgency.[65] While it is appropriate to engage in military operations to stamp out armed insurgents from the forests, intelligence gathering with regard to knowing how they convey and plan their actions in the forests is very significant.

Ever since the forest-based conflict began in Nigeria; it has become very difficult to easily identify locations of the insurgents through accelerated intelligence gatherings. This is because the forests often cover very large areas and are thick. Part of the reason for the failure is poor intelligence gathering management and intelligence compromise by the intelligence community. Victory against insurgents rarely comes from destruction of troops on a battlefield and, as they typically blend into the population, the enemy is often more difficult to find than to neutralize.[66]

Lack of technology like drones is hindering the capacity of the military forces to accurately identify at a stipulated time the bases of the armed groups. With low knowledge of technology such as drones, and other spy devices, adequate knowledge of the enemies' base has remained palatable. According to Adamu Sagir, "security has to be based on human intelligence and intelligence gathering through the use of technological devices capable of rescuing Nigeria from its current security challenges."[67]

There is a disconnection in the relationship between the public and the intelligence officers. This explains the apparent apathy on the part of the populace to cooperate with security agencies in furnishing them with security intelligence. The technology for intelligence gathering through the community locals can now be synchronized with other digital technologies such as device tracking, security cameras, and other key tools in crime fighting.[68]

The reason why intelligence operations failed to meet the expected outcome amid the military confrontation between the armed forces and the insurgency in the forest is because of an acute shortage of funding to boost the morale of the intelligence experts. The intelligence officers in both the army and police, including other intelligence agencies, do not get the required funds to acquire necessary equipment, which would aid accelerated intelligence operations in the forests.

Furthermore, the persistence in fighting among security agencies has created serious lacuna in intelligence operations and counterinsurgency. Internal wrangling among them slows down the war against criminalities and insurgency in Nigeria.[69] That is why forest-based conflict is very complex and difficult to stop. As the armed forces intensify their counterattacks against armed groups in the forest, the more the insurgents continue to expand their tentacles in the forests across strategic locations in the country.

Struggle for distinct roles and responsibilities within the intelligence sector has scaled down the operational capacities of intelligence experts. Instead of maintaining the role of assigning responsibilities to the appropriate institutions, some of these roles have been diverted to suit certain interests at the expense of the others.[70]

Shortage of trained personnel is hindering intelligence gathering in rural areas. Criminals and terrorists that commit violent crimes in Nigeria often run to rural communities and villages for hiding.[71] As a result, security agencies lack enough men that would go round the whole rural areas to conduct intelligence search and to catch the culprits perpetrating criminal activities. Despite these enormous challenges, the intelligence institutions in the country in collaboration with the international intelligence agencies have intensified efforts in unraveling the activities of armed non-state actors operating across the country. This intelligence information is shared with other security forces for adequate strategy formulation and counterattack methods suitable for forest-situated security operations.

CONCLUSION

Warfare has changed considerably. Forests have become the theater of military operations due to urbanization and growth of human population. Insurgents are now uncomfortable living in the cities and towns; instead, they found refuge in the forests where they feel safer to plan and execute their atrocious operations. These armed groups do not occupy and operate in just any just type of forests; they inhabit thick forests adjacent to highways for hit and run attacks, kidnapping, and armed robbery. Through these forests, road users are robbed, kidnapped, and placed into forest detention. The implications of the forest-prone internal security operation are huge, including the disruption of agricultural production, deforestation, displacement of communities and wild animals, loss of revenue from forest tourism, and overstretching of the security architecture of the country. While the Nigerian State has been recording success in the joint military efforts of downgrading the insurgents, the security situations in the country appear to have remained precarious as different parts of the country continue to face direct attacks and violence. Given that the forest-based security operations are new in the security calculation of the country's military, developing the right strategy to counter it has remained the major challenge of the security agencies. In all these, military technology for effective execution of the forest security operation has become paramount more than ever before. With technology, insurgents can easily be located and dealt with in the forests. Forests-based conflict is now a reality and never-ending anytime soon. The more armed groups regroup, the more they expand their tentacles in the forests. It, therefore, behooves the national government to consistently put into serious consideration the emergence of this kind of warfare in their defense planning and expand their military calculation along that direction for the safety and security of her citizens.

NOTES

1. Chukwuma, C.C. Osakwe and Bem, Japhet Audu. 2017. "Nigeria's Military Operations in the Lake Chad Basin." *Journal of Defense Management* 7, no. 1 (2017): 1–7. DOI:10.4172/2167-0374.1000162.
2. African Natural Resource Centre. *How Forestry Contributes to the African Development Bank's High 5 Priorities: Challenges and Opportunities* (Algeria: African Natural Resources Centre, 2018), 1–2.
3. Forest Peoples Programme. *Forest Peoples: Numbers across the World* (United Kingdom: Forest Peoples Programme Fosseway Business Centre, 2012), 2.
4. Okoli, Chukwuma A. and Ochim, I. Fidelis, "Forestlands and National Security in Nigeria: A Threat-Import Analysis." *Journal of Political Science and Leadership Research* 2, no. 2 (2016): 44. Accessed on November 11, 2022. www.iiardpub.org.
5. Jackie, Crafford; Joseph, Mulders; Nuveshen, Naidoo; Ms. Dineo, Maila; Valmak, Mathebula and Kyle, Harris. *Full Report: Forestry and Macroeconomic Accounts of Nigeria: The Importance of Linking Ecosystem Services to Macroeconomics* (New York: United Nations Environment Programme, 2017), 13.
6. Oyetunji, P.O. "The Effects of Forests on Economic Growth in Nigeria, 1990 – 2015." *World of Natural Sciences an International Scientific Journal* 23 (2019): 256.
7. Christian, Yakubu Oche and Ahmad, Abu Bawa. "Constraints To Forest Management Practices in Kebbi State, Nigeria." *International Journal of Environmental Design & Construction Management* 17, no. 4 (2019): 83.
8. Suleiman, Iguda Ladan. "Forests and Forest Reserves as Security Threats in Northern Nigeria." *European Scientific Journal* 10, no. 35 (2014): 1–2.
9. Harwell, Emily. Forests *in Fragile and Conflict-affected States* (Washington, DC: Program on Forests (PROFOR), 2010).
10. Suleiman, "Forests And Forest Reserves," 4.
11. Okoli and Ochim, "Forestlands and National Security," 1–3.
12. Okoli and Ochim, "Forestlands and National Security," 5–6.
13. Solomon, Odeniyi. "Bandits, Terrorists Occupy 1,129 sq m of Forest Reserves in Nigeria –NPS." *Punch*, September 15, 2021. Accessed on November 11, 2022. https://punchng.com/bandits-terrorists-occupy-1129sqm-of-forest-reserves-in-nigeria-nps/?amp.
14. Ismail, Mudashir; Idowu, Isamotu; Lami, Sadiq; Shehu, Umar; Habibu, Umar Aminu and Misbahu, Bashir. "Things You Need to Know About Forests Governed By Bandits, Boko Haram." *Daily Trust*, February 27, 2021. Accessed on November 11, 2022. https://dailytrust.com/9-forests-governed-by-bandits-b-haram,2.
15. Ismail et al., "Things you need to know about Forests."
16. Okoli and Ochim, "Forestlands and National Security."
17. Uche, Igwe. "Nigeria's Conflict and Insecurity Makes us Rethink Authority in 'Ungoverned' Spaces." *Ise Blogs*, March 12, 2021. Accessed on November 11, 2022.

https://blogs.lse.ac.uk/africaatlse/2021/03/12/nigeria-conflict-insecurity-rethink-how-authority-policy-works-ungoverned-governance-spaces/.

18. Uche, "Nigeria's conflict and insecurity."
19. Uche, "Nigeria's conflict and insecurity."
20. Dapo, Akinrefon. "South-West Forests Now Preferred Crime Scenes?" *Vanguard*, June 18, 2015. Accessed on November 11, 2022. https://www.vanguardngr.com/2015/06/south-west-forests-now-preferred-crime-scenes/amp/.
21. Okoli and Ochim, "Forestlands and National Security."
22. Uche, "Nigeria's conflict and insecurity."
23. Obinna, Odogwu, Jude, Chinedu and Okey, Sampson. "Southeast Forests of Kidnappers: Governors of Region in Frantic Search for Solution." *The Sun Online*, November 6, 2022.
24. "Nigerian Government Admits Bandits, Kidnappers have Taken over Forests Nationwide." *Sahara Reporters*, January 11, 2024. Accessed on January 29, 2024. https://saharareporters.com/2024/01/11/nigerian-govenrment-admits-bandits-kidnappers-have-taken-over-forests-nationwide.
25. Voke, Charles Mgbonyenbi and Frank, C.A. "Emeni. Militancy and Sustainable Development in the Niger Delta: Excerpts from the Fourth Republic." *UJAH* 21, no. 4 (2020). Accessed on November 11 2022. *http://dx.doi./org/10.4314/ujah.v21i4.*
26. Chris, I. Nwagboso. "The Nature of Internal Security Problems in African States: The Nigerian Experience." *International Journal of Academic Research in Business and Social Sciences* 6, no. 4 (2016). Accessed on November 11, 2022. http://dx.doi.org/10.6007/IJARBSS/v6-i4/2079.
27. Mgbonyenbi and Emeni, "Militancy and Sustainable Development."
28. Transparency International. *Discussion Paper: Military Involvement in Oil Theft in the Niger Delta* (United Kingdom: Transparency International, 2019), 1.
29. Lynn, L. Taylor. "Boko Haram Terrorism: Reaching Across International Boundaries to Aid Nigeria In The Humanitarian Crisis." *ILSA Journal of International & Comparative Law* 21, no. 1 (2014): 4.
30. Azeez, Olaniyan. "Once Upon a Game Reserve: Sambisa and the Tragedy of a Forested Landscape." *Arcadia* Spring 2 (2018): 2.
31. Kingsley, Omonobi. "Desert Sanity: Troops rescue 550 Women, Children in Sambisa." *Vanguard*, April 29, 2022. Accessed on November 11, 2022. https://www.vanguardngr.com/2022/04/desert-sanity-troops-rescue-550-women-children-in-sambisa/amp/.
32. "Nigerian Troops Invade Alleged IPOB's Eastern Security Network Operational Camp, Kill Scores." *Sahara Reporters,* May 29, 2022. Accessed on November 11, 2022. https://saharareporters.com/2022/05/29/nigerian-troops-invade-alleged-ipobs-eastern-security-network-operational-camp-kill.
33. "Troops Clear Ipob/Esn Enclave, Capture Cache of Arms, Ammunition In Imo." *Scrip*, May 23, 2022, Accessed on November 11, 2022. https://www.scirp.org/journal/paperinformation.aspx?paperid=73006#ref39.
34. Ekene, Lionel. "Silent and Deadly: Increasing Activities of Nigeria's Special Forces." *Military Africa*, January 19, 2022. Accessed on November 11, 2022. https://

www.military.africa/2022/01/silent-and-deadly-increasing-activities-of-nigerias-special-forces/.

35. John, Kennedy Uzoma. "Sit-At- Home: Troops Dislodge IPOB/ESN Stronghold at Imo/Anambra Border Forest." *Nigerian Tribune Online,* January 18, 2022. Accessed on November 11, 2022. https://tribuneonlineng.com/sit-at-home-troops-dislodge-ipob-esn-stronghold-at-imo-anambra-border-forest/.

36. Lawrence, Njoku. "Soldiers bombard IPOB Camps in Enugu." *Guardian,* October 16, 2022. Accessed on November 11, 2022. https://guardian.ng/news/soldiers-bombard-ipob-camps-in-enugu/.

37. "Army Dislodges Criminal Hideouts, Arrest Three in Imo, Anambra – Official." *Vanguard*, February 24, 2022. Accessed on November 11, 2022. https://www.vanguardngr.com/2022/02/army-dislodge-criminal-hideouts-arrest-three-in-imo-anambra-official/amp/.

38. Oghebnahogie, Oshoko. "Troops Liquidate 4 IPOB/ESN Elements in Fierce Encounter in Imo State." *Security King*, May 16, 2022. Accessed on November 11, 2022. https://securitykingng.org/security/troops-liquidate-ipob-esn/.

39. "ESN Flee from Forests in Orsu Local Government Area of Imo State after Military Reinforcements Arrive." *Nigerian Watch*, February 22, 2022. Accessed on November 11, 2022. https://nigerianwatch.com/esn-flee-from-forests-in-orsu-local-government-area-of-imo-state-after-military-reinforcements-arrive/, 5.

40. "Army Combs Ekiti Forests, Discovers Decayed Body." *The Nation*, June 3, 2018. Accessed on November 11, 2022. https://thenationonlineng.net/army-combs-ekiti-forests-discovers-decayed-body/amp/.

41. Isaac, Shobayo. "29-Year-Old Lady Disarms Kidnapper of AK-47 Frees Three Victims in Plateau." *Nigerian Tribune Online*, December 19, 2022. Accessed on November 11, 2022. https://tribuneonlineng.com/29-year-old-lady-disarms-kidnapper-of-ak-47-frees-victims-in-plateau?

42. "El-Rufai Restates Call for Bombardment of Forests Inhabited by Terrorists." *Guardian*, March 31, 2022. Accessed on November 11, 2022. https://guardian.ng/news/el-rufai-restates-call-for-bombardment-of-forests-inhabited-by-terrorists/

43. Kingsley, Nwezeh. "Air Strikes Hit Terrorists' Hospitals in Sambisa Forest." *Thisdaylive*, August 13, 2022. Accessed on November 11, 2022. https://www.thisdaylive.com/index.php/2022/09/13/air-strikes-hit-terrorists-hospitals-in-sambisa-forest/amp/.

44. "How we Dislodged Boko Haram from Sambisa Forest: Nigerian Army." *Premium Times* December 29, 2016. Accessed on November 11, 2022. https://www.premiumtimesng.com/news/headlines/219153-dislodged-boko-haram-sambisa-forest-nigerian-army.html.

45. Kingsley, Omonobi. "Bombardment of Bandit's Camps: Residents Jubilate in Zamfara, Hold Appreciation Rally for Military." *Vanguard*, July 22, 2021. Accessed on November 11, 2022. https://www.vanguardngr.com/2021/07/bombardment-of-bandits-camps-residents-jubilate-in-zamfara-hold-appreciation-rally-for-military/amp/.

46. Tarkaa, David. "Scores Killed As Military Jet Raids ISWAP Hospital In Sambisa Forest." *Leadership*, February 22, 2022. Accessed on November 11, 2022.

https://leadership.ng/scores-killed-as-military-jet-raids-iswap-hospital-in-sambisa-forest/.

47. Oghenahogie, Oshoko. "Military Air Strikes Target Terrorist's Hideouts at Alhaji Isiaka Location & Kurudu in Kaduna State; Liquidate 14 Terrorists." *Security King*, September 1, 2022. Accessed on November 11, 2022. https://securityking.org/security/military-air-strikes-target/, 1.

48. "High Value Terrorists' Killed as Nigerian Air Force Intensifies Bombardment of Sambisa Forest." *Premium Times*, January 30, 2016. Accessed on November 11, 2022. https://www.premiumtimesng.com/news/top-news/197690-high-value-terrorists-killed-nigerian-air-force-intensifies-bombardment-sambisa-forest.html

49. "Military Kills 5 ISWAP Commanders, over 200 Terrorists in Borno." *Daily Trust*, September 5, 2022. Accessed on November 11, 2022. https://dailytrust.com/military-kills-5-iswap-commanders-over-200-terrorists-in-borno/.

50. Adelani, Adepegba. "NAF Bombs Bandits' Hideout at Kuduru Forest in Kaduna." *Punch*, December, 2020. Accessed on November 11, 2022. https://punchng.com/naf-bombs-bandits-hideout-at-kuduru-forest-in-kaduna/?amp.

51. Abdulrazaq, Adebayo. "Insecurity: NAF, Sister Agencies Launch Joint Operations in Kwara." *Daily Post*, December 17, 2022. Accessed on November 11, 2022. https://dailypost.ng/2022/12/17/insecurity-naf-sister-agencies-launch-joint-operations-in-kwara/.

52. Solomon, Odeniyi. "NAF Continues Bandits' Bombardment; Military Acquires More Jets." *Punch*, November 11, 2022. Accessed on November 11, 2022. https://punchng.com/naf-continues-bandits-bombardment-military-acquires-more-jets/?utm_source=auto-read-also&utm_medium=web&amp.

53. Oluchukwu, Ignatus Onianwa. 2022. "International Media and Counterterrorism Operations in Nigeria: Issues and Challenges." In Ismayil Elnur and Karadogan Ismayil Ebru (Eds.), *Media and Terrorism in the 21st Century*, IGI Global, 144.

54. Oikhala, Gabriel Izokpu. "Survival Strategies of Internal Security in Nigeria: The Amotekun Option." *Journal of Administrative Science* 19, no. 1 (2022). Accessed on November 11, 2022. http://jas.uitm.edu.my,114.

55. Chijioke, Kelechi Iwuamadi; Elias, C. Ngwu and Magnus, Chijindu Onyemaobi. "Regional Security Outfits and the Challenges of Insecurity in Nigeria." *University of Nigeria Journal of Political Economy* 11, no. 2 (2021): 480–491.

56. Friedrich Ebert Stiftung and European Union. *Policy Brief: Evolution of Regional Security Outifts in Nigeria: Ethnic or Community Policing* (Nigeria Office: Friedrich Ebert Stiftung, 2020), 14. Accessed on November 11, 2022. https://pscc.fes.de/fileadmin/user_upload/images/publications/Pub_Evolution_of_Regional_Security_Outfits_in_Nigeria_Ethnic_or_Community_Policing.pdf.

57. Nwagboso, "The Nature of Internal Security, 5."

58. Odewale, Ayotunde David and Lamidi, Kazeem Oyedele. "Regionalization of Non-State Security Agencies in Southwest Nigeria: Prospects and Challenges of Amotekun." *European Scientific Journal* 116, no. 20 (2020). Accessed on November 11, 2022. URL:http://dx.doi.org/10.19044/esj.2020.v16n20p103.

59. "Bandits, Criminals Have Taken Over Forests In Osun State – South-West Corps, Amotekun Commander, Adewimbi." *Sahara Reporters*. August 27,

2022. Accessed on November 11, 2022. https://saharareporters.com/2022/08/27/bandits-criminals-have-taken-over-forests-osun-state-south-west-corps-amotekun-commander.

60. Wale, Oyewale. "Amotekun Denies Clash with Herdsmen in Oyo." *Punch*, January 10, 2021. Accessed on November 11, 2022. https://punchng.com/amotekun-denies-clash-with-herdsmen-in-oyo/?amp.

61. Daniel, Ayantoye, Abiodun, Nejo, Peter, Dada, Bola, Bamigbola, Daud, Olatunji and Olufemi, Olaniyi. "Kidnappings: Police, Amotekun Comb South-West Forests, 48 Arrested." *Punch*, November 12, 2022. Accessed on November 13, 2022. https://punchng.com/kidnappings-police-amotekun-comb-south-west-forests-48-arrested/?amp.

62. Abiodun, Nejo and Bola, Bamigbola. "Police, Amotekun Comb Ekiti, Osun Forests for Killer Herdsmen, Others." *Punch*, April 4, 2021. Accessed on November 11, 2022. https://punchng.com/police-amotekun-comb-ekiti-osun-forests-for-killer-herdsmen-others/?amp.

63. "Operation Burst, Amotekun Comb Forests in Ogbomoso For Kidnappers." *Inside Oyo*, August 8, 2022. Accessed on November 11, 2022. https://www.insideoyo.com/operation-burst-amotekun-comb-forests-in-ogbomoso-for-kidnappers/.

64. Oladayo, Ogunbowale. Intelligence Gathering: Strategy for Security Of Lives And Property. Accessed on December 1, 2022. https://oladayobowaleoluju-won.medium.com/intelligence-gathering-strategy-for-security-of-lives-and-property-a1e78091c209/4/November/2018.

65. Zakary, Bidoma. "Nigeria's DIA: Traducers and Needless Attacks." *Vanguard*, March 14 2022. Accessed on November 11, 2022. https://www.vanguardngr.com/2022/03/nigerias-dia-traducers-and-needless-attacks/amp/.

66. Ngboawaji, Daniel Nte, Paul, Eke and Kinikanwo, Anele. "Rural Intelligence Gathering and the Challenges of Counter Insurgency: Views from the Niger Delta." *Bangladesh e-Journal of Sociology* 7, no. 1 (2010): 28–29.

67. Fatteh, Hamid. "Modern Military Technology Recommended Halting Nigeria's Festering Security Challenges." *Punch*, August 7, 2022. Accessed on November 11, 2022. https://punchng.com/modern-military-technology-recommended-to-halt-nigerias-festering-security-challenges/?amp.

68. Solomon, M. Ekolama. "Citizen Participation in Intelligence Gathering: A Tool for Crime Prevention and Control in Nigeria." *Thisdaylive*, August 22, 2017. Accessed on November 11, 2022. https://www.thisdaylive.com/index.php/2017/09/22/citizen-participation-in-intelligence-gathering-a-tool-for-crime-prevention-and-control-in-nigeria/amp/.

69. Zakary, "Nigeria's DIA,"

70. Joseph, Undu. "Criminality: Security Experts Canvass Better Intelligence Gathering." *Vanguard*, November 28, 2016. Accessed on November 11, 2022. https://www.vanguardngr.com/2016/11/criminality-security-experts-canvass-better-intelligence-gathering/amp/.

71. Chris, I. Nwagboso and Nwagboso, Nnenna Salome. "The Challenges of Intelligence Gathering in the Third World: A Critique of the Nexus between Poverty and

Public Policy Process." *International Journal of Public Administration and Management Research* 6, no. 2 (2021). Accessed on November 11, 2022. URL:https://doi.org /10.36758/ijpamr/v6n2.2021/05, 67.

REFERENCES

Abdulrazaq, Adebayo. "Insecurity: NAF, Sister Agencies Launch Joint Operations in Kwara." *Daily Post*, December 17, 2022. Accessed on December 1, 2022. https://dailypost.ng/2022/12/17/insecurity-naf-sister-agencies-launch-joint-operations-in-kwara/.

Abiodun, Nejo and Bola, Bamigbola. "Police, Amotekun Comb Ekiti, Osun Forests for Killer Herdsmen, Others." *Punch*, April 4, 2021. Accessed on November 11, 2022. https://punchng.com/police-amotekun-comb-ekiti-osun-forests-for-killer-herdsmen-others/?amp.

Adelani, Adepegba. "NAF Bombs Bandits' Hideout at Kuduru Forest in Kaduna." *Punch*, December 7 2020. Accessed on December 1, 2022. https://punchng.com/naf-bombs-bandits-hideout-at-kuduru-forest-in-kaduna/?amp.

Adewale, Akingbade. "Confronting the Challenges of Insecurity in Ondo State: The Role of Geospatial Surveillance and Mapping." *Global Journal of Arts, Humanities and Social Sciences* 10, no. 5 (2022): 25–38.

African Natural Resource Centre. *How Forestry contributes to the African Development Bank's High 5 Priorities: challenges and opportunities* (Algeria: African Natural Resources Centre, 2018).

"Army Combs Ekiti Forests, Discovers Decayed Body." *The Nation*, June 3, 2018. Accessed on November 11, 2022. https://thenationonlineng.net/army-combs-ekiti-forests-discovers-decayed-body/amp/.

"Army Dislodges Criminal Hideouts, Arrest Three in Imo, Anambra –Official." *Vanguard*. February 24, 2022. Accessed on November 11, 2022. https://www.vanguardngr.com/2022/02/army-dislodge-criminal-hideouts-arrest-three-in-imo-anambra-official/amp/ .

Azeez, Olaniyan. "Once Upon a Game Reserve: Sambisa and the Tragedy of a Forested Landscape." *Arcadia* Spring 2 (2018).

"Bandits, Criminals Have Taken Over Forests In Osun State – South-West Corps, Amotekun Commander, Adewimbi." *Sahara Reporters*, August 27, 2022. Accessed on November 11, 2022. https://saharareporters.com/2022/08/27/bandits-criminals-have-taken-over-forests-osun-state-south-west-corps-amotekun-commander.

Chijioke, Kelechi Iwuamadi; Elias, C. Ngwu and Magnus, Chijindu Onyemaobi. "Regional Security Outfits and the Challenges of Insecurity in Nigeria." *University of Nigeria Journal of Political Economy* 11, no. 2 (2021): 480–491.

Chris, I. Nwagboso and Nwagboso, Nnenna Salome. "The Challenges of Intelligence Gathering in the Third World- A Critique of the Nexus between Poverty and Public Policy Process." *International Journal of Public Administration and Management Research* 6, no. 2 (2021). Accessed on November 11, 2022. URL:https://doi.org

/10.36758/ijpamr/v6n2.2021/05 http://journals.rcmss.com/index.php/ijpamr. www.academix.

Christian, Yakubu Oche and Ahmad, Abu Bawa. "Constraints to Forest Management Practices in Kebbi State, Nigeria." *International Journal of Environmental Design & Construction Management* 17, no. 4 (2019).

Chukwuma, C. C. Osakwe and Bem, Japhet Audu. "Nigeria's Military Operations in the Lake Chad Basin." *Journal of Defense Management* 7, no. 1 (2017): DOI:10.4172/2167-0374.1000162.

Daniel, Ayantoye; Abiodun, Nejo; Peter, Dada; Bola, Bamigbola; Daud, Olatunji and Olufemi, Olaniyi. "Kidnappings: Police, Amotekun Comb South-West Forests, 48 Arrested." *Punch*, November 12, 2022. Accessed on November 11, 2022. https://punchng.com/kidnappings-police-amotekun-comb-south-west-forests-48-arrested/?amp.

Dapo, Akinrefon. "South-West Forests Now Preferred Crime Scenes?" *Vanguard*, June 18, 2015. Accessed on November 11, 2022. https://www.vanguardngr.com/2015/06/south-west-forests-now-preferred-crime-scenes/amp/.

"El-Rufai Restates Call for Bombardment of Forests Inhabited by Terrorists." *Guardian*. March 31, 2022. Accessed on November 11, 2022. https://guardian.ng/news/el-rufai-restates-call-for-bombardment-of-forests-inhabitedbyterrorists/.

"ESN Flee from Forests in Orsu Local Government Area of Imo State after military reinforcements arrive." *Nigerian Watch*, February 22, 2021. Accessed on November 11, 2022. https://nigerianwatch.com/esn-flee-from-forests-in-orsu-local-government-area-of-imo-state-after-military-reinforcements-arrive/.

Fatteh Hamid. "Modern Military Technology Recommended Halting Nigeria's Festering Security Challenges." *Punch*, August 7, 2022. Accessed on November 11, 2022. https://punchng.com/modern-military-technology-recommended-to-halt-nigerias-festering-security-challenges/?amp.

Forest Peoples Programme. *Forest Peoples: Numbers across the World* (United Kingdom: Forest Peoples Programme Fosseway Business Centre, 2012).

Friedrich Ebert Stiftung and European Union. *Policy Brief: Evolution of Regional Security Outfits in Nigeria: Ethnic or Community Policing* (2020): 14.

Harwell, Emily. *Forests in Fragile and Conflict-affected States* (Washington DC: Program on Forests (PROFOR), 2010).

"High Value Terrorists' Killed as Nigerian Air Force Intensifies Bombardment of Sambisa Forest." *Premium Times,* January 30, 2016. Accessed on November 11, 2022. https://www.premiumtimesng.com/news/top-news/197690-high-value-terrorists-killed-nigerian-air-force-intensifies-bombardment-sambisa-forest.html.

"How we Dislodged Boko Haram from Sambisa Forest: Nigerian Army." *Premium Times*, December 29, 2016. Accessed on November 11, 2022. https://www.premiumtimesng.com/news/headlines/219153-dislodged-boko-haram-sambisa-forest-nigerian-army.html.

Isaac, Shobayo. "29-Year-Old Lady Disarms Kidnapper of AK-47 Frees Three Victims in Plateau." *Nigerian Tribune Online*, December 19, 2022. Accessed on November 11, 2022. https://tribuneonlineng.com/29-year-old-lady-disarms-kidnapper-of-ak-47-frees-victims-in-plateau?

Ismail Mudashir, Idowu Isamotu; Lami Sadiq; Shehu Umar; Habibu Umar Aminu and Misbahu Bashir. "Things You Need to Know About Forests Governed By Bandits, Boko Haram." *Daily Trust*, February 27, 2021. https://dailytrust.com/9-forests-governed-by-bandits-b-haram. Accessed on November 11, 2022.

Jackie, Crafford; Joseph, Mulders; Nuveshen, Naidoo; Ms. Dineo, Maila; Valmak, Mathebula and Kyle, Harris. *Full Report: Forestry and Macroeconomic Accounts of Nigeria: The Importance of Linking Ecosystem Services to Macroeconomics* (New York: United Nations Environment Programme, 2017).

John Kennedy, Uzoma. "Sit-At- Home: Troops Dislodge IPOB/ESN Stronghold at Imo/Anambra Border Forest." *Nigerian Tribune Online*, June, 2022. Accessed on November 11, 2022. https://tribuneonlineng.com/sit-at-home-troops-dislodge-ipob-esn-stronghold-at-imo-anambra-border-forest/.

Joseph, Undu. "Criminality: Security Experts Canvass Better Intelligence Gathering." *Vanguard.* November 28 2016. Accessed on November 11, 2022. https://www.vanguardngr.com/2016/11/criminality-security-experts-canvass-better-intelligence-gathering/amp/.

Kingsley, Nwezeh. "Air Strikes Hit Terrorists' Hospitals in Sambisa Forest." *Thisdaylive*, August 13, 2022. https://www.thisdaylive.com/index.php/2022/09/13/air-strikes-hit-terrorists-hospitals-insambisa forest/amp/.

Kingsley, Omonobi. "Bombardment of Bandit's Camps: Residents Jubilate in Zamfara, hold Appreciation Rally for Military." *Vanguard.* July 22, 2021. Accessed on November 11, 2022. https://www.vanguardngr.com/2021/07/bombardment-of-bandits-camps-residents-jubilate-in-zamfara-hold-appreciation-rally-for-military/amp/.

Kingsley, Omonobi. "Desert Sanity: Troops Rescue 550 Women, Children in Sambisa." *Vanguard*, April 29, 2022. https://www.vanguardngr.com/2022/04/desert-sanity-troops-rescue-550-women-children-in-sambisa/amp/. Accessed on November 11, 2022.

Lawrence, Njoku. "Soldiers bombard IPOB camps in Enugu." *Guardian.* October 16, 2022. Accessed on November 11, 2022. https://guardian.ng/news/soldiers-bombard-ipob-camps-in-enugu/.

Lynn, L. Taylor. "Boko Haram Terrorism: Reaching Across International Boundaries to Aid Nigeria In The Humanitarian Crisis." *ILSA Journal of International & Comparative Law* 21, no. 1 (2014).

"Military Kills 5 ISWAP Commanders, over 200 Terrorists in Borno." *Daily Trust.* September 5, 2022. Accessed on November 11, 2022. https://dailytrust.com/military-kills-5-iswap-commanders-over-200-terrorists-in-borno/.

"NAF to Nigerians: With Intelligence Gathering, we're Able to Bomb Boko Haram Supply Lines." *Vanguard*, December 26, 2017. Accessed on November 11, 2022. https://www.vanguardngr.com/2017/12/naf-nigerians-intelligence-gathering-able-bomb-boko-haram-supply-lines/amp/.

Ngboawaji Daniel Nte, Paul Eke and Kinikanwo Anele. "Rural Intelligence Gathering and the Challenges of Counter Insurgency: Views from the Niger Delta." *Bangladesh e-Journal of Sociology* 7, no. 1 (2010): 28–29.

"Nigerian Troops Invade Alleged IPOB's Eastern Security Network Operational Camp, Kill Scores." *Sahara Reporters*, May 29, 2022. Accessed on November

11, 2022. https://saharareporters.com/2022/05/29/nigerian-troops-invade-alleged-ipobs-eastern-security network-operational-camp-kill.

Obinna, Odogwu; Jude, Chinedu and Okey, Sampson. "Southeast Forests of Kidnappers: Governors of Region in Frantic Search for Solution." *The Sun*, November 6, 2022. Accessed on November 11, 2022. https://www.sunnewsonline.com.

Odewale, Ayotunde David and Lamidi, Kazeem Oyedele. "Regionalization of Non-State Security Agencies in Southwest Nigeria: Prospects and Challenges of Amotekun." *European Scientific Journal* 16, no. 20 (2020). Doi:10.19044/esj.2020.v16n20p103 Accessed on November 11, 2022. URL:http://dx.doi.org/10.19044/esj.2020.v16n20p103.

Oghenahogie, Oshoko. "Troops Liquidate 4 IPOB/ESN Elements in Fierce Encounter in Imo State." *Security King*, May 16, 2022. Accessed on November 11, 2022. https://securitykingng.org/security/troops-liquidate-ipob-esn/.

Oikhala, Gabriel Izokpu. "Survival Strategies of Internal Security in Nigeria: The Amotekun Option." *Journal of Administrative Science* 19, no. 1 (2022): 99–135 http://jas.uitm.edu.my.

Okoli, Chukwuma A. and Ochim, Fidelis I. "Forestlands and National Security in Nigeria: A Threat-Import Analysis." *Journal of Political Science and Leadership Research* 2, no. 2 (2016). www.iiardpub.org.

Oladayo, Ogunbowale. "Intelligence Gathering: Strategy for Security of Lives And Property." November 4, 2018. Accessed on December 1, 2022. https://oladayobowaleolujuwon.medium.com/intelligence-gathering-strategy-for-security-of-lives-and-property-a1e78091c209/4/November/2018.

Oluchukwu, Ignatus Onianwa. "International Media and Counterterrorism Operations in Nigeria: Issues and Challenges." In Ismayil Elnur and Karadogan Ismayil Ebru (Eds.), *Media and Terrorism in the 21st Century* (United States: IGI Global, 2022).

"Operation Burst, Amotekun Comb Forests In Ogbomoso For Kidnappers." *Inside Oyo*, August 8, 2022. Accessed on November 11, 2022. https://www.insideoyo.com/operation-burst-amotekun-comb-forests-in-ogbomoso-for-kidnappers/.

Oyetunji, P. O. "The Effects of Forests on Economic Growth in Nigeria, 1990 – 2015." *World of Natural Sciences an International Scientific Journal* 23 (2019): 254–265.

Solomon, M. Ekolama. "Citizen Participation in Intelligence Gathering: A Tool for Crime Prevention and Control in Nigeria." *Thisdaylive*. August 22, 2017. Accessed on November 11, 2022. https://www.thisdaylive.com/index.php/2017/09/22/citizen-participation-in-intelligence-bgathering-a-tool-for-crime-prevention-and-control-in-nigeria/amp/.

Solomon, Odeniyi. "Bandits, Terrorists Occupy 1,129 sqm of Forest Reserves in Nigeria –NPS." *Punch*, September 15, 2021. Accessed on November 11, 2022. https://punchng.com/bandits-terrorists-occupy-1129sqm-of-forest-reserves-in-nigeria-nps/?amp.

Solomon, Odeniyi. "NAF Continues Bandits' Bombardment; Military Acquires More Jets." *Punch*, November 11, 2022. Accessed on November 11, 2022. https://punchng.com/naf-continues-bandits-bombardment-military-acquires-more-jets/?utm_source=auto-read-also&utm_medium=web&amp.

Suleiman, Iguda Ladan. "Forests And Forest Reserves as Security Threats in Northern Nigeria." *European Scientific Journal* 10, no. 35 (2014): 10–35.

Tarkaa, David. "Scores Killed as Military Jet Raids ISWAP Hospital In Sambisa Forest." *Leadership*, February 22, 2022. Accessed on November 11, 2022. https://leadership.ng/scores-killed-as-military-jet-raids-iswap-hospital-in-sambisa-forest/.

Transparency International. *Discussion Paper: Military Involvement in Oil Theft in the Niger Delta* (United Kingdom: Transparency International, 2019).

Troops Clear Ipob/Esn Enclave, Capture Cache of Arms, Ammunition In Imo. *Scrip*, May 23, 2022. Accessed on November 11, 2022. https://www.scirp.org/journal/paperinformation.aspx?paperid=73006#ref39.

Uche, Igwe. "Nigeria's Conflict and Insecurity Makes us Rethink Authority in 'Ungoverned' Spaces'." *Ise Blogs*, March 12, 2021. . Accessed on November 11, 2022. https://blogs.lse.ac.uk/africaatlse/2021/03/12/nigeria-conflict-insecurity-rethink-how-authority-policy-works-ungoverned-governance-spaces/.

Voke, Charles Mgbonyenbi and Frank, C.A. Emeni. "Militancy and Sustainable Development in the Niger Delta: Excerpts from the Fourth Republic." *UJAH* 21, no. 4 (2020). http://dx.doi./org/10.4314/ujah.v21i4.

Wale, Oyewale. "Amotekun Denies Clash with Herdsmen in Oyo." *Punch*, January 10, 2021. Accessed on November 11, 2022. https://punchng.com/amotekun-denies-clash-with-herdsmen-in-oyo/?amp.

Zakary, Bidoma. "Nigeria's DIA: Traducers and Needless Attacks." *Vanguard*, March 14, 2022. Accessed on November 11, 2022. https://www.vanguardngr.com/2022/03/nigerias-dia-traducers-and-needless-attacks/amp/.

Part III

ENVIRONMENTAL SECURITY AND DEVELOPMENT

Chapter 13

Socioeconomic Impact of Land Swamp Reclamation

Implications for Environmental Security in the Niger Delta Region, Nigeria

Michael Uzodianya Didia and
Temple Probyne Abali

Land reclamation is the process of returning lands to an improved state and may involve acquiring new land from oceans, riverbeds, or lake beds. Land reclamation's primary goal is to rehabilitate degraded property. Reclamation of land is done to mine sand, build roads, build homes, explore for crude oil, and expand urban cities.[1] Government agencies as well as private developers are involved in reclamation. To construct initiatives that benefit the public good, the government occasionally takes coastal regions forcibly from the native community, clears the mangrove forest, and fills the area with sand. This poses a severe threat to the coastal environment, especially in regard to biodiversity, as this is severely affected. The flora and fauna of the mangrove area so forcefully acquired by the government and sand filled, often for urban development, are destabilized or destroyed, and the ecosystem balance is disturbed irretrievably. These and other land expansion activities are to blame for the trend in loss of biodiversity.

Government and culture in Nigeria have an impact on land policy. This is reflected in the Land Use Act, which was enacted in 1978 and declares that the Federal Government owns all lands in Nigeria, including underwater territory that is 200 nautical miles from the coast. The governor of each state in the country holds the land and water territories in trust for the federal government and is authorized to be the only signatory on all Certificates of Occupancy documents. As a result of the immense power granted to each state's government, it is simple for them to acquire or confiscate land in the name of the public good and then reclaim it as private property for their own personal gain.

Generally, Environmental Impact Assessment (EIA) studies need to be completed before coastal properties are purchased in order to design an effective management strategy aimed at preventing the loss and extinction of common and uncommon species of fauna and flora. The continuous development of reclaimed land for urban purposes has endangered the coastal environment, especially species of both flora and fauna. Unfortunately, a proper EIA or landscape assessment is rarely performed. Although the government is saddled with the responsibility of protecting the coastal areas and properties from unauthorized landscape reclamations, the locals of coastal lands traditionally safeguard their communities, especially when the coastal lands, flora, and fauna are considered important to their culture.

There is often a tight balance to tread between protecting the coastal lands and reclaiming the lands for other purposes. Even though reclamation of coastal land harms aquatic species, if it is used for nonintrusive agricultural uses, it may bring some economic benefits to the affected communities. In Nigeria, reclaimed land is utilized to construct housing, a road system, farms for livestock grazing, recreation areas, commercial centers, and landfills.

In the Niger Delta, reclaimed land is also bought from coastal communities to increase the area available for the construction of homes for human occupancy. The estimated population of the Niger Delta increased from 28,856,000 in 2005 to 45,715,000 in 2020.[2] It is projected to increase to over 60,000,000 in 2030.[3] The coastal regions are in dire need of expansion and, therefore, the lands are being reclaimed for urban development to cater for increased population and economic growth. There are five stages or phases of land reclamation in the Niger Delta, and each stage has an impact on the ecosystem. These stages include: (a) clearing mangrove forests; (b) using bulldozers to compact swamps; (c) pumping white sand from the ocean floor to shore; (d) filling land to be reclaimed with sand; and (e) erecting roads, structures, businesses, and parks.

Private company owners reclaim coastal regions to carry out private ventures that will improve their financial circumstances. Oil firms cut down coastal forests and install pipes and oil wells in wetlands. Small-town population growth had also resulted in the reclamation of coastal regions to make place for the building of homes to house more people. However, a lot of land reclamation efforts are business-focused rather than development-focused. The reason for this is the expanding sand mining industry that has engulfed most coastal communities. After a few years of operation, sand mines are frequently shut down, leaving a devastated ecosystem behind. Reclamation is carried out without an adequate environmental impact assessment having been conducted and no remediation is carried out on disturbed areas of the environment. Numerous species have perished as a result of this circumstance. Examples of such species are the African elephant, chimpanzee, crested genet, and the Niger Delta pygmy hippopotamus.

Significantly, the presence of extensive swamps and marshland in the Niger Delta implies that the topography is not firm. The existence of a maze of creeks simply points to the fact that the topography is also compartmentalized.[4] Land reclamation of this area has both socioeconomic and environmental impacts that require examination. The major issues to grapple with, therefore, are: what are the socioeconomic impacts of land reclamation in the Niger Delta region of Nigeria and what are their implications for environmental security?

This paper, therefore, examines land reclamation techniques being used in the region and the socioeconomic impact of land reclamation and makes recommendations for sustainable land reclamation for environmental security in the region.

METHODS OF LAND RECLAMATION

Several techniques of land reclamation are employed in the Niger Delta region.[5] The most common methods include but are not restricted to hydraulic sand filling, use of chikoko mud, use of concrete breakwater, and landfilling or dumping of solid wastes in areas to be reclaimed.

HYDRAULIC SAND FILLING

This involves the use of dredging equipment to scoop and pump out mainly sand (or mud or other materials) from the traction load on the riverbed. With pay-loaders, the sand is packed into heavy trucks and transported to where it will be used to raise and stabilize the land. The necessary bounding concrete materials are first erected after which the sand is used to fill/raise up the ground level and allowed to stabilize before actual construction work commences. Hydraulic sand filling is the process of reclaiming land by first dredging fill material in a dredging area using floating equipment (dredgers), then transporting the fill material by dredger, barge, or pipeline from the dredging area to the reclamation site, and finally placing the fill material as a mixture of fill material and (process) water in the reclamation area. A typical example where this activity is carried out is Chokocho in the Niger Delta. This practice is very expensive with higher-level technology and a heavier capital outlay to operationalize but more durable than other methods of reclamation.

THE USE OF CHIKOKO MUD (SOFT ORGANIC MARINE CLAY)

Unlike the hydraulic sand-filling technique, which involves a much higher level of technology and a heavier capital outlay to operationalize, reclamation with the aid of chikoko mud can be easily carried out even by the poorest people. The basic material used in this process is chikoko mud that can be, traditionally, scooped off in blocks from the mangrove swamps. Soft organic marine clay is noticeably and strongly bound by intertwining and adventitious mangrove plant roots and since it is made of fine particles, it inhibits permeability. Chicoco mud is a very soft and extremely compressible organic marine mud found extensively and to considerable depths within the saline tidal flat or mangrove swamp of the Niger delta region in southern Nigeria. Soft organic marine clay is highly compressible, barely able to support an average human weight but air-dried chicoco has been used successfully by the natives for shore protection, especially if placed above water. Soft organic marine clay with improvement proves a viable engineering material. For instance, air-dried chikoko mud has been used extensively by natives for landscaping, shore and slope protection, and land reclamation. However, it is prone to flooding if improperly used and well-managed.

THE USE OF CONCRETE BREAKWATER

This method of reclamation mainly involves the erection of concrete walls some distance into the river channel. The space between the wall and the riverbank is filled up and sealed with concrete. The platform at the Port Harcourt harbor of Nigeria Ports Authority (NPA) is a good example of structure preceded by this type of reclamation.

DUMPING OF SOLID WASTES

A very subtle way of reclaiming land, especially near the waterfronts, is through the dumping of solid waste. Over time, such wastes fill up low-lying areas adjacent to swamps and rivers. When the land becomes relatively strong, some structures begin to emerge on such previously swampy or floodable areas. Diobu waterfronts in Port Harcourt are typical examples where this method of reclamation is carried out.

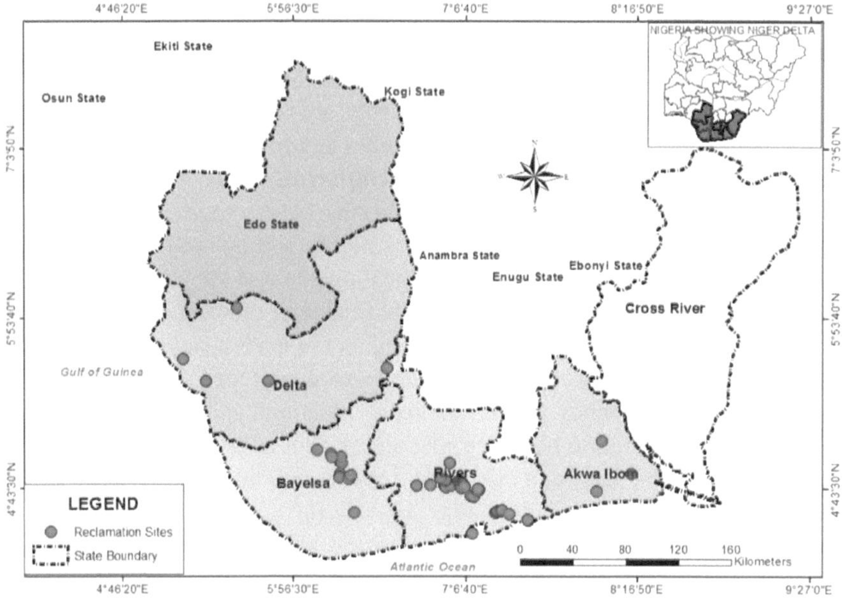

Figure 13.1 Sites Show the Trend of Land Reclamation in the Niger Delta. *Source*: Created by the author.

DISTRIBUTION OF LAND RECLAMATION SITES IN THE NIGER DELTA

The distribution of land reclamation sites across the Niger Delta region reveals hotspots of reclamation activities across the study areas as shown in figure 13.1. Rivers State shows a dominance of reclamation activities with forty-two cases of reclamation sites, which is the highest recorded so far in the Niger Delta. Bayelsa State is the second highest with sixteen cases, followed by Delta with four reclamation sites, and finally Akwa-Ibom with only three sites.

SOCIOECONOMIC IMPACT OF LAND RECLAMATION IN THE NIGER DELTA

From the foregoing, the Niger Delta has a steadily growing population estimated at more than 45 million people in 2020 and accounts for more than 23 percent of Nigerian total population. The population density is also among the highest in the world, with 265 people per square kilometer, according to the Niger Delta Development Commission. The people of the area are predominantly fishermen, crop farmers, petty traders, civil servants, and other paid employments. These occupations are dominantly practiced by all age groups and sex in the study area. Farming as practiced in the area shows to be lucrative as major crops cultivated include cassava, plantain, cocoyam, vegetables, coconut, palm fruit, plantain. These fruit crops are cultivated bountifully. The region is rich in crude oil, which has played a key role in attracting investors to the region and has also increased the employment rate in the region but has resulted in environmental devastations that have important implications for environmental security in Nigeria. Gas flaring and oil spills have resulted in environmental conflicts between the local people and the Nigerian state on the one hand and between local communities among themselves on the other hand. Part of the problem, as already stated, is the inadequacy in land supply due to the swampy nature of much of the coastal land area. There is a shortfall in land for occupancy as community populations and areas expand and communities that hitherto had been spatially distant come in closer proximity to one another. This has been one of the reasons behind the land reclamation efforts.

It can be deduced that Rivers State has the highest number of land reclamation sites in the Niger Delta. This is because the region requires more land area to be reclaimed as it is witnessing greater population and economic growth, is more riverine than others, and has less availability of land to accommodate the increasing population. Significantly, the people are more open to land reclamation out of necessity and are indifferent to environmental concerns.

The types of reclamation methods prevalent in these states carried out in small scales of reclamation by local communities are dumping of solid wastes and soft organic marine clay (chikoko mud fill) and the use of concrete breakwater owing to the inability to afford more sophisticated technology. The land areas reclaimed using these methods are, therefore, generally small and scattered. However, hydraulic sand filling is majorly carried out by the government and multinationals based on affordability.

Significantly, the purposes of land reclamation in the Niger Delta are grouped into two: (a) direct and (b) indirect purposes. The direct purposes include land expansion, construction activities, land acquisition, and succession (land-forming activities of mangroves). The indirect purposes include sand mining, exploratory activities, stream expansion/canalization, disturbance limitation, and agriculture. The positive socioeconomic benefit derived from land reclamation is that the lives of people may be improved, and their standard of living and quality of life may be enhanced as more land becomes available for several socioeconomic activities. Land reclamation aims at transforming wasted land into useful land; hence, it helps to increase land area, create habitat for wildlife, improve the quality of life of the environment, as well as turn the wasted land into residential, commercial, and industrial areas. Following reclamation, more structures and infrastructure can be built for economic growth; the availability of housing and more road networks, thereby easing traffic congestion; and the presence of industries, factories, or businesses, as applicable, thereby increasing employment opportunities for residents and lowering the unemployment rate.

Similarly, land reclamation impacts the coastal environment in several ways as follows:

Site Vegetation Clearing

Before reclamation, the site is cleared, which leads to the destruction of native plant species within the reclaimed area. This could lead to permanent loss of mangroves or other plant species of economic/medicinal importance and habitat for marine organisms and wildlife and their emigration to unaffected areas, thereby upsetting the ecological balance. During clearing, heavy machinery fell trees. This further denigrates the soil structure and converts the aquatic environment to a marshy environment. The swampy environment is a mixture of mud and water. After a long period of perturbation, the swampy soil gets solidified and changes to a terrestrial environment. But the original flora and fauna may have been lost.

Increased Erosion of the Cleared Areas/Riverbanks

Once the bulldozer rolls in to mow down the mangroves, it creates depressions for tidal pressure to wash ashore. This is because the adventitious roots of the mangrove forest serve as tidal breaks. The presence of mangroves along the coast stabilizes and reinforces the soil against erosion. Mangrove litter decomposes to form manure, which further consolidates the soil structure. Thus, the removal of vegetation loosens the soil and makes it susceptible to the force of erosion. In the same vein, the wheel of the bulldozer fragments the soil particles and makes it porous and prone to leaching or wind erosion.

Increase Access for Hunting and Logging

Logging activities destroy plant covers along the coast and create a passage for people to enter to hunt rare animals or harvest wild plant species. Similarly, reckless and indiscriminate hunting in open forests reduce species population and thus reduce species abundance and diversity, which may lead to the extinction of some species.

Changes in Topography of Sand-filled Area/Riverbed and Dredged Areas

Landfilling and dredging generally lead to changes in topography. In mangrove ecosystems, landfilling and dredging change the land topography and the entire aquatic system. Dredging could lead to changes in River channel morphology and stream morphometric properties, such as an increase in the length and breadth of the river channel, which affects the flow dynamics of the hydrology of the stream network. Dredging and filling activities and processes change stream bedform morphology and alter benthic ecosystems and the benthic species that depend on them. In addition, dredging could increase the seabed's depth, decrease the water level, and negatively impact intertidal aquatic communities. This may result in the death or migration of amphibious species along the coast as dehydrated intertidal areas experience a decrease in dissolved oxygen levels and aquatic food resources.[6]

Increased Turbidity

This could lead to the reduction in species composition and diversity of aquatic resources. It could also lead to death of fish and smothering of aquatic organisms. Increased turbidity leads to the blockage of light beyond the water surface, thus hampering photosynthetic activities below the water surface such as the benthic region.

Disturbance of Aquatic Life

This includes zooplankton, phytoplankton, benthic organisms, and fisheries. It also affects soil-dwelling organisms from adverse impact on water and soil quality. The felling of trees and the uprooting of their stumps lead to soil fragmentation. Evacuation of the tree stumps leads to the loss of many soil organisms from their habitat. Sand filling of the coasts leads to the burial of millions of species. Loss of trees leads to increased soil erosion and increased sedimentation of rivers.

GROUND WATER/SOIL QUALITY COULD BE IMPAIRED BY LEACHATES FROM GENERATED DREDGED SOIL

The dumping of dredged soil on mangrove vegetation smothers the plants and increases the heavy metal concentration of the soil.[7] This affects the survival of soil-dwelling organisms by increasing the acidity and alkalinity of the soil. Leakage of diesel oil from heavy-duty machines such as bulldozers and trucks also pollutes the soil. Similarly, the contamination of surface water and groundwater with used engine oil can pollute the groundwater aquifer, thus affecting neighboring communities that drink water from boreholes and hand-dug wells.

Impairment of Environmental Quality

The bulldozers that fell the trees and the trucks that evacuate the logs all generate smoke, which pollutes the surrounding air, leading to poor air quality and resulting in acid rain. This leads to the impairment of the health of aquatic and terrestrial organisms around the reclaimed area.

Improper Disposal of Solid Waste

This could lead to the contamination of soil surface and groundwater, disruption of fishing activities, and decrease in the aesthetic value of the environment.

IMPLICATIONS OF LAND RECLAMATION ON ENVIRONMENTAL SECURITY

Land reclamation, though with its many benefits, has certain disadvantages. Land reclamation is associated with some dangers, such as flooding and soil liquefaction. Reclaimed lands are expensive and can be damaging to corals and marine life. Common indirect impacts include influx of surface water and sediments, fragmentation of a wetland from a contiguous wetland complex, loss of recharge area, or changes in local drainage patterns.

Land reclamation often results in flooding, human displacement, building collapse, large displacement of the marine sediments, and the development of mud waves as well as the disruption of the ocean's ecosystem. Reclamation usually leads to the decline of biological diversity, the decrease of natural wetlands, and the extinction of habitats for animals and plants.[8] For migratory species, the living environments of marine plants and marine animals are seriously affected.[9]

Reclamation has been found to change shoreline evolution and wetland hydrology; deteriorate soil and water quality; alter vegetation succession, benthic animal and microbial communities, and fisheries; and impair ecosystem functioning and services.[10] Reclamation of water bodies into urban land use has many environmental consequences such as damaged ecology. Urban land transformation leads to the creation of residential and commercial buildings around water bodies, causing degradation of water ecology and an influx of nutrients.[11]

The processes of land reclamation can cause long- and short-term problems, which are harmful to habitats. There is an obvious change where the land is created, but habitats can be disturbed or destroyed in the process of gathering the earth or dredging for materials to add to the land. Reclamation has been observed to impact fisheries, benthic animal and microbial populations, vegetation succession, shoreline evolution, wetland hydrology, soil and water quality, and ecosystem services.[12] Floods and flooding are encouraged by land reclamation, erosion has been enhanced through poor reclamation practices, and fauna and flora have been lost. Reclamation takes meandering coastlines and replaces them with straight lines, destroying mangrove forests and shallows in the process.[13] It may appear to bring benefits in the short term, but in the long term, it can cause ecological disaster. For example, the loss of wetlands from land reclamation can worsen drought.[14]

Reclamation of water bodies into urban land use has many environmental consequences such as creation of residential and commercial buildings around water bodies, causing degradation of water ecology and an influx of nutrients.[15] Environmental problems caused by reclamation of wetlands are the release of organic pollutants, changes in salt and freshwater systems, changes in groundwater supply, air pollution, and fires. Ultimately, reclamation usually leads to the decline of biological diversity, the decrease of natural wetlands, and the extinction of habitats for animals and plants.[16] For migratory species, the living environment of marine plants and marine animals has been seriously affected. Significantly, if this persists without intervention, there would be an important environmental security issue.

CONCLUSION

It is easy to argue that most of the local people profit from land reclamation efforts. Examples of these advantages include the improvement of infrastructure and higher wages; accessibility to professional or unqualified employment; the development of recreation areas; job opportunities; the expansion of land capacity; higher citizen living standards; the creation of new land; and the mainstay of human development and the survival of any thriving community. Basically, land reclamation has boosted the socioeconomic benefits in terms of infrastructural development, economic/employment opportunities, land capacity expansion, and human development. For instance, reclaimed land has increased the number of roads and housing estates that have been constructed in the region. This may not have been possible as it is now. However, the reclamation of land has exposed areas to flood, erosion, environmental degradation, loss of livelihood from fishing activities, and loss of biodiversity.

NOTES

1. Numbere, A. "The Impact of Landscape Reclamation on Mangrove Forest and Coastal Areas in the Niger Delta, Nigeria." In *Landscape Reclamation*, Loures, L. O. (Ed.), (2020). DOI:10.5772/intechopen.82053.

2. Didia, M. U. "Investigation of the Impact of Land Reclamation Activities in the Niger Delta. Unpublished Ph.D Thesis. Institute of Geosciences and Environmental Management, Rivers State University, Port Harcourt." (2020). See also, Didia, M. U., and Ede, P. N. "Impact of Land Reclamation on the Vegetal Cover of Bayelsa State, Nigeria." *Journal of Geographical Research* 4, no. 2 (2021): 17–21.

3. Didia, U. M., and Owukio, S. S. "GIS-Based Assessment of Socio-Economic Impacts of Swamp Reclamation in Abonnema, Rivers State, Southern Nigeria." *International Journal of Information Systems and Informatics* 3, no. 2 (2022): 67–80.

4. Abali, T. P. "Hydraulic Geometry as a Key Indicator of Channel Reclamation in the Niger Delta Region of Nigeria. A Case Study of Dockyard Channel in Port Harcourt." *EVP Multidisciplinary Journal of Public Policy and Contemporary Issues* 1, no. 1 (2008): 89–94. See also, Abali, T. P and Aminigbo, L. M. O. "Morphological Implication of Land Reclamation in Port Harcourt, Southern Nigeria." A Paper Presented at the 4th International "Başkent" Congress on Physical, Social and Health Sciences, Istanbul, Turkey, February 26–27, 2022

5. Umeuduji, J. and Abali, T. P. "Land Reclamation and River Channel Morphology in Port Harcourt, Nigeria." *Journal of Geographic Thought (JOGET)* 8, no. 1&2 (2007): 28–39.

6. Numbere, O. Aroloye. "Mangrove Restoration Under Different Disturbances Regimes in the Niger Delta, Nigeria." In *Mangrove Ecosystem Restoration*, edited by Sahadev Sharma (Intechopen, 2021). Accessed on February 23, 2021, from Mangrove Restoration under Different Disturbances Regime in the Niger Delta, Nigeria | IntechOpen.

7. Didia, M. U., and Ede, P. N. "Impact of Land Reclamation on the Vegetal Cover of Bayelsa State, Nigeria." *Journal of Geographical Research* 4, no. 2, (2021): 17–21.

8. Ibama, B., and Nengi, D. I. "There is a Price to Pay for Every Wetland Reclamation and Conversion: Experiences from Port Harcourt Municipality." *MOJ Ecology & Environmental Sciences* 5, no. 5 (2020): 221–228

9. Ibid.

10. Newton, Alice, Icely, John, Cristina, Sonia, Perillo, Gerardo M. E., Eugene Turner, R., Ashan, Dewan, Cragg, Simon et al. "Anthropogenic, Direct Pressures on Coastal Wetlands." *Frontiers in Ecology and Evolution* 8 (2020): 144.

11. Adegboyega, S. A., Oloukoi, J., Olajuyigbe, A. E., and Ajibade, O. E. "Evaluation of Unsustainable Land Use/Land Cover Change on Ecosystem Services in Coastal Area of Lagos State, Nigeria." *Applied Geomatics* 11 (2019): 97–110.

12. Newton et al., 2020.

13. Perillo, Gerardo, Eric Wolanski, Donald R. Cahoon, and Charles S. Hopkinson, eds. *Coastal Wetlands: An Integrated Ecosystem Approach*. Elsevier, 2018.

14. Ardón, Marcelo, Jennifer L. Morse, Benjamin P. Colman, and Emily S. Bernhardt. "Drought-induced Saltwater Incursion Leads to Increased Wetland Nitrogen Export." *Global Change Biology* 19, no. 10 (2013): 2976–2985.

15. Akhtar, Naseem, Syakir Ishak, Muhammad Izzuddin, Bhawani, Showkat Ahmad, and Umar, Khalid. "Various Natural and Anthropogenic Factors Responsible for Water Quality Degradation: A Review." *Water* 13, no. 19 (2021): 2660.

16. Deane, David C., Fordham, Damien A. , He, Fangliang, and Bradshaw, Corey J. A. "Future Extinction Risk of Wetland Plants is Higher from Individual Patch Loss than Total Area Reduction." *Biological Conservation* 209 (2017): 27–33.

REFERENCES

Abali, T. P., and Aminigbo, L. M. O. "Morphological implication of land reclamation in Port Harcourt, Southern Nigeria." *A Paper Presented at the 4th International "Başkent" Congress on Physical, Social and Health Sciences,* Istanbul, Turkey, February 26–27, 2022.

Abali, T. P. "Hydraulic geometry as a key indicator of channel reclamation in the Niger Delta region of Nigeria. A case study of Dockyard channel in Port Harcourt." *EVP Multidisciplinary Journal of Public Policy and Contemporary Issues* 1, no. 1 (2008): 89–94.

Adegboyega, S. A., J. Oloukoi, A. E. Olajuyigbe, and O. E. Ajibade. "Evaluation of unsustainable land use/land cover change on ecosystem services in coastal area of Lagos state, Nigeria." *Applied Geomatics* 11 (2019): 97–110.

Akhtar, Naseem, Muhammad Izzuddin Syakir Ishak, Showkat Ahmad Bhawani, and Khalid Umar. "Various natural and anthropogenic factors responsible for water quality degradation: A review." *Water* 13, no. 19 (2021): 2660.

Ardón, Marcelo, Jennifer L. Morse, Benjamin P. Colman, and Emily S. Bernhardt. "Drought-induced saltwater incursion leads to increased wetland nitrogen export." *Global change biology* 19, no. 10 (2013): 2976–2985.

Deane, David C., Damien A. Fordham, Fangliang He, and Corey JA Bradshaw. "Future extinction risk of wetland plants is higher from individual patch loss than total area reduction." *Biological Conservation* 209, (2017): 27–33.

Didia, M. U. "Investigation of the impact of land reclamation activities in the Niger Delta. Unpublished Ph.D Thesis. Institute Of Geosciences and Environmental Management, Rivers State University, Port Harcourt." (2020).

Didia, M. U., and Ede, P. N. "Impact of land reclamation on the vegetal cover of Bayelsa State, Nigeria." *Journal of Geographical Research* 4, no. 2 (2021): 17–21.

Didia, U. M., and Owukio, S. S. "GIS-Based assessment of socio-economic impacts of swamp reclamation in Abonnema, Rivers State, Southern Nigeria." *International Journal of Information Systems and Informatics* 3, no. 2 (2022): 67–80.

Ibama, B., and D. I. Nengi. "There is a price to pay for every wetland reclamation and conversion: Experiences from Port Harcourt Municipality." *MOJ Eco Environ Sci* 5, no. 5 (2020): 221–228.

Newton, Alice, John Icely, Sonia Cristina, Gerardo ME Perillo, R. Eugene Turner, Dewan Ashan, Simon Cragg et al. "Anthropogenic, direct pressures on coastal wetlands." *Frontiers in Ecology and Evolution* 8, (2020): 144.

Numbere, O. Aroloye. "Mangrove restoration under different disturbances regimes in the Niger Delta, Nigeria." In *Mangrove ecosystem restoration*, edited by Sahadev Sharma. Intechopen, 2021. Accessed on February 23, 2021, from Mangrove Restoration under Different Disturbances Regime in the Niger Delta, Nigeria | IntechOpen.

Perillo, Gerardo, Eric Wolanski, Donald R. Cahoon, and Charles S. Hopkinson, eds. *Coastal wetlands: An integrated ecosystem approach*. Elsevier, 2018.

Umeuduji, J., and Abali, T. P. "Land reclamation and river channel morphology in Port Harcourt, Nigeria." *Journal of Geographic Thought (JOGET)* 8, no. 1&2 (2007): 28–39.

Chapter 14

Environmental Security in Kenya

A Case of Geothermal Projects in Samburu and Narok Counties

Clifford Collins Omondi Okwany, Evelyn Atieno Owino, and Brian Ocheing Okoth

Environmental security, which generally refers to removing threat/fear through restoration and management of natural resources, has been growing since the 1990s.[1] Concerns about unsustainable development practices threatening indigenous communities have also been a point of debate.[2] These debates have increasingly focused on changes in climate regimes and have produced global actionable plans such as the famous Paris Agreement of 2015, a treaty on climate action to achieve a zero-emission world through sustainable development. This Agreement put renewable energy at the center of financial institutions such as the European Investment Bank (EIB) and World Bank to meet the envisioned 2030 SDGs through goal number 13—climate action[3] and establishing institutions to address the growing Africa's renewable energy.[4]

Kenya is one of the leading countries investing in such climate action, developing geothermal energy with the partnership of the World Bank, EIB,[5] *Kreditanstalt für Wiederaufbau* (KfW)—a German state development and investment bank,[6] and Japan's International Cooperation Agency (JICA).[7] Geothermal energy is recognized as the most significant energy potential,[8] with the ambition and potential of nuclear energy.[9] Kenya's geothermal project is the largest source of energy in the country,[10] amounting to more than 40 percent,[11] or 810 megawatt (MW; see table 14.1). In 2018, the geothermal grid was at 31.6 percent and produced between 5,033 gigawatt hours (GWh)[12] to 5,127 GWh.[13] However, despite the promising growth in renewable energy, the pastoral communities have been affected by the geothermal developments as evidenced by forceful evictions, noise pollution, and environmental health

Table 14.1 The Capacity of Olkaria Geothermal and Estimated Grid

KenGen-Owned Olkaria Developments	*Units*	*Year Established*	*Capacity in Megawatts*
Oklaria I	I	1981	15
	II	1982	15
	III	1985	15
	IV	2014	70
	V	2015	70
	VI	2022	83
Olkaria II	I	2003	35
	II	2003	35
	III	2010	35
Olkaria III	I	2000	48
	II	2009	36
	III	2014	26
	IV	2016	29
Olkaria IV	I	2014	70
	II	2015	70
Olkaria V	I	July 2019	79
	II	October 2019	79
Olkaria VI	Planned for development		-
Oklaria VII	Planned for development		-
TOTAL GRID			810

hazards, and they risk losing their livelihoods.[14] Such facts have highly focused on the Olkaria geothermal development in Naivasha Nakuru County.

At the same time, there is limited research focusing on the proposed great geothermal project in Baringo County, Suswa in Narok County, and the anticipated Suguta Trough in Samburu County and how these potential developments have an impact on the communities around them. More so, such limitation is also evident in the case of Suguta Trough showing how such resources along borderlands escalate boundary violent conflict among ethnic communities and contestation among local governments since the implementation of devolution—transfer of power, functions, and services to the local/county governments in Kenya.

The connection between indigenous people's land and renewable energy is evident in Kenya. Whether these projects will uplift the indigenous communities' cultural, economic, and social rights is a contested debate. The Kenyan case proves that such rights have been infringed on. Policy formulated with a focus on the indigenous people's concerns and rights is not active, public participation is limited, the adverse effects of renewable energy on the communities around the project are never addressed, and land tenure challenges exist.[15] As such, the challenges have led to evictions, threatening indigenous communities' livelihoods.[16] Despite such investments, the rate of implementing the Community Land Act 2016, which protects the pastoralists' land rights, is slow. The land tenure stipulates customary rights, and it provides community rights for their adjudication, documentation through registration and ownership. A bill was passed in parliament in the year 2022 and gazetted to amend the Community Land Act 2016 and express the provisions that prohibit unlawful occupation of community, private, and public land, and it also set out the procedure for evicting the illegal occupants.[17] Despite this, geothermal energy developments are ongoing and threatening pastoral livelihoods and land rights without protection being accorded to them as enshrined in the law.

We analyze the geothermal energy project in Suswa, bordering Narok and Kajiado Counties in Kenya, showing how this potential renewable energy project has already created grievances between government agencies such as the GDC and the pastoral communities in the southern Rift Valley region (see, figure 14.1 showing geothermal potential in Kenya). We also analyze the potential Suguta Trough in Samburu County, showing how the *Emuruangogolak* volcano on the border between Samburu and Turkana Counties has the potential to uplift Kenyan renewable energy ambition but causes environmental risks to frontier inhabitants in the northern Rift Valley region, risking ethnic community borderland violent conflicts, and county contestation over natural resources. The analysis is based on empirical data collected between October 2021 and February 2023 from interviews and focus group

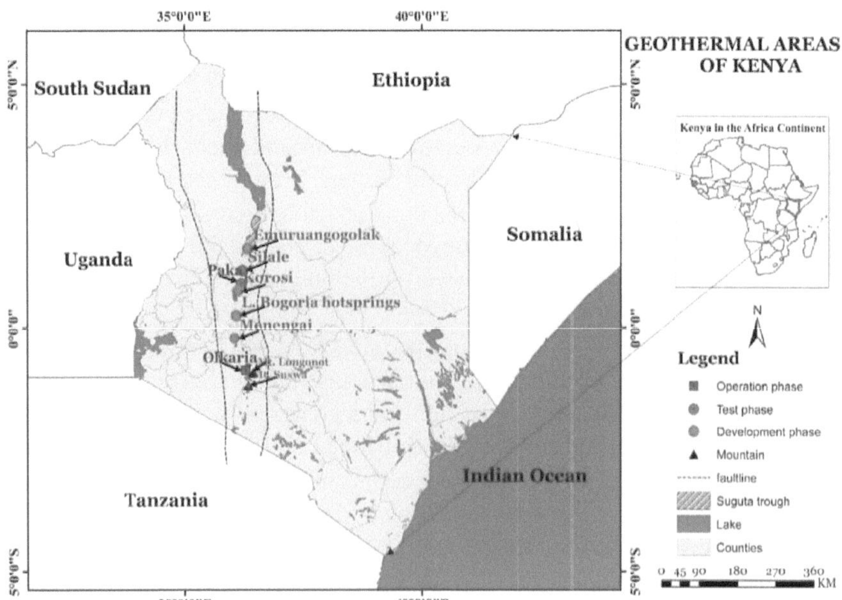

Figure 14.1 Geothermal Areas of Kenya. *Source*: Created by the author.

discussions from communities in Baragoi, Samburu North, Suswa in Narok, and GDC offices in Nairobi and Nakuru Counties.

We first analyze environmental security and then examine the historical perspective of the frontier, the development of geothermal potential in Olkaria, and the risks the plants pose to the indigenous communities. Lastly, we analyze cases of the potential geothermal developments of Suguta Trough and Suswa. We point out that environmental security should focus on not only sustaining prosperous economies but also social inclusion, a situation where a government practices subsidiarity—the use of a top-down approach but also a bottom-up approach to remove any necessary threats that may affect the pastoral livelihood in areas with potential geothermal energy for national development. In addition, we show that geothermal energy in Kenya risks losing its renewable status and being a burdensome project as its upkeep risks surpassing its value because of water scarcity in the project area. Geothermal energy becomes renewable when there is a surplus of water supply, and Kenya's Rift Valley region faces water challenges (see water points along the geothermal lines in figure 14.1), and thus, the burdensome status of the project despite having caused environmental insecurity.

THEORETICAL CONCERNS IN ENVIRONMENTAL SECURITY

Environmental security has been defined as the ultimate security, considering the consequences of global warming and climate change on the environment, which threaten human survival.[18] Others looked at it through military and political lenses.[19] Although with a focus on international relations, the concept is a garden of different cases.[20] Environmental concerns have been constructed from political perspectives to explain political considerations, especially in the 1990s, when serious discussions around changes in environmental regimes were in their infancy.[21]

These concerns in the 1990s took a global perspective, linking the international political economy to the conflict between the poor global south and the choices of the prosperous global north, which were feared to lead to a confrontation.[22] Natural resources are often a motivation to conflict due to greed.[23] Greed is destructive and has environmental consequences, leading to grievances contributing to warfare. Some work on environmental security[24] focused on how security policies are shaped to address environmental issues—linking national security, conflict, and environmental degradation. Thus, environmental security is concerned with removing the threat and fear caused by ecological problems, and it connects ecology to security.[25] Such concern focuses on actions of removal of external threats—a nation-centric scenario. However, environmental security can also take a different perspective, such as a biocentric scenario—removal of a threat to the existence of the ecosystem,[26] which is not of concern in this chapter.

The concept has been described within the framework of Anthropocene—human historical activities and their impact on ecological and geological changes and how such impact threatens life on earth.[27] These concerns are significant for understanding environmental security, which is still a growing field. However, our concern in this chapter does not consider international relations per se. Still, it is concerned with government policies that focus on sustainable energy development but have an effect on communities around such projects, which is our focus. Jon Barnett[28] points out different frameworks that explain the concept of environmental security: (1) normative nature—it is concerned with ecological security; (2) it is political in the sense that it connotes securitization—an actor pointing at a referent object as a threat; (3) it is theoretical in the sense that it identifies environmental issues in conflict, violence, threats, and fears; and (4) it is conceptual because it redefines the concept of security. We concentrate on these normative and theoretical concerns, giving explanations for the contestation between the government's focus on sustainable energy and indigenous communities' security.

Environmental security is a contested concept—it is complex and sometimes misleadingly dichotomous.[29] Thus, linking environment and security depends on the meaning of security and the research method one uses.[30] As such, the antagonistic nature and complex meaning of security make environmental security complex and contested. Some view security as the absence of threat, and others view security as the absence of fear.[31] It is considered the absence of threat in the sense that if the state removes all possible threats such as poverty, external enemy, corruption, and so on, the citizens might feel safe; the absence of fear may refer to the survival of the fittest.[32] Societies experience security differently. To some, a threat might be floods, drought, the police, or projects that emerge with increased global capitalism. To others, such threats might not exist due to different experiences and conditions created by the actor referring to a threat—securitization. Security can be military, environmental, or geothermal projects we analyze in this chapter; it depends on what one wants to designate as a threat. In other words, security depends on who speaks it or projects it. In the case of this chapter, the referent object is energy security with a focus on geothermal energy. We do not focus on biocentric and nation-centric security, as already explained. However, we emphasize human-centric—securitizing people from environmental threats, showing how geothermal discoveries benefit the state but threaten indigenous livelihood.

The concept of environmental security attempts to grapple with understanding the relationship between modernity and the environment. To some extent, the concept is concerned with the exploitative nature of capitalism and the free world and how such ideology leads to destruction. For example, environmental security is concerned with how colonialism and imperialism impact the indigenous people in relation to their environment, looking at the "risk societies"—frontier inhabitants and how modern production endangers and threatens their livelihood through natural resource extraction.[33] In other words, we analyze how civilization consumes scarce natural resources and threatens ecological existence.[34] For example, energy security has been a global concern, and just like any other security concept, "the fundamental questions to ask are; 'for whom'; 'from what threats'; and 'by what means'."[35] As such, we concentrate on this view, explaining how exploiting natural resources to generate electricity for the growing population affects frontier societies. We define environmental security as the removal of threat/fear through restoration and management of natural resources, and environmental insecurity as the threat/fear that comes with exploiting natural resources, leading to conflict and social disorder. Thus, it explains the impact of civilization's demands on energy consumption and the exploitation of geothermal energy to generate electricity.

HISTORICAL PERSPECTIVE OF GEOTHERMAL DEVELOPMENT IN KENYA: A CASE OF OLKARIA

The pastoral lands, mainly arid and semiarid lands (ASALs), were neglected by the British colonial government, and the communities living in these areas, predominantly pastoralists, were mere subjects on their communally owned land.[36] The postcolonial government did the same, neglecting these lands by formulating and conducting an executive order—Sessional Paper No. 10 of 1965, stipulating African socialism on economic development. The executive order focused on developing high-potential agricultural districts, anticipating a trickle-down effect on the ASALs.[37] Such neglect continued with Sessional Paper No. 1 of 1986, another strategic economic order to save the country from the world recession and address some of the economic problems created by the oil crisis of the 1970s.[38] Today, these areas have become frontiers for development. For example, Turkana 100 percent ASAL, Samburu 85–100 percent, Baringo 50–85 percent, and Narok 30–50 percent ASAL[39] have renewable energy potential. These ASALs have attracted geothermal energy development and were prioritized in Vision 2030—an economic blueprint for development.[40] In 2012, Sessional Paper No. 8 was developed to prioritize the ASALs with a sustainable development focus.[41] As such, government neglect and interest in these frontier areas reflect what Giorgio Agamben referred to as the state of exception[42]—the state ignoring the frontier spaces but firmly providing security, mainly when there are some economic interests.

Estimates indicate that 42 percent of Kenya's GDP is from natural resources, and most sectors that depend on natural resources offer about 70 percent of employment in Kenya.[43] Exploiting such resources would push Kenya toward becoming a middle-income country.[44] With such a trajectory on natural resource exploitation, Kenya's legal texts have environmental security principles such as public participation in natural resource management, the "polluter pays" principle of the Rio Declaration of 1992, "Climate Action" adopted from the Paris Agreement of 2015,[45] sustainability from the SDGs, and intergenerational equity.[46] In addition, the National Policy for Energy 2018, the Energy Act of 2019, and Kenya's Vision 2030 stipulate community engagement to ensure security and manage community expectations and any threat that might result from energy development. These legal texts appeal to the environmental security debates.[47] However, the conduct of these legal frameworks on community protection and environmental security is still wanting. They face criticism, mainly focusing on indigenous people's rights,[48] corruption, poor policy implementation, and governance issues, leading to energy poverty.[49] The new geothermal potentials in Suswa in Narok County and Suguta Trough and Baragoi in Samburu County pose mixed reactions—the push for energy security; however, they pose further environmental insecurity to the frontier inhabitants and risk their livelihood.

Geothermal energy development aligns with the Sessional Paper No. 10 of 2012 on Kenya's Vision 2030, stipulating socioeconomic transformation through infrastructure development and energy foundation. In 2005, Kenya's installed geothermal energy was at 129 MW, producing 1088 GWh. However, in 2010, there was an increase in installation of 167 MW, which produced 1430 GWh, and (a forecast of installing) 530 MW in 2015,[50] and in 2020 the five established Olkaria plants were at 726.6 MW.[51] In 2018, the total electricity generation capacity from all energy sources was at 2,670 MW with a demand of 1,841 MW, and geothermal energy accounted for more than 40 percent of the total grid, making Kenya a global leader in renewable energy with low cost.[52] In the region, Kenya takes the lead with energy production in the Horn of Africa, particularly in thermal, geothermal, and wind.[53]

Some of the earlier explorers of geothermal energy were Italy starting in 1904, Iceland in 1930, Japan in 1951, New Zealand in 1958, and the United States in 1960. Kenya began exploring geothermal opportunities in 1956 and joined the club with the first supply of 15 MW to the national grid.[54] Despite the exploration in 1956, the development started after the gazette in 1971, and Olkaria 1 was commissioned in 1981 as the first African geothermal energy project.[55] The Olkaria geothermal development is situated in Hell's Gate National Park, designated as one of the UNESCO World Heritage sites.[56] In

2012, the project had 29 wells; in 2018, there were 290 wells.[57] The developments of renewable energy projects have since created forceful eviction of the Maa communities,[58] the Maa continue to be pushed away, and their grazing rights are denied due to the expansion of the Olkaria plants.[59] Currently, there are five plants with new Olkaria VI and VII envisioned.[60] The development of Olkaria geothermal initiatives has highly demonstrated adverse effects on the Maa communities.[61] For example, the development of Olkaria IV in January 2012 and June 2014 came with a Resettlement Action Plan (RAP), which led to the relocation of four villages of the Maa community, namely Oloosinyat, Oloolkarian, Oloonong'ot, and OloMayiana.[62] The displacement affected the cultural and ancestral attachment to the land and pushed them away from the national park, which they depended on due to its tourism attraction.[63]

Implementing public–private partnership (PPP) increased the participation of the private sector in Kenya's electrification projects, leading to 43 percent of private sector-generated energy. Such privatization risks energy security in the country due to private entities monopolizing energy pricing policy.[64] However, the involvement of PPP has increased Kenya's electricity from 2,295 MW in 2015 to 5,040 MW in 2020,[65] a capacity that is enough to light up 75 percent of Kenya's population.[66] Geothermal energy development has been envisioned to grow Kenya into an industrial economy.[67] An estimate of 1,984 MW production was projected to increase the geothermal grid in 2020,[68] and the ambition under vision 2030 is to develop geothermal energy production alone to 5000 MW.[69] Such prediction has envisaged the production of more plants in Olkaria or other parts of the rift valley region with the potential for geothermal energy, despite the evident environmental threats accompanying the Olkarai geothermal development in the indigenous Maa community.[70] In addition, further geothermal developments are taking place in Bogoria-Salali in Baringo County and Suswa in Narok, and there is potential geothermal energy in Suguta Trough in Samburu County. However, studies on such envisioned renewable energy developments in the frontier and their impact on the pastoral communities are limited in the academic space.

Result and Analysis: A Mixed Reaction to the Energy Development and Security

The government-hyped geothermal energy and its development are advantageous to the country's development and also have serious consequences for energy poverty but are also depriving the pastoral communities of their livelihood, thereby creating conflicts over land. The areas with geothermal developments and discoveries in Kenya are predominantly occupied by pastoral communities. These communities have existed under a communal land tenure system for centuries, because it is suitable for pastoral livelihood.

However, the discoveries and developments of geothermal resources lead to anticipation and privatization of land, and such changes lead to land grabbing, corruption, conflict, and violence.

The developments, such as Olkaria geothermal units, have already proved the project's impact on health hazards such as noise and waste pollution in Hell's Gate National Park—environmental insecurity and forceful evictions from ancestral community land are also evident. Interviews with land experts indicated that despite these projects' discoveries and development, the Community Land Act 2016, which protects pastoral communities from land grabbing and corruption, has not been implemented in the area.[71] The law stipulates community ownership in the land tenure system. Investments such as geothermal energy projects must consult, involve the community, and negotiate on the land before implementing a project because the law makes the community the custodian of the land. Despite the lack of an implementation process, the Community Land Act 2016 was put up in parliament for amendments.[72] In addition, Kenya has a great framework from legal text stipulating environmental security and community protection. However, these legal frameworks are eroded by corruption, political interests, and patronage,[73] blocking implementation and leading to environmental insecurity and community eviction from their ancestral land.

THE CASE OF SUSWA GEOTHERMAL POTENTIAL

Suswa is located west of the capital city Nairobi, and it is south of Kenya's rift valley region, bordering Nakuru County to the south and Kajiado County to the north, and east of Narok County. The area covers 308.6 square kilometers and is part of the Narok east administrative units.[74] The area has a low-lying Quaternary volcano named Mount Suswa, a famous Ol Doinyi Nyukie crater built by post-caldera lava.[75] In addition, the mountain is a protected conservancy, and it attracts tourists due to its rich vegetation, including its two craters and famous caldera. The inhabitants are predominantly the pastoral Maa community, and the potential geothermal discoveries and development are advantageous to the exchequer.[76] However, geothermal projects have environmental risks to pastoral livelihood, Mount Suswa conservancy, wildlife, and tourist attraction.

The case of Suswa geothermal development in Narok, just like other new geothermal developments in Baringo, proves the intent of Kenya's government to exploit and manage geothermal natural energy resources, which risks the pastoral communities' livelihood. The GDC, a government's special purposed vehicle for generating geothermal energy,[77] is envisioned to add renewable energy to the total electricity grid. The Narok County Suswa three-phase

Table 14.2 The GDC's Potential Geothermal Production

Geothermal Development Plan	*Units*	*Units*	*Units*
Narok, Suswa Potential	Phase I—100 MW	Phase II—100 MW	Phase III—100 MW
Baringo, Bogoria-Salali	Paka—100 MW	Salali—100 MW	Korosi—100 MW
Naruku Menegai	Phase I—105 MW	Phase II—60 MW	Phase III—300 MW

projects are envisioned to add 300 MW each, the Nakuru County's Menengai three-phase projects are estimated to add 465 MW, and Baringo County's Bogoria-Silali project also has three sites, Paka, Korosi, and Silali, each envisioned to add 100 MW (see table 14.2). These estimates will add 1,065 MW to the anticipated 5,000 MW;[78] with an environmental impact assessment done in 2013, geothermal development in Suswa was licensed to GDC.[79] Such figures pose a promising future for energy security in Kenya. However, as we emphasize below, there are reservations about environmental insecurity, such as health hazards to the habitat around the geothermal projects and forceful evictions of the frontier inhabitants.

The Menengai Phase One project is in operation and producing 105 MW, while the potency of Suswa is delayed due to technical challenges facing drilling and lack of water availability.[80] These exploitations and development of natural resources pushed Kenyan development and electricity security, proving that exploiting natural resources could be beneficial.[81] However, removing threat/fear through restoration and protection of the ecosystem and communities around the project, its environmental security is questionable. Thus, reservations exist on how such developments impact the pastoral communities' livelihood, wildlife, parks, and conservation. Our interviews with geothermal experts prove these reservations:

> The geothermal resource is deep in the subsurface, say 5km or more. Rainwater takes about 10000 years to drip down to those depths. Kenya, in general, is animal friendly towards the wildlife on the surface. Power plants are designed to allow animal traffic high and low. There is environmental staff addressing any animal concerns. There are some isolated accidents for sure. You have to understand that nature is cruel, and mistakes happen. Olkaria's development exemplifies such scenarios. For sure, Suswa should learn from Olkaria's mistakes.[82]

The geothermal potential is situated in Mount Suswa, "comprising a 12 km diameter double caldera system, a vast network of obsidian caves and habitat to a variety of wild animals and plants."[83] The GDC geothermal plant developments in Suswa threaten the ecosystem; it is a risk to economic activity in the Suswa conservancy, where the caldera and caves are important tourist attractions. The plight of Hell's Gate Mational Park in Naivasha predicts Suswa's future.[84] The Olkaria drilling in Naivasha has created noise and waste pollution, reducing the number of wild animals in Hell's Gate National Park.[85] Such allegations were confirmed in media interviews.[86] Simon Thomset, a bird expert in Hell's Gate, affirmed that the noise and waste pollution endangers rare bird species such as the rare lammergeyer vultures. However, the Geothermal Director at KenGen, Geoffrey Muchemi, denied the allegations affirming that environmental assessment was done and the geothermal development is committed to environmental security and sustainability in energy production.[87] Such denial of environmental insecurity overlaps with the economic vision of increasing the electricity grid to make Kenya an industrial economy by 2030.

Land in Narok and Kajiado Counties have experienced rapid change. Neighboring the capital city Nairobi, Suswa is at a distance of about fifty kilometers; increased privatization of land has attracted many outsiders and the development of the Suswa geothermal plant and the Standard Gauge Railway (SGR) has attracted speculation of land, thereby resulting in high land prices. As such, the anticipation of development is attracting the selling and buying of land at a market rate of USD 3,000 for an eighth of an acre or 50 feet by 100 feet, despite pastoral community land being communally owned for grazing purposes.[88] Suswa is a grazing corridor and the migration route between Kajiado and Narok pastoralists.[89] However, the development ventures have attracted private land ownership, leading to the loss of pastoral livelihood.

The education level in Suswa is deficient, and this hinders the pastoral communities' power to negotiate their benefits from the emerging projects in Suswa;[90] this high illiteracy rate was caused by the government's historical neglect of these communities through development and economic plan.[91] However, changes emerged with Kenya's Vision 2030 economic blueprint, stipulating the development and exploration of natural resources in these areas;[92] thus, the neglect and the shift of exploitation reflect Agamben's state of exception.[93] In addition, the weak tenure system of the pastoral land, which is communally owned, and limited information encourage land grabbing, the anticipation of new developments, and threaten indigenous livelihood.

The Olkaria geothermal development experience that led to environmental health hazards and forceful displacements of the Maa community proves a likelihood of a similar scenario in Suswa's future. The former geothermal development led to KenGen's water pollution, the dumping of hydrogen sulfide with a rotten egg smell and brine and other industrial wastes in rivers, and has negatively impacted the overall health of those living nearby. The industrial wastes have poisoned the *Ricinus communis*—castor beans or oil plant, impacting pastoral livestock health.[94] The GDC has a department for CSR, and they have engaged and sensitized communities concerning the potential geothermal project in Suswa. Still, there is a delay due to water scarcity and technical problems.[95] As such, the water scarcity is likely to drain the existing wells, and the pastoral communities in Suswa are likely to feel the negative impacts, such as forceful eviction and health hazards (noise and water pollution).

Essentially, both GDC and KenGen hyped the development of geothermal as renewable energy for Kenya's future. However, geothermal energy is only naturally renewable and sustainable in island countries such as Japan, Ireland, Indonesia, and the Philippines because there is plenty of water in these islands. Water is the taxi for the energy produced by the geothermal, and lack of it means geothermal is not renewable.[96] Kenya is not an island—water is a challenge, and Olkaria development started losing their wells, and there are suggestions to use the water from Lake Naivasha,[97] which further risks the ecosystem of the lake, and Suswa geothermal energy development is likely to face the same challenges due to water scarcity.[98] As such, developing renewable geothermal energy in Kenya is challenging, and it takes a long time and requires resources (money and sustainability of water) to make it renewable.[99] The Suswa project not only risks the pastoral livelihood but also risks being a burdensome project—upkeep surpassing the value of the project. The project has limited water available for capturing the geothermal energy from the subsurface and transporting it to the surface for use; thus, the project risks its renewable status.

POTENTIAL GEOTHERMAL DISCOVERY: A CASE OF SUGUTA TROUGH

There have been discoveries of geothermal energy resources in Suguta Trough, located in the Emuruangogolak volcano about 100 kilometers south of Lake Turkana.[100] The area borders Baragoi[101] town, some 35–40kilometers from Nachola in Samburu North;[102] it is a semi-territoriality scenario, characterized by insecurity due to government neglect.[103] The area is dominated by Turkana and Pokot community armed groups taking advantage of limited government presence, raiding and rustling cattle, and conducting roadside banditry.

The Suguta Valley is famously known as the "valley of death" after the killing of the forty-two General Service Unit police who traced stolen cattle to the Turkana community armed groups and were ambushed.[104] The Kenyan government's anticipation of the Suguta Trough's potential geothermal discoveries will open the area, and security forces will create a territoriality scenario—monopolizing control and influence.[105] The discovery is advantageous to the communities around the Suguta Trough because it is also an area demarcated for the Lamu–South Sudan–Ethiopia Transport Corridor (LAPSSET), an anticipated mega road infrastructure project. However, the discoveries and feasibility study of the Suguta Trough have already created tension among the ethnic communities living around the Suguta Valley.

The Turkana pastoralists dominate the Emuruangogolak area bordering Samburu and Turkana Counties; however, the Samburu Ethnic communities are claiming Suguta Trough is located in Samburu County.[106] Turkana communities in Nachola Ward and Suguta Valley also claim that the trough is in their territory.[107] However, both communities are skeptical of the potential geothermal discovery in the belief that claiming it might lead to displacements and health hazards. Some media reports indicated standoffs, and the Samburu County government has been urged to intervene.[108] With the government's anticipation and sending of investors to the Suguta Trough, the communities complained that the investor and the government did not involve them. The Turkana community standoffs indicated demands for a memorandum of understanding and community participation in the potential geothermal project.

In addition, Suguta Trough lies between Turkana and Samburu Counties border area, both county/local governments claim the land, leading to county border conflict due to the natural resource endowment in the Suguta Valley.[109] Such county contestation began with the development of devolved governance in Kenya, and natural resources endowment is a driver for such county boundary conflicts, and more so, violent conflict among communities along these borders—Samburu and Turkana in the case of Suguta Valley shift from cattle raiding to borderland contestation due to rent-seeking politicians fueling borderland violent conflict,[110] and the Suguta Trough is a driver for such conflicts.

CONCLUSION

Kenya's legal texts have shown an ecologically friendly framework; however, the discovery and development of geothermal energy overlap between environmental security and the reverse of it. Renewable energy developments in Olkaria, Menengai, and Bogoria-Salali and the new potential sites in Suswa and Suguta Trough are promising. However, the Olkaria development proves environmental insecurity, such as health hazards and water and noise pollution, affecting wildlife.[111] The Olkaria plant also risks being a sustainable energy project due to water scarcity to make it renewable for a more extended period. In addition, forceful evictions demonstrate energy poverty, poor policy implementation, environmental governance problems, and corruption.[112] The framework is an overlap between renewable energy focus and risking pastoral livelihoods rights.

The Suswa geothermal development is likely to face negative experiences similar to those of Olkaria, and the communities around the Suswa geothermal project have pointed out different complaints on environmental risks and health hazards; they have reported potential cases of poisoning of animals, cancer, and miscarriages, which their neighboring communities experience in Olkaria Naivasha.[113] These cases robustly demonstrate the effect of environmental insecurity created by geothermal development, despite the existing legal texts and frameworks stipulating community protection and engagement. In addition, the emerging geothermal potential in Suswa already proves threats to pastoral livelihood and environmental risk to Mount Suswa conservation, risking tourism economy[114] and health hazards to Suswa residents.

The geothermal project in Suswa has been delayed due to drilling and water availability challenges.[115] As such, drilling would risk the shared water wells, and waste production also threatens the tourist attraction of Mount Suswa, which has been the case in Olkaria's effect on Hell's Gate National Park.[116] The lack of sustainable water means that the Suswa geothermal will risk being renewable because geothermal energy is available, but the water to transport it from the subsurface to the surface for use is not available. The Olkaria wells have dried up, and there is an ongoing negotiation to use the water from Lake Naivasha, which further risks the lake's ecosystem. Furthermore, apart from the pollution and environmental insecurity risks, the pastoral communities face threats to their land from anticipation and privatization of land despite the legal framework, the Community Land Act 2016, stipulating land rights through communities' registration and decision-making.

The implementation of the Community Land Act 2016 has been slow, and in 2022 the law was subject to amendments; yet, the developments and discoveries of the geothermal potential are ongoing. The land in Suswa's case differs from the Suguta Trough politically; the former is closer to the capital city. Thus, anticipation and privatization of the communal land are high. Suguta Trough faces insecurity challenges and still suffers from the state of exception. However, geothermal developments will attract development ventures, leading to privatization which risks pastoral livelihood. Thus, geothermal potential in these frontier spaces proves environmental insecurity as the threat that comes with the exploitation of natural resources, leading to conflict and social disorder.

The Suguta Trough geothermal potential is advantageous to the communities in Suguta Valley. The road infrastructure anticipated to pass through Suguta Valley will open the rural frontier area to development. However, limited community participation already escalates conflict due to ethnic boundaries. The feasibility study of the Suguta Trough has already created ethnic tension between the Samburu and Turkana communities. In addition, the devolved governance in Kenya creates tension and fuels boundary violent conflict due to potential resources within county boundaries such as the Suguta Trough. Thus, the already developing geothermal projects have proven to be environmentally insecure and are risks to pastoral livelihood, and the potential ones, Suswa and Suguta Trough, will risk the same. These developments attest to the government's focus on sustainable economic development with limited social inclusion; the subsidiary practice concentrates on a top-down approach to environmental security but ignores the inclusive, bottom-up approaches that point to environmental risks likely to affect the ecosystem.

NOTES

1. S Buzan, Barry, Ole Wæver, and Jaap De Wilde. *Security: a new framework for analysis*: Lynne Rienner Publishers. 1998.

2. Luthra, Sunil, Sanjay Kumar, Dixit Garg, and Abid Haleem. "Barriers to renewable/sustainable energy technologies adoption:Indian perspective." *Renewable and Sustainable Energy Reviews* 41 (2015): 762–776. doi: https://doi.org/10.1016 /j.rser.2014.08.077; see also, Greaves, Wilfrid. "Arctic (in)security and indigenous peoples: comparing Inuit in Canada and Sámi in Norway." *Security Dialogue* 47, no. 6 (2016): 461–480. doi: 10.1177/0967010616665957.

3. Renkens, Ilse. "The impact of renewable energy projects on indigenous communities in Kenya: The cases of the Lake Turkana Wind Power project and the Olkaria Geothermal Power plants." *International Work Group for Indigenous Affairs (IWGIA)*, 2019.

4. Kazimierczuk, Agnieszka H. "Wind energy in Kenya: A status and policy framework review." Renewable and Sustainable Energy Reviews, 107, (2019): 434–445. https://doi .org /10 .1016 /j .rser .2018 .12 .061.

5. Mwangi, Martin N. "Financing geothermal projects." *Short Course on Geothermal Project Management Development*, 2008.

6. GDC. *Anual report and financial statement*. Geothermal Development Company, 2015.

7. JICA. "Promoting development in Kenya: the growth Path of Japanese ODA." *Japan International Cooperation Agency*, 2021.Accessed February 27, 2023. https://www.jica.go.jp/kenya/english/office/others/c8h0vm0000f0mtzd-att/brochure _01.pdf.

8. Richter, Alexander. "KenGen brings the first unit of its 173 MW Olkaria V geothermal power plant online, adding 86.6 MW of geothermal power generation to the grid in Kenya. The second unit of the same size is to come online soon as well." *Think Geoenergy*, 2019. Accessed February 24, 2023. https://www.thinkgeoenergy.com/kengen-announces-86-6-mw-of-olkaria-v-unit-1-geothermal-plant-to -be-online/.

9. Mbali, Kenneth Kivisi, and Okonda J. Justus. "Kenya's nuclear energy dilemma." *International Journal of Nuclear Safety and Security* 1, no. 2 (2022): 137–149. https://doi.org/10.1504/IJNSS.2022.127979.

10. GoK. *Report of the Auditor General on the Kenya power and lightening company*, edited by Office of the Auditor General. Government of Kenya, 2020, 55..

11. Ministry of Energy. *Kenya national electrification strategy: Key highlights.* Nairobi Ministry of Energy, 2018.

12. GoK (2020).

13. KNBS. *Economic Survey 2019.* Kenya National Bureau of Statistics, 2019.

14. Schade, Jeanette. "Kenya 'Olkaria IV' case study report: human rights analysis of the resettlement process, 2017." https://nbn-resolving.org/urn:nbn:de:0168-ssoar-51409-6.

15. Sena, Kanyinke. "Renewable energy projects and the rights of marginalised/indigenous communities in Kenya." *International Working Group for Indigenous Affairs Report* 21, 2015.

16. Renkens (2019).

17. GoK. *Land (Amendment bill) 2022. In 165,* edited by Government of Kenya Kenya Gazette Suppliment. Government printers, Nairobi, 2022..

18. Myers, Norman. "Ultimate security: The environmental basis of political stability." *New York* 8, 1993.

19. Homer-Dixon, Thomas F. "On the threshold:environmental changes as causes of acute conflict." *International Security* 16, no. 2 (1991): 76–116. http://www.jstor.org/stable/2539061.

20. Haas, Peter M., Robert Owen Keohane, Marc A.Levy, and Les Gasser. *Institutions for the earth: sources of effective international environmental protection.* MIT Press, 1993.

21. Buzan et al.. (1998).

22. Heilbroner, Robert L. *An inquiry into the human prospect: looked at again for the 1990s*:WW Norton & Company, 1991; See also, Homer-Dixon (1991).

23. Homer-Dixon, T. "Environment, scarcity,and violence Princeton University Press." *Princeton NJ*, 1999.

24. Deudney, Daniel. "The case against linking environmental degradation and national security." *Millennium* 19, no. 3 (1990): 461–476. https://doi.org/10.1177/03058298900190031001. See also, Homer-Dixon (1999), and Barnett, Jon. *The meaning of environmental security: ecological politics and policy in the new security era.* Zed Books, 2001..

25. Floyd, Rita. *Security and the environment: .securitisation theory and US environmental security policy* Cambridge University Press, 2010.

26. See, for example, Taylor, Paul W. *Respect for nature: a theory of environmental ethics* 51. Princeton University Press, 2011.

27. Hardt, Judith Nora. *Environmental security in the anthropocene: assessing theory and practice*: Routledge, 2017.

28. Barnett (2001).

29. Dalby, Simon. *Environmental security.* Vol. 20, U of Minnesota Press, 2002, Barnett (2001).

30. Busby, Joshua. *Environmental security.* Oxford University Press, Oxford, 2018.

31. Buzan et al. (1998).

32. Buzan, Barry. *People, states & fear: an agenda for international security studies in the post-cold war era.* ECPR Press 2008..

33. Dalby (2002).

34. Makarieva, Anastassia M., Victor G. Gorshkov, and Bai-Lian Li. "Energy budget of the biosphere and civilization: rethinking environmental security of global renewable and non-renewable resources." *Ecological Complexity* 5, no. 4 (2008): 281–288. https://doi.org/10.1016/j.ecocom.2008.05.005.

35. Floyd, Rita, and Richard A. Matthew. *Environmental security.* Routledge, London,2013, 249.

36. Hughes, Lotte. "Moving the Maasai: a colonial misadventure." *St Antony's Series. Basingstoke, UK: Palgrave Macmillan,* 2006. https://doi.org/10.1057/9780230246638.

37. Zeleza, Tiyambe. "Economic policy and performance in Kenya since independence." *Transafrican Journal of History* 20 (1991): 35–76. https://www.jstor.org/stable/24520302.

38. Kennedy, Osoro. 2013. "Determinants of balance of payments in Kenya." *European Scientific Journal* 9, no. 16 (2013): 1857–7431.

39. See, ASALs in Njoka, Jesse T., Pius Yanda, Faustin Maganga, Emma Liwenga, Adolphine Kateka, Abdallah Henku, Edmund Mabhuye, Nico Malik, and Cynthia Bavo. "Kenya: country situation assessment." *Cilia Schubert,* 2016. Accessed February 24, 2023.

40. See, Government of Kenya. "Kenya's vision 2030, famous version." (2007). https://vision2030.go.ke/publication/kenya-vision-2030-popular-version/.

41. See, GoK. "Sessional Paper No. 8 of 2012 on national policy for the sustainable development of Northern Kenya and other Arid lands." Ministry of State for Development of Northern Kenya and other Arid Lands. Accessed February 24, 2023.

42. Agamben, Giorgio. "State of exception." In *State of Exception.* University of Chicago Press, Chicago, 2008.

43. UNEP. Green economy assessment report. edited by United Nation Environmental Programme (UNEP). United Nations Environment Programme, 2014.

44. World Bank. "World Bank Group Country Survey 2015." *World Bank* (2015): 1–150. https://doi.org/10.48529/a768-z546.

45. GoK. *National climate change action plan (NCCAP) 2018–2022*, edited by Ministry of Environment and Forestry. Government of Kenya, Nairobi, 2018.

46. See, for example, Environmental Management and Co-ordination Act 1999 and the Climate Change Act 2016, stipulating Climate action and sustainable development.

47. Mallowah, Stephen, and Christopher Oyier."The Environment and Climate Change Law Review: Kenya." *The Law Review*, 2023. Accessed February 27, 2023. https://thelawreviews.co.uk/title/the-environment-and-climate-change-law-review/kenya#:~:text=The%20main%20object%20of%20the,the%20sustainable%20development%20of%20Kenya..

48. Sena (2015), Koissaba, Ben R. Ole. "Geothermal energy and indigenous communities." 2017. https://eu.boell.org/sites/default/files/geothermal-energy-and-indigenous-communities-olkariaproject-kenya.pdf, Koissaba, Ben R. Ole. "Geothermal energy and indigenous communities: the Olkaria Projects in Kenya." *Report for Heinrich Böll Stiftung the European Union.* Brussels, 2018a., Renkens (2019).

49. Leal Filho, Walter, Zeus Guevara, Ayyoob Sharifi, Dinesh Surroop, Salil K. Sen, Carelle Mang-Benza, Kevin Ovita Teddy,Samuel Awoniyi, Usman Muhammad, and Cosmos Nike Nwedu. "Energy poverty in African countries: an assessment of trends and policies." *SSNR,* 2022.http://dx.doi.org/10.2139/ssrn.4229621.

50. Bertani, Ruggero. "Geothermal power generation in the world 2005–2010 update report." *Geothermics* 41 (2012): 1–29. https://doi.org/10.1016/j.geothermics.2011.10.001.

51. Richter (2019).

52. Kenya National Electrification Strategic plan, 2018–2022.

53. Kombe, Emmanuel Yeri, and Joseph Muguthu. "Geothermal energy development in East Africa: barriers and strategies." *Journal of Energy Research and Reviews* 2 (2019): 1–6. doi:10.9734/JENRR/2019/45278.

54. Bw'Obuya, Nicholas Martin. "The socio-economic and environmental impact of eothermal energy on the rural poor in Kenya: the impact of a geothermal power plant on a poor rural community in Kenya." Sida/SAREC and AFREPREN/FWD, 2002.

55. Hughes, Lotte, and Daniel Rogei. "Feeling the heat: responses to geothermal development in Kenya's Rift Valley." *Journal of Eastern African Studies* 14, no. 2 (2020): 165–184. https://doi.org/10.1080/17531055.2020.1716292.

56. UNESCO. "The African Great Rift Valley - Hell's Gate National Park." UNESCO World Heritage Convention, 2010. Accessed February 27, 2023. https://whc.unesco.org/en/tentativelists/5511/.

57. Dayan, George Moshe. *Geothermal Drilling at Olkaria Fields Lake Bogoia and Lake Naivasha.* Kenya Electricity Generating Company, Kenya, 2018.

58. Koissaba, Ben R. Ole. Campaign Update: Kenya-Maasai protest against new land concessions for geothermal extraction in Kenya, 2019.

59. Rogei, Daniel. "Mediating maendeleo: examining the nexus between geothermal extraction, wildlife conservation and community well-being in Olkaria-Suswa, Southern Kenya." *Carleton University,* 2021.

60. Renkens (2019).

61. Schade (2017).

62. Rogei (2021).

63. Renkens (2019); see also Kong'ani, Lilian Namuma S., Raphael G. Wahome,and Thuita Thenya. "Variety and management of developmental conflicts: the case of the Olkaria IV geothermal energy project in Kenya." *Conflict, Security & Development* 21, no. 6 (2021): 781–804. https://doi.org/10.1080/14678802.2021.2000806.

64. Ibid, Kazimierczuk (2019).

65. Koissaba, Ben R. Ole. *The Olkaria projects: A case study of geothermal energy and indigenous communities in Kenya.* Heinrich Boell Stiftung, Brussels, 2018b.

66. ITA. "Kenya -country commercial guide: energy -electricity power systems." *International Trade Administration*, 2022. Accessed February 27, 2023. https://www.trade.gov/country-commercial-guides/kenya-energy-electrical-power-systems#:~:text=The%20access%20rate%20for%20urban,rural%20Kenya%20stands%20at%2065%25.

67. Fetzek, Shiloh. "Geothermal expansion and Maasai land conflicts in Kenya." *Resillience Compass*, 2015.

68. Koissaba (2018a).

69. ITA (2022).

70. Fetzek (2015); see also, Sena (2015); Koissaba (2017, 2018a, 2018b, 2019); and Renkens (2019).

71. Conference on Implementation of the Community Land Act 2016 Sponsored by Ministry of Lands and Physical Planning (MoLPP), French Institute for Research in Africa (IFRA) and Food and Agriculture Organization of the United Nations (FAO), June 13–14, 2022.

72. Interviews conducted on Land registry, February 2023.

73. Interview in Samburu, February 2023.

74. Narok County Government. *Narok county intergrated development plan 2018-2023*. Narok County Government, 2018.

75. Nash, W. P., I. S. E. Carmichael, and R. W. Johnson. "The mineralogy and petrology of Mount Suswa, Kenya." *Journal of Petrology* 10, no. 3 (1969): 409–439. https://doi.org/10.1093/petrology/10.3.409; Johnshon, RW. "Volcanic Geology of Mount Suswa, Kenya." Philosophical Transactions of the Royal Society of London. Series A, Mathematical and Physical Sciences 265, no. 1164 (1969): 383–412. https://doi.org/https://doi.org/10.1098/rsta.1969.0061.; and Skilling, I. P. "Incremental caldera collapse of Suswa volcano, Gregory rift valley, Kenya." *Journal of the Geological Society* 150, no. 5 (1993): 885–896. https://doi.org/10.1144/gsjgs.150.5.0885.

76. Narok County Government (2018).

77. The Sessional Paper No. 4, 2004, divided the energy sector to different subsectors, and GDC was formulated under the Energy Act No. 12 of 2006, https://www.gdc.co.ke/.

78. GDC. *Annual report and financial statement*. Geothermal Development Company, Nairobi, 2019.

79. GDC (2019).

80. Interviews at GDC offices, February 2023.

81. Interviews at GDC offices, February 2023.

82. Interview with a geothermal expert and engineer, February 28, 2023.

83. REDPLAN. Environmental and Social impact Assessment (ESIA) Study report; Suswa Geothermal Development Project in Narok and Kajiado Counties. Research on Environment and Development Planning (REDPLAN) Consultants Ltd (n.d.).

84. Interviews at Mount Suswa Conservancy, August 2021.

85. Interviews at Naiburta campsite, Hell's Gate, September 2022.

86. Gitonga, Antony, and Kibiwott Koross. "Is hells gate geo-power a blessing or curse?" *The Standard Newspaper*, 2015. Accessed on February 28, 2023. https://www.standardmedia.co.ke/features/article/2000119744/is-hells-gate-geo-power-a-blessing-or-curse.

87. Gitonga and Koross (2015).

88. Interviews in Ngong town, May 2022.

89. Interviews with communities in Suswa, June to August 2022.

90. Interviews with teachers, August 2022.

91. See, for example, Sessional Paper No. 10 of 1965, and Sessional Paper No. 1 of 1986, stipulating economic development in highlands and neglecting pastoral lands in Kenya.

92. Sessional Paper No. 8 of 2012.

93. Agamben (2008).

94. Mwebe, John, and Elias Jika. "Communities in Kenya demand better engagement on geothermal power plant project." *IPA*, 2018. Accessed February 26, 2023. https://accountability.medium.com/communities-in-kenya-demand-better-engagement-on-geothermal-power-plant-project-4290f7f32194.

95. Interviews at Kisharu village in Suswa, June 2022; Interviews with GDC offices, February 2023.

96. Interview with a geothermal expert and engineer, February 28, 2023.

97. Interview with a geothermal expert and engineer, February 28, 2023.

98. Interviews with community members in Kisharu village in Suswa, June 2022; Interviews with GDC officials in Nairobi, February 2023.

99. Interview with a geothermal expert and engineer, February 28, 2023.

100. Dunkley, P. N., M. Smith, D. J. Allen, and W. G. Darling. "The geothermal activities and geology of the northern sector of the Kenya Rift Valley." In *Research report SC/93/1*. British Geological Survey, Nottingham. 1993; See also, JICA (2010).

101. Dunkley et al. (1993).

102. Fieldwork experience in Nachola and Suguta Valley, October 2022.

103. Okwany, Clifford Collins Omondi, and Evelyne Atieno Owino. "Kapedo deathtrap: a semi-territoriality advantage to the bandits." *The Star,* October 15, 2022; , Okwany, Clifford Collins Omondi."Territoriality as a method for understanding armed groups in Kenya and strengthening policy responses." *United States Institute of Peace,* 2023. https://doi.org/10.37805/pn2023.1.lpbi.

104. Manga, Beauttah. "Valley of death that claimes lives of top officers." *The Standard Newspaper*, 2012. Accessed February 23, 2023. https://www.standardmedia.co.ke/rift-valley/article/2000070537/valley-of-death-that-claims-lives-of-top-officers.

105. Okwany and Owino (2022), Okwany (2023), Okwany, Clifford Collins Omondi, Evelyne Atieno Owino, and Zedekiah Opondo Sidha. "The nature of armed groups in Kenya in the context of territoriality." *Amani Journal*, 2023.

106. Interviews in Tuum, Masikita, and Bendera, October 2022.

107. Interviews in Nachola and Suguta Valley, October 2022.

108. K24TV. "Baragoi geothermal standoff: county government urged to intervene." *K24TV*, 2018. Accessed April 2, 2022. https://www.youtube.com/watch?v=fr2Vwooyn64.

109. Interviews in Turkana and Samburu County offices conducted in October 2022 and January 2023.

110. Fieldwork experience from 2021 to 2024 in both Turkana, Baringo, West-Pokot, and Samburu Counties.

111. Schade (2017), Koissaba (2018a, 2018b), Renkens (2019), Kong'ani, Wahome, and Thenya (2021) and Sena (2015).

112. Leal Filho et al. (2022).

113. Interviews in Kisharu Village in Suswa, June 2022.

114. Interviews with tour guides at Mount Suswa Conservancy, August 2021.

115. Interviews at GDC offices, February 2023.

116. Renkens (2019) and Kong'ani, Wahome, and Thenya (2021); Interviews with tour guides at Naiburta campsite, Hell's Gate, September 2022.

REFERENCES

Agamben, Giorgio. "State of Exception." In State of Exception, 1–95. Chicago: University of Chicago Press., 2008. https://doi.org/10.7208/9780226009261

Barnett, Jon. *The meaning of environmental security: Ecological politics and policy in the new security era,* London, UK:Zed Books, 2001.

Bertani, Ruggero. 2012. "Geothermal power generation in the world 2005–2010 update report." *Geothermics* 41, (2012):1–29. https://doi.org/10.1016/j.geothermics.2011.10.001.

Busby, Joshua. *Environmental security,* Oxford, UK: Oxford University Press, 2018.

Buzan, Barry. *People, states & fear: An agenda for international security studies in the post-cold war era,* Brighton, Sussex: European Consortium for Political Research (ECPR) Press, 2008.

Buzan, Barry, Ole Wæver, and Jaap De Wilde. *Security: A new framework for analysis*, Boulder, CO: Lynne Rienner Publishers 1998.

Bw'Obuya, Nicholas Martin. *The socio-economic and environmental impact of geothermal energy on the rural poor in Kenya*: AFREPREN/FWD, 2002.

Dalby, Simon. *Environmental security.* Vol. 20, Minneapolis, MN: University of Minnesota Press, 2002.

David-Chavez, Dominique M, and Michael C Gavin. "A global assessment of Indigenous community engagement in climate research." *Environmental Research Letters* 13, no.12 (2018): 123005. https://doi.org/10.1088/1748-9326/aaf300.

Dayan, George Moshe. *Geothermal drilling at Olkaria fields Lake Bogoia and Lake Naivasha*, Kenya: Kenya Electricity Generating Company, 2018.

Deudney, Daniel 1990. "The case against linking environmental degradation and national security." *Millennium* 19, no. 3 (1990): 461–476. https://doi.org/10.1177/03058298900190031001.

Dunkley, P. N., M. Smith, D. J. Allen, and W. G. Darling. "The geothermal activities and geology of the northern sector of the Kenya rift valley." In *Research report SC/93/1*. Nottingham: British Geological Survey, 1993.

"Geothermal Expansion and Maasai Lan D Conflicts in Kenya." Resillience Compass, Climate Diplomacy, 2015, accessed December 12, 2023.

Floyd, Rita. *Security and the environment: Securitisation theory and US environmental security policy,* Cambridge: Cambridge University Press, 2010.

Floyd, Rita, and Richard A Matthew. *Environmental security,* London: Routledge, 2013.

GDC. *Anual Report and Financial Statement,* Nairobi, Kenya: Geothermal Development Company, 2015.

GDC. *Annual report and Financial Statement,* Nairobi, Kenya: Geothermal Development Company, 2019.

Gitonga, Antony, and Kibiwott Koross. 2015. "Is Hells Gate geo-power a blessing or curse?." *The Standard Newspaper*, accessed on February 28, 2023. https://

www.standardmedia.co.ke/features/article/2000119744/is-hells-gate-geo-power-a-blessing-or-curse.
GoK. "Sessional Paper No. 8 of 2012 on National Policy for the Sustainable Development of Northern Kenya and other arid lands," 2012. Minitry of State for Development of Northern Kenya and other Arid Lands, accessed on February 24, 2023. https://www.ndma.go.ke/index.php/resource-center/policy-documents/send/44-policy-documents/4305-sessional-paper-on-national-policy-for-development-of-asals.
GoK. *National Climate Change Action Plan (NCCAP) 2018–2022 edited by Ministry of Environment and Forestry*, Nairobi: Government of Kenya, 2018.
GoK. *Report of the Auditor General on the Kenya Power and Lightening Company edited by Office of the Auditor General*, Government of Kenya, 2020.
GoK. "Land (Amendment Bill) 2022." In *165*, edited by Government of Kenya Kenya Gazette Suppliment, Nairobi: Government Printers, 2022.
Greaves, Wilfrid. "Arctic (in)security and Indigenous peoples: Comparing Inuit in Canada and Sámi in Norway." *Security Dialogue* 47, no. 6 (2016): 461–480. https://doi.org/10.1177/0967010616665957.
Haas, Peter M, Robert Owen Keohane, Marc A Levy, and Les Gasser. *Institutions for the earth: Sources of effective international environmental protection,* MIT Press, 1993.
Hardt, Judith Nora. *Environmental security in the Anthropocene: Assessing theory and practice,* Routledge, 2017.
Heilbroner, Robert L. *An inquiry into the human prospect: Looked at again for the 1990s*, WW Norton & Company, 1991.
Homer-Dixon, T. *Environment, scarcity, and violence Princeton University Press*, Princeton NJ., 1999.
Homer-Dixon, Thomas F 1991. "On the threshold: Environmental changes as causes of acute conflict." *International security* 16, no. 2: 76–116. http://www.jstor.org/stable/2539061.
Hughes, Lotte. *Moving the Maasai: A colonial misadventure,* St Antony's Series. Basingstoke, UK: Palgrave Macmillan, 2006. https://doi.org/10.1057/9780230246638.
Hughes, Lotte, and Daniel Rogei. "Feeling the heat: Responses to geothermal development in Kenya's Rift Valley." *Journal of Eastern African Studies* 14, no. 2 (2020): 165–184. https://doi.org/10.1080/17531055.2020.1716292.
ITA. "Kenya -country commercial guide: Energy -electricity power systems." I*nternational Trade Administration*, 2022. Accessed on February 27, 2023. https://www.trade.gov/country-commercial-guides/kenya-energy-electrical-power-systems#:~:text=The%20access%20rate%20for%20urban,rural%20Kenya%20stands%20at%2065%25.
JICA. *Situation analysis study on geothermal development in Africa*, edited by Japan International Cooperation Agency: West Japan Engineering Consultants, 2010.
JICA. "Promoting development in Kenya: The growth path of Japanese ODA." *Japan International Cooperation Agency*, 2021. Accessed on February 27, 2023. https://www.jica.go.jp/kenya/english/office/others/c8h0vm0000f0mtzd-att/brochure_01.pdf.

Johnshon, R. W. "Volcanic geology of Mount Suswa, Kenya." *Philosophical Transactions of the Royal Society of London. Series A, Mathematical and Physical Sciences* 265, no. 1164 (1969): 383–412. https://doi.org/10.1098/rsta.1969.0061.

K24TV. "Baragoi geothermal standoff: County government urged to Intervene." *K24TV*, 2018. Accessed on April 2, 2022. https://www.youtube.com/watch?v=fr2Vwooyn64.

Kazimierczuk, Agnieszka H. "Wind energy in Kenya: A status and policy framework review." *Renewable and Sustainable Energy Reviews* 107, (2019): 434–445. https://doi.org/10.1016/j.rser.2018.12.061.

Kennedy, Osoro. "Determinants of balance of payments in Kenya." *European Scientific Journal* 9, no. 16 (2013): 1857–7431.

KNBS. "Economic Survey," 1–324. Herufi House, Lt. Tumbo Avenue , Nairobi: Kenya National Bureau of Statistics, 2019.

Koissaba, Ben R. Ole. "Geothermal energy and indigenous communities," 2017. https://eu.boell.org/sites/default/files/geothermal-energy-and-indigenous-communities-olkariaproject-kenya.pdf.

Koissaba, Ben R. Ole. "Geothermal Energy and Indigenous Communities: The Olkaria Projects in Kenya." *Report for Heinrich Böll Stiftung and the European Union. Brussels* (2018).

Koissaba, Ben R. Ole. *The Olkaria projects: A case study of geothermal energy and indigenous communities in Kenya*. Brussels: Heinrich Boell Stiftung, 2018b. Retrieved from https://eu.boell.org.

"Campaign Update: Kenya-Maasai Protest against New Land Concessions for Geothermal Extraction in Kenya." Cultural Survival 2019, accessed November 5, 2023.

Kombe, Emmanuel Yeri, and Joseph Muguthu. "Geothermal energy development in East Africa: Barriers and strategies." *Journal of Energy Research and Reviews* 2, (2019): 1–6. https://doi.org/10.9734/JENRR/2019/45278.

Kong'ani, Lilian Namuma S, Raphael G Wahome, and Thuita Thenya. 2021. "Variety and management of developmental conflicts: The case of the Olkaria IV geothermal energy project in Kenya." *Conflict, Security & Development* 21, no. 6 (2021): 781–804. https://doi.org/10.1080/14678802.2021.2000806.

Leal Filho, Walter, Zeus Guevara, Ayyoob Sharifi, Dinesh Surroop, Salil K Sen, Carelle Mang-Benza, Kevin Ovita Teddy, Samuel Awoniyi, Usman Muhammad, and Cosmos Nike Nwedu. "Energy Poverty in African Countries: An Assessment of Trends and Policies." *SSNR* (2022). http://dx.doi.org/10.2139/ssrn.4229621.

Luthra, Sunil, Sanjay Kumar, Dixit Garg, and Abid Haleem. "Barriers to renewable/sustainable energy technologies adoption: Indian perspective." *Renewable and sustainable energy reviews* 41, (2015): 762–776. https://doi.org/10.1016/j.rser.2014.08.077.

Makarieva, Anastassia M, Victor G Gorshkov, and Bai-Lian Li. "Energy budget of the biosphere and civilization: Rethinking environmental security of global renewable and non-renewable resources." *Ecological complexity* 5, no. 4 (2008): 281–288. https://doi.org/10.1016/j.ecocom.2008.05.005.

Mallowah, Stephen, and Christopher Oyier. "The environment and climate change law review: Kenya," 2023. *The Law Review*. Accessed on February 27, 2023.

https://thelawreviews.co.uk/title/the-environment-and-climate-change-law-review/kenya#:~:text=The%20main%20object%20of%20the,the%20sustainable%20development%20of%20Kenya.

Manga, Beauttah. "Valley of death that claimes lives of top officers." *The Standard Newspaper*, 2012. Accessed on February 23, 2023. https://www.standardmedia.co.ke/rift-valley/article/2000070537/valley-of-death-that-claims-lives-of-top-officers.

Mbali, Kenneth Kivisi, and Okonda J Justus. "Kenya's nuclear energy dilemma." *International Journal of Nuclear Safety and Security* 1, no. 2 (2022): 137–149. https://doi.org/10.1504/IJNSS.2022.127979.

Ministry of Energy. *Kenya national electrification strategy: Key highlights,* Nairobi Ministry of Energy, 2018.

Mwangi, Martin N "Financing Geothermal Projects." In Short Course on Geothermal Project Management Development. Imperial Botanical Beach Hotel, Entebbe, Uganda: UNU-GTP, KenGen and MEMD-DGSM, 2008.

Mwebe, John, and Elias Jika. "Communities in Kenya demand better engagement on geothermal power plant project." *IPA*, 2018. Accessed on February 26, 2023. https://accountability.medium.com/communities-in-kenya-demand-better-engagement-on-geothermal-power-plant-project-4290f7f32194.

Myers, Norman. "Ultimate security: The environmental basis of political stability." *New York* 8, 1993.

Narok County Government. *Narok County Intergrated Development Plan 2018–2023*, Narok County Government, 2018.

Nash, WP, ISE Carmichael, and RW Johnson. "The mineralogy and petrology of Mount Suswa, Kenya." *Journal of Petrology* 10, no. 3 (1969): 409–439. https://doi.org/10.1093/petrology/10.3.409.

Njoka, Jesse T., Pius Yanda, Faustin Maganga, Emma Liwenga, Adolphine Kateka, Abdallah Henku, Edmund Mabhuye, Nico Malik, and Cynthia Bavo. 2016. "Kenya: Country situation assessment." *Cilia Schubert*, accessed on February 24, 2023. https://idl-bnc-idrc.dspacedirect.org/bitstream/handle/10625/58566/IDL-58566.pdf.

Odhiambo, Allan. "US firm injects 29MW geothermal power into grid." *Nation Newspaper*, 2016. Accessed on February 2, 2023. https://www.businessdailyafrica.com/US-firm-injects-29MW-geothermal-power-into-grid/-/539546/3067644/-/90fl3tz/-/index.html.

Okwany, Clifford Collins Omondi. "Territoriality as a method for understanding armed groups in Kenya and strengthening policy responses." *United States Institute of Peace,* 2023. https://doi.org/10.37805/pn2023.1.lpbi.

Okwany, Clifford Collins Omondi , Evelyne Atieno Owino, and Zedekiah Opondo Sidha. "The Nature of Armed Groups in Kenya in the Context of Territoriality." In Peace and Security in Africa's Borderlands 73–94. Nairobi Amani Journal, International Peace Support Training Centre, 2023.

Okwany, Clifford Collins Omondi, and Evelyne Atieno Owino. 2022. "Kapedo Deathtrap: A semi-territoriality advantage to the bandits." *The Star,* 15th October 2022.

Patel, Sonal. "Olkaria Geothermal Expansion Project, rift valley province, Kenya." *Power*, 2015. Accessed on February 15, 2023. https://www.powermag.com/olkaria-geothermal-expansion-project-rift-valley-province-kenya/.

PowerTech. "Kenya electricity generating company Olkaria II expansion." *Power Technology*, 2009. Accessed on May 27, 2022. https://www.power-technology.com/projects/olkariageothermal/.

REDPLAN. Environmental and social impact assessment (ESIA) study report; Suswa Geothermal Development Project in Narok and Kajiado counties. Research on Environment and Development Planning (REDPLAN) Consultants Ltd.

Renkens, Ilse. *The impact of renewable energy projects on indigenous communities in Kenya: The cases of the Lake Turkana Wind Power project and the Olkaria Geothermal Power plants*. International Work Group for Indigenous Affairs (IWGIA), 2019.

Richter, Alexander. "KenGen brings the first unit of its 173 MW Olkaria V geothermal power plant online, adding 86.6 MW of geothermal power generation to the grid in Kenya. The second unit of the same size is to come online soon as well." *Think Geoenergy*, 2019. Accessed February 24, 2023. https://www.thinkgeoenergy.com/kengen-announces-86-6-mw-of-olkaria-v-unit-1-geothermal-plant-to-be-online/.

Rogei, Daniel. "Mediating maendeleo: Examining the nexus between geothermal extraction, wildlife conservation and community well-being in Olkaria-Suswa, Southern Kenya." *Carleton University*, 2021.

Schade, Jeanette. "Kenya 'Olkaria IV' case study report: Human rights analysis of the resettlement process," 2017. https://nbn-resolving.org/urn:nbn:de:0168-ssoar-51409-6.

Sena, Kanyinke. "Renewable Energy Projects and the Rights of Marginalised/Indigenous Communities in Kenya," 1–24. Eks-Skolens Trykkeri, Copenhagen, Denmark: International Working Group for Indigenous Affairs (IWGIA) and Indigenous Peoples National Steering Committee on Climate Change (IPNSCCC), 2015.

Skilling, IP. "Incremental caldera collapse of Suswa volcano, Gregory rift valley, Kenya." *Journal of the Geological Society* 150, no. 5 (1993): 885–896. https://doi.org/10.1144/gsjgs.150.5.0885.

Taylor, Paul W. *Respect for nature: A theory of environmental ethics* 51, Princeton University Press, 2011.

UNEP. *Green economy assessment report edited by United Nation Environmental Programme (UNEP)*, United Nations Environment Programme, 2014.

UNESCO. "The African great rift valley - Hell's Gate National Park." *UNESCO World Heritage Convention, 2010*. Accessed February 27, 2023. https://whc.unesco.org/en/tentativelists/5511/.

World Bank. "World Bank Group Country Survey 2015." *World Bank* (2015): 1–150. https://doi.org/10.48529/a768-z546.

Zeleza, Tiyambe. "Economic policy and performance in Kenya since independence." *Transafrican Journal of History* (1991): 35–76. https://www.jstor.org/stable/24520302.

Chapter 15

Mangrove Forest Management in Africa

Ufuoma Efe Akpovwovwo

Mangroves are distinct from other coastal vegetation types with a multiplicity of ecological services. Their uniqueness is depicted basically in their physiology, adaptability, and general life cycle. Globally, they are known to thrive basically in the tropical and subtropical regions, within latitudes of 32⁰N and 38⁰S.[1] They occur usually in coastal regions, within locations which are receptive to freshwater from the rivers in the hinterlands and salt water from the sea. This culminates in brackish water, with sediments being the underlying substrate. Tidal influence, emanating from daily tidal-originated inundations, is a determinant of behavior and activities of these waters. These activities characterize the mangrove terrain as the intertidal zone. This unique positioning is a basis for its nutrient-rich status owing to nutrient influx both from the sea and hinterland via the rivers.

Mangrove forests can be categorized on the basis of environmental geomorphology and structure, in the following manner: (i) basin mangroves—usually occur as a zone behind the riverine and fringing mangroves, being connected indirectly to the sea; (ii) dwarf mangroves—appear stunted and are characterized by regular tidal inundations; (iii) hammock mangroves—develop on substrate of prior developed peat; (iv) fringing mangroves—characteristic location adjoining the sea serves as an impetus for consistent and regular tidal impact; (v) riverine mangroves—occur usually in close proximity with major rivers; and (vi) overwash mangroves—consist mainly of secluded mangrove stands, however they are restricted to a few mangrove species.[2]

Mangrove forests can also be zoned based on the degree/manner of tidal immersion, thereby serving as a determinant of plant and animal occurrence within the area of the mangrove forest in question. Sections of mangrove forests characterized by low tidal inundation enjoy consistent immersion in water. Rhizophoraceae species are adapted to consistent immersion in water and, therefore, occur more within this zone. Other sections are, however, only immersed during high tide. *Avicennia spp.* are adapted to immersion only during high tide.[3] This further influences the unique adaptations attributed to the various species, which include salt-excreting leaves, stilt roots, viviparity, and aerial roots.

MANGROVE FOREST LOCATIONS IN AFRICA

The total mangrove-covered area in Africa is put at 25,960 square kilometers.[4] Mangroves occur on the Atlantic Coast (Western and Central Africa) and the Coast of the Indian Ocean (Eastern Africa, South Africa, and Madagascar) of Africa. In 1999/2000, the mangrove-covered area in Nigeria ranked the largest with an area of 8,572 square kilometers and Mauritania the lowest, with an area of 0.4 square kilometers. Mangrove forests in Africa occupy an area of about 3.2 million hectares, representing approximately 22 percent of the world's mangroves. Africa is rated as having the second largest mangrove area, after Asia. Furthermore, about 19 percent of global mangrove forests occur in Africa.[5] Mangrove forests in Nigeria form about 7 percent of the total mangrove forests in the world, ranking third after Indonesia and Brazil.[6] The western Atlantic region harbors 37 percent of the continental mangrove forests, while the central Atlantic and the Eastern Indian Ocean mangrove forests harbor about 14 percent each, respectively.

In West Africa, mangrove forests occur majorly in the following countries: Gabon, Cameroon, Guinea, Guinea Bissau, and Nigeria.[7] Mangroves also occur within the coastlines of Liberia in West Africa and Angola in Central Africa.[8] Furthermore, the presence of mangroves is also recorded in Senegal, Gambia, Sierra Leone, Liberia, Cote D'Ivoire, Ghana, and Benin.[9] In addition to Nigeria's mangrove forest area ranking the largest, its canopy height is relatively higher than those of Benin and Togo. It has also been established that the mangroves of the Niger Delta of Nigeria rank second in area coverage globally, after that of Bangladesh.[10]

The mangrove forests in Guinea, Guinea Bissau, Gambia, and Senegal are observed to, however, extend further inland than those of the aforementioned.[11] Mangroves are also known to occur on some islands, off the coast of western Africa, such as Tristão islands in the north of Guinea and the Bijagós Archipelago.[12] Mangroves also occur in the northern part of West Africa in Mauritania. The mangrove forests of areas in North Africa, including Somalia, Djibouti, and Eritrea, are generally scanty.[13]

In East Africa, which is also referred to as the Western Indian Ocean zone, mangroves occur in Kenya, Tanzania, Madagascar, and Mozambique. The

Western Indian Ocean zone comprises of the following countries: Comoros Island, Kenya, Madagascar, Mauritius, Mozambique, Seychelles, Somalia, South Africa, and Tanzania.[14]

A major portion of mangroves in Madagascar is reported to occur more within its west coast than the east coast, owing to the relatively drier climate governing the east coast.[15] In Tanzania, the Rufiji Delta contains the most extensive mangrove forest, with other portions occurring in Kilwa, Mtwara, and Mnazi bay.[16] The largest expanse of mangrove forest in Kenya occurs in Lamu district, while other patches can be found in Kwale, Kilifi, Tana River, and Mombasa districts. The largest mangrove forest area in Mozambique occurs within the Zanzibar River basin. Other mangrove forest patches occur within the creeks of Lumbo, Mecuti, Ibo island, and between the areas of the banks of the Beira and Save Rivers.[17] Mangrove forests in Mozambique are the third largest in the area, and first along the coastline countries of East Africa, having an area of about 3000 square kilometers.[18] The mangrove forest area in Zambia is one of the largest in East Africa.[19] Mangrove regeneration of the extreme locations of Mauritania[20] and South Africa is generally slow, culminating in stunted growth. The largest expanse of mangroves in Seychelles occurs in Astove, Cosmoledo, and Aldabra.

The northern limit of mangroves in Africa occurs in the Parc National du Banc d'Arguin in Mauritania, forming a discontinuous distribution comprising mainly of *Avicennia germinans.* In Somalia, mangroves occur along the Kisimayo coast, and around the Caanoole and Bushbush estuaries.[21] Mangrove forests in South Africa are more or less subtropical, with their largest expanse occurring within the creeks of Mhlathuze and St Lucia.[22] Mangroves within the northern extremes and the southern extremes are characterized by relatively low levels of regeneration.

MANGROVE SPECIES COMPOSITION IN AFRICA

The uniqueness of mangroves stemming majorly from their adaptability properties characterizes the following species in the following manner: Stilt roots (*Rhizophora, Bruguiera*, and *Ceriops* spp.), pneumatophores (*Avicennia* and *Sonneratia* spp.) and others such as viviparity and salt-excreting leaves.[23] The mangroves of the Western and Eastern African coasts vary significantly in composition and structure.[24] Mangrove species composition in East Africa comprises *Sonneratia alba, Rhizophora mucronata, Bruguiera gymnorrhiza, Ceriops tagal, Heritiera littoralis, Xylocarpus granatum, Avicennia marina*, and *Lumnitzera racemosa.*[25] *Avicennia marina* has been observed to be the most dominant within the region.[26] *Rhizophora mucronata* and *C. tagal* are dominant in Kenya, constituting about 70 percent of the total mangrove forest species.[27] Mangrove species in Mozambique include *A. marina, B. gymnorrhiza, C. tagal, H. littoralis, L. racemosa, R. mucronata, S. alba,* and *X. granatum.*[28] *Rhizophora, Avicennia,* and *Acrosticum* spp. were observed to occur in both Eastern and Western regions of Africa.[29]

Mangrove species in West Africa include Rhizophoraceae *(Rhizophora mangle, Rhizophora harrisonii, Rhizophora racemosa, Avicennia germinans, Laguncularia racemosa*, and *Conocarpus erectus. Laguncularia* sp. and *Conocarpus* sp. are, however, less abundant.[30] The occurrence of the exotic species, *Nypa fruticans,* has also been reported in West Africa.[31] In literature, however, there is a contention on whether *N. fruticans* should be awarded the exotic species status.[32] *Nypa fruticans* is usually referred to as a Mangrove palm or simply as Nipa palm. *Rhizophora* spp. are the most dominant in Nigeria and Cameroon, with relatively few occurrences of *Avicennia* spp.[33] The mangroves in Mauritania comprise majorly of *A. germinans and R. racemosa.* The South African mangrove species include *B. gymnorrhiza and R. mucronata.*[34]

MANGROVE RESOURCES

The resourcefulness of the mangrove forests is myriad and multifaceted, being determined in most cases by nature and the local inhabitants of the host communities. These resources range from income-generating activities, to ecotourism to general ecosystem services such as coastal erosion prevention. Mangroves have the capability of reducing wind strength by approximately 70 percent.[35] Mangroves also serve as a large carbon reserve, thereby ensuring efficiency in carbon sequestration. Mangroves constitute the largest carbon reservoirs on earth.[36]

As a result of the high nutrient influx from both sea and hinterland water sources, mangroves serve as a food source and a habitat and nursery site for a large variety of marine life such as fish species and several crustaceans (crabs, shrimps, periwinkles, etc.). Several other organisms such as migratory birds and sea turtles find temporary solace within the mangrove forest. Other animals include manatee, crocodile, and hippopotamus among several others. Other faunal species occurring within the East African mangrove ecosystem include: *Littorina scabra, Uca* spp., *Sesarma* spp., *Cerithidea decollata*, nematodes, bivalves, copepods, and molluscs.[37] For instance, the Niger Delta of Nigeria has been adjudged a biodiversity hotspot, of which a major part comprises of the mangrove forest.[38]

The indigenous people who serve as hosts and have lived within and around the mangrove forests for generations derive their livelihoods from these same forests. They have been able to strike a balance and have lived sustainably within the mangrove ecosystem. Several resources derived from the mangrove forest by the local people include food (crabs, fish, etc.), wood, tannin (for fishing traps/nets and dyes), medicine, and ecotourism, among many others.

MANGROVE LOSS/DEGRADATION

Population explosion is a growing trend globally, with urbanization and industrialization being close allies. The consequences of these phenomena are evident in every environment, with the mangrove forest not being an exception. Degradation, destruction, and climate change are ongoing consequences. The basic need for sustenance of the growing population, in addition to the need for development, presents an impetus for environmental degradation. The need for environmentally friendly energy, as a replacement for the use of fuelwood, charcoal, and crude oil cannot be overemphasized. The search for marginal lands, agriculture, and urban development in the face of land scarcity is a growing concern and is a driver of mangrove forest degradation. Over time, large portions of mangrove forest ecosystems have been cleared, felled, destroyed, and undergone reclamation, for the purpose of ensuring developmental progress. Mangrove forest destruction, degradation, and species extinction have been the consequence, bearing grave implications for the sustainability of the earth's ecosystem.

Mangrove forest loss in Nigeria and Guinea has been ranked the largest while that occurring in Mauritania is the least.[39] Mangrove loss of about 980 hectares has been reported in Cameroon for the period between 1975 and 2007, resulting in several cases of outright mangrove forest extinction on some mainland areas and on the offshore island of Kwelekwele. Mangrove forests in Tanzania have been observed to have been depleted at a rate of about 0.25 percent per annum for the period between 2010 and 2016.[40] In Kenya, a mangrove forest loss of about 0.15 percent for the period of 2010–2016 has also been recorded. This rate of loss was observed to have decreased from about 0.28 percent during the 2000–2010 period. In Kenya, the implementation of a mangrove ban corresponded with the period of decrease in mangrove loss rate. The losses were reported to be occurring majorly in Kwale and Lamu. Some gains were observed in Mombasa, Kilifi, and Tana River areas, likely due to success in management efforts during that period. The incidence of loss has been attributed to population increase and inefficiency in implementation and administration of governmental policies.[41]

Destruction of large expanses of mangrove forests in East Africa is attributed particularly to the growing practice of aquaculture.[42] Mangrove forest loss in South Africa, occurring in only fourteen out of the seventeen originally mangrove-occupied estuaries, has been observed, representing about 7 percent loss over a period of 17 years.[43]

Generally, across the mangrove forest in Africa, the timber trade has been a major driver for degradation. Energy sources for domestic and industrial purposes, through charcoal production and building construction, have had serious negative impact on the mangrove forest. Sedimentation and subsequent ecosystem alteration, emanating from increased deforestation or selective harvesting, have also been factors. Overexploitation of resources such as crude oil, consequent on increased industrialization, is also worthy of note. This is particularly evident in the Niger Delta of Nigeria, where the crude oil exploitation has resulted in oil pollution and creek channel dredging for exploitation activities. The associated development of residential areas for the oil workers has further contributed to the negative impact on the mangrove forests. Some other factors contributing to mangrove loss across Africa comprise salt making, rice farming, and aquaculture.[44]

Nipa palm colonization has been observed over time to be actively occurring within the mangrove forests in Africa. *Nypa fruticans*, commonly called nipa palm, occurs naturally within the Indo-Pacific mangrove forest block. It was introduced into Africa in the 1900s, particularly to Western and Central Africa. As a result, nipa palm is found occurring within the mangrove forests of countries spanning between Guinea Bissau and Angola, usually termed the Guinea current large marine ecosystem countries (GCLMEC).[45] Over the years, the invasive nature of nipa palm has been observed. Its invasive nature is borne from the versatility of its seeds, owing to the factors of seed resilience and ability to alter soil structure and characteristics.[46] Increasing trend of urbanization and associated large-scale mangrove destruction has further served as a driver for further invasion of nipa palm. Success in eliminating nipa palm has been limited. Nipa palm invasion is counterproductive as it lacks the capability of prosecuting the numerous ecosystem services attributed to the native mangrove species. Mangrove loss of about 21,000 hectares in the Niger Delta has been reported over a period of seventeen years, terminating in 2003.[47] This is attributed to the activities of the oil and gas industry, urbanization, and the spread of the exotic nipa Palm. During the period of 2002–2016, a mangrove forest loss of about 8 percent in the Western section of the Niger Delta was detected.[48] Consequently, mangrove forest loss has been recorded to be actively occurring.

All these losses in area coverage contribute to the decrease in the level of resilience of the global environment in tackling the vicissitudes of climate change. Climate change effects are likely to result in further mangrove loss emanating from the occurrence of acute overbearing weather events such as tropical cyclones, droughts, sea level rise, and associated flooding.[49] In charting a course toward a sustainable solution, concerted efforts need to be made in carrying out interdisciplinary efforts with a focus on attaining conservation and/or sustainable management.

It, therefore, becomes necessary for engagement in collaborations with international organizations, and in being signatories to several international environmental conventions such as Ramsar Convention on Wetlands. Furthermore, the use of advanced technology/strategies (e.g., remote sensing), for the purpose of improved mangrove forest and monitoring is critical. Remote sensing provides a possibility of a more informed, time-efficient, and accurate insight into mangrove ecosystem characteristics for the purpose of tackling current and future issues.

Losses of Mangrove Forest and Environmental Security: Implications for Africa

Environmental degradation and destruction, especially pertaining to the mangrove forest ecosystem, is already an established fact playing out on the global scale. The characteristic fragmented nature and consequent smaller size relative to other ecosystems put the mangrove ecosystem at a disadvantage. Its coastal location, making it a foremost contact for natural hazards, is a further disadvantage. In addition, human pressure is a major driver of mangrove forest degradation, being a consequence of overpopulation in coastal areas. This is in association with the attendant issues of urbanization and industrialization. The majority of economic activities occur within these coastal cities. For instance, coastal locations account for more than 75 percent of fishing landings within West Africa. The input of mangrove fish nursery services in the West African economy comes to about 400 million dollars yearly.[50] In East Africa, the tourism industry thrives more within the coastal areas.

The rich mangrove forest biodiversity serves as a source of food, energy, medicines, and livelihoods for the local people and the entire continent. The implication of mangrove forest loss, therefore, translates into millions of dollars at the risk of being wasted without hope of replenishment. The case of endemic species, characteristic of mangrove forest ecosystems, is at the brink of extinction and requires much-needed attention. Mangrove loss could, therefore, translate to loss of species types and environmental services, running into millions of dollars. These losses are potentially irreversible or are likely to take a relatively long period of restoration time. *Avicennia germinans,* for instance, has been discovered to be gradually going extinct from the mangrove forests of Nigeria in Western Africa.[51] The potential extinction of these key mangrove species could result in a chain reaction such that other components of the mangrove ecosystem, comprising non-mangrove associates and fauna, stand the risk of being lost.

Consequently, mangrove loss results in the disruption of the delicate balance between the ecosystem and the rural host communities. This eventually culminates in poverty and consequently the operation of unsustainable practices.

The role of climate change on global mangrove forest degradation cannot be denied. Increased occurrence of storms and rising sea and temperature levels are a constant reminder of the effect of climate change. Temperature levels are forecasted to rise by about 4.8°C by 2081–2100.[52] The character, functioning, and zonation of mangrove forests globally are likely to be adversely impacted. For instance, the photosynthetic process will be stalled, as carbon dioxide uptake for this process will be limited.[53] Rising temperature levels could result in increased salinity, thereby limiting the growth and productivity of the mangroves.[54] Mangroves are, however, equipped with the ability to alleviate climate change effects. One of which is the ability to carry out carbon sequestration. Of all the vegetation types, mangroves have proven to be the most efficient in carbon sequestration. Carbon in mangroves is stored in a combination of above and below-ground biomass, as well as in the underlying soil. Global mangrove forest loss translates to about 0.08 gigatonnes of carbon in the above-ground biomass and 0.4 gigatonnes in the soil.[55] The amount of carbon sequestered every year comes to about twenty-six million tonnes of carbon every year.[56] The implications of the continued removal and degradation of mangroves on climate change, as it pertains to carbon storage, are dire. Excessive carbon dioxide amounts in the atmosphere will be a result, consequently worsening the already bad climate change situation. In the final analysis, rural communities ultimately will bear the brunt of these consequences.

The consequence of environmental degradation is the diminished quality and value of ecological products and services. This further results in the emergence of a different set of situations and conditions which are alien and thus detrimental to the sustainability of the mangrove forest. In a bid to addressing the imbalance and subsequent malfunctioning arising from habitat alteration occasioned by mangrove forests loss, several situations usually arise. They include, but are not limited, to the following: migration, conflicts, and rising poverty levels. All these result in a situation of insecurity, induced by the environment's deficiencies and inability to carry out efficient functioning. A case in point is the situation in the Niger Delta of Nigeria, whereby issues of pollution, migration, food insecurity, and acid rain have reached a critical level. Complaints made by fishermen linked fish catch decline to continued oil spillage, resulting in a reduction of economic power and a rise in the poverty level. This situation further contributed to the increased incidence of militancy in the region.[57] This scenario replicates itself across the mangrove forest areas in Africa.

Furthermore, the unexpected nature and increased occurrence of natural disasters occasioned by climate change have taken a great toll on the environment. The destruction of life and property occasioned by the increased occurrence of natural disasters, arising from climate change and human interference, has culminated in the rise in the vulnerability level of the mangrove forest environment. In West Africa, for instance, sea level rise has resulted in increased erosion, placing the rural communities at risk of losing land, food, and livelihoods.[58] The central and western African coastal nations are more vulnerable owing to their relatively low-lying coasts.

It has become pertinent to value the resources and services of mangroves, as a basis for ensuring their protection, sustainable use, and sustainability of the people and their environment. The value attributed to ecosystem services basically emanates from the perception of the people mostly in contact with the mangrove forest environment.[59] The water filtering service carried out by mangrove roots ensures the purification of water flowing from the hinterland en route the ocean. This service is especially beneficial to downstream communities, as it pertains to domestic and several other purposes. Research in Sri Lanka estimated the value of the ecological service of water purification of the mangrove forest at about 1.8 million dollars per year.

Man derives more benefits from mangroves than any other living organism.[60] The fishing industry, for instance, benefits about six million dollars per year from about 12,000 hectares of mangroves.[61] It has also been discovered that about 60 percent of the fishes in the vicinity of the Gulf of Guinea breed in the mangrove forests of the Niger Delta of Nigeria.[62] Current mangrove cover contributes about 1,000 trillion fish of economic value and other marine life yearly.[63] The sheer structure of the mangroves is a resource in itself such that it has been recorded that the tourism industry generates a lot of income from the mangrove forest environment.

Ecosystem functions (services) such as hydrological flux, biological productivity, biogeochemical recycling, protection from natural forces, community, and wildlife habitat also serve as a source of mangrove forest valuing. Reduction in the ability to deliver these services in any way or form results in the reduction of the quality of life for both flora and fauna. This also poses the risk of eventual extinction of the mangrove ecosystem. For instance, reduced coastal protection from storms, flooding, and other natural hazards correspond to the loss of property of high financial worth. The loss of life, occasioned by the destruction of local communities, is a serious consequence which should be put in perspective. The importance of the ecological services of mangrove forests cannot therefore be overemphasized. In addition, the typhoon-related waves have been observed to reduce by about 90 percent after going through 1 kilometer of mangrove forest.[64]

Generally, the value of mangrove-related ecosystem services is valued at between 200,000 and 900,000 USD per year globally.[65] These figures reveal the magnitude of value accruing from mangroves globally. The situation in Africa cannot be far-fetched. This clearly shows that mangrove forests hold massive economic potentials for the mangrove regions in the tropics.[66] Globally, the benefits emanating from mangroves are put at about 1.6 billion USD per year.[67]

MANGROVE RESTORATION STRATEGIES/WAY FORWARD

Effective mangrove management, in the face of increasing degradation, is extremely necessary. Mangrove management involves steps or strategies taken toward a major goal of ensuring sustainability of the mangrove forest ecosystem. The scope of mangrove management basically comprises of mangrove regeneration, restoration, and conservation/protection. Mangrove restoration is a process involving steps taken toward ensuring the attainment of the original conditions of the mangrove forest. These conditions should comprise intrinsically of species composition and structure. Specific location conditions are likely to influence the choice of strategy. A very important component at the crux of any strategy is the choice of specie composition to be adapted. For instance, the adoption of *R. racemosa* for reafforestation/restoration programs in the Niger Delta in Nigeria has been recommended. On the other hand, *A. germinans* was observed to be limited in occurrence and was also recommended for further studies and restoration efforts to prevent its extinction.[68]

The need for the adoption of natural recovery for effective management is therefore being emphasized.[69] For this process to be time efficient, natural and artificial seedling recruitment and regeneration methods are critical.[70]

In East Africa, saplings, or propagules of *R. mucronata, C. tagal, B. gymnorhiza,* and *A. marina,* and in rare cases, *S. alba*, have been choice species for restoration projects. Several of these strategies have been carried out in the East African countries, but with few verifiable successes. This situation has been attributed mainly to insufficient monitoring over a sufficiently long time, culminating in inadequate site planning and assessment.[71] Basically, insufficient information for effective planning is lacking in most cases. For instance, information relating to critical points in the course of the mangrove life cycle, such as the fruiting/propagule peak production season, which vary across mangrove species, is not readily available for planning purposes. This places great importance on data inventory, and by consequence, collection.

Data should be all-encompassing, inclusive of field data and (for the purpose of tackling the characteristic inaccessibility of the mangrove forest) satellite data. However, mangrove forest in situ measurements are scarce and limited. This limitation further poses a challenge of engagement for the purpose of rigorous research and comparability.[72] The advent of remote sensing has brought about an improvement in mangrove inventory and monitoring, working in concert with other conventional in situ methods.

Lessons can be learned from other regions of the world, which have registered some level of success. Indonesia, for instance, recorded some successes in mangrove forest restoration by basically incorporating the component of collaborations with indigenous communities, government interventions, and collaborations with research bodies. In addition, these strategies were largely dependent on well-developed information systems linked to robust multidisciplinary research.[73] In several other parts of the world where attempts have been made at achieving effective mangrove management (such as Malaysia, South East Asia, the Phillipines, Florida, etc.), the following recommendations were made: (1) the planting site surface should be stabilized before the beginning of planting; (2) depending on the level of exposure of the planting site, seedlings should be fenced; (3) tall saplings of 1–1.3 meters are the most appropriate for planting; (4) monitoring of restoration programs is necessary; (5) educational/environmental awareness on the importance of the mangrove forest is important; (6) host communities should be incorporated; (7) indigenous knowledge is critical to the survival of any strategy; (8) alternative sustainable livelihoods should be provided for the local people; (8) fast-growing mangroves should pioneer the restoration process in order to counter the development of invasive grasses; (9) mangrove forest management should be holistic, incorporating all other components of the ecosystem; (10) sources of degradation should be determined and examined before the commencement of restoration; (11) replanting is best done in areas which sustained mangrove growth sometime in the past; (12) protection of thriving mangrove forests; and (13) restoration of disturbed hydrological systems.

In East Africa generally and in Cote D'ivoire in West Africa, the following lessons and recommendations were made: (1) choice of mangrove recruits for replanting should be taken directly from the mangrove forest floor; (2) choice of mangrove recruits from seedlings harvested from water is also beneficial; and (3) clternative sources of local wood supply should be identified or reafforested.[74]

It is imperative to note that collaborations with international bodies should be sought for to ensure lasting solutions. In addition, the standpoint of mangrove forest conservation/protection should be improved upon.

CONCLUSION

The importance of mangrove forests to the people and to the global environment cannot be overemphasized. It is a known fact that Africa has continually been characterized by population explosion, the consequences of which have been evidenced by increases in environmental degradation, especially of the mangrove forest. More attempts and research need to be conducted to ensure the stalling of the degradation process and eventual improvement of environmental conditions. The lessons learned are therefore timely and instructive. All hands need to be on deck, ranging from host communities to policymakers to researchers, to ensure effective mangrove forest management in Africa.

NOTES

1. Saenger, Peter G. *Mangrove Ecology, Silviculture and Conservation*. Netherlands: Springer, 2003.
2. "Guidelines on Mangrove Ecosystem Restoration for the Western Indian Ocean Region Western Indian Ocean Ecosystem Guidelines and Toolkits." 2020. www.wiomn.org.
3. Institute for Biodiversity and Protected Areas. *A la découverte des Mangroves en Afrique de l'Ouest – Manuel d'Education à l'Environnement* (Guinea -Bissau: The Restoration Initiative (TRI) – IBAP/UICN/GEF, 2021).
4. Fatoyinbo, Temilola E. and Marc Simard. "Height and Biomass of Mangroves in Africa from ICESat/GLAS and SRTM." *International Journal of Remote Sensing* 34 (2), 2013: 668–81. doi:10.1080/01431161.2012.712224.
5. Ajonina, Gordon, Abdoulaye Diamé, and James Kairo. "Current Status and Conservation of Mangroves in Africa: An Overview." *World Rainforest Movement Bulletin* 133, 2008: 1–6.
6. Food and Agriculture Organization. Main Report: *Global Forest Resources Assessment* 2020. Rome: FAO, 2020.
7. . United Nations Environmental Programme. *Mangroves of Western and Central Africa* United Kingdom: United Nations Environment Programme – Regional Seas Programme, 2007.
8. FAO 2020
9. Feka, Zebedee N. and Isaac Morrison. "Managing Mangroves for Coastal Ecosystems Change: A Decade and beyond of Conservation Experiences and Lessons for and from West-Central Africa." *Journal of Ecology and The Natural Environment* 9(6), 2017: 99–123. doi:10.5897/JENE2017.0636.
10. Itzkin, Adela et al. "Approaches to Enhance Integration and Monitoring for Social-Ecological Systems." *Land,* 11(10), 2022. MDPI. doi:10.3390/land11101848.
11. Fatoyinbo and Simard, 2013.
12. "GEF-Manuel-MangroveFR-Lowdef." n.d.
13. Ajonina, Diame and Kairo, 2008.
14. "Guidelines on Mangrove Ecosystem Restoration for the Western Indian Ocean Region Western Indian Ocean Ecosystem Guidelines and Toolkits." 2020. www.wiomn.org.
15. Taylor, Michelle, Corinna Ravilious, and Edmund P. Green. "Mangroves." UNEP WCMC. Cambridge: UNEP World Conservation Monitoring Centre, 2003.

16. Wang, Haitao et al, "Temporal and Spatial Variations of Greenhouse Gas Fluxes from a Tidal Mangrove Wetland in Southeast China." *Environmental Science and Pollution Research* 23(2), 2016: 1873–1885.

17. Itzkin et al. 2022.

18. Giri, Chandra "Status and Distribution of Mangrove Forests of the World Using Earth Observation Satellite Data." *Global Ecology and Biogeography,* 20(1), 2011: 154–159.

19. Shapiro, Aurélie et al. "The Mangroves of the Zambezi Delta: Increase in Extent Observed via Satellite from 1994 to 2013." *Remote Sensing,* 7(12), 2015. MDPI AG: 1671504–18. doi:10.3390/rs 215838.

20. Otero, Viviana, et al. "Mangroves at Their Limits: Detection and Area Estimation of Mangroves along the Sahara Desert Coast." *Remote Sensing,* 8(6), 2016. MDPI AG. doi:10.3390/rs8060512.

21. Ajonina, Diame and Kairo, 2008.

22. Taylor, Ravilious, and Green, 2003.

23. Kuenzer, Claudia et al. "Remote Sensing of Mangrove Ecosystems: A Review." *Remote Sensing,* 3(5), 2011: 878–928.

24. Machava-António, Vilma et al. "A Comparison of Mangrove Forest Structure and Ecosystem Services in Maputo Bay (Eastern Africa) and Príncipe Island (Western Africa)." *Forests,* 13(9), 2022: 1466. 17 J. O. Bosire et al., *Mangroves of the Western Indian Ocean: Status and Management*, 2015.

25. J. O Bosire et al. 2015.

26. Machava-António et al, 2022.

27. Riungu, Purity M. et al. "Value Chain and Sustainability of Mangrove Wood Harvesting in Lamu, Kenya." *Trees, Forests and People,* 9, 2022: 100322.

28. Aurélie Shapiro et al, 2013.

29. Duke, Norman C., "Mangrove floristics and biogeography." *Tropical Mangrove Ecosystems,* 41, 1992: 63–100.

30. "GEF-Manuel-MangroveFR-Lowdef." n.d.

31. Spalding, Mark, Francois Blasco, and Colin Field. *World Mangrove Atlas.* Okinawa: International Society for Mangrove Ecosystems, 1997.

32. Ajonina, Diamé, and Kairo, 2008.

33. Ellison, J. C. and I. Zouh. "Vulnerability to Climate Change of Mangroves: Assessment from Cameroon, Central Africa." *Biology,* 1, 2012: 617–638.

34. Kuenzer et al., 2011.

35. McIvor, A. L. et al. "Reduction of Wind and Swell Waves by Mangroves." Natural Coastal Protection Series: Report 1. Cambridge Coastal Research Unit Working Paper 40. ISSN 2050-7941. (2012).

36. Donato, Daniel C. et al. "Mangroves among the Most Carbon-rich Forests in the Tropics." *Nature Geoscience,* 4(5), 2011: 293–297.

37. Taylor, Ravilious, and Green, 2003.

38. Moudingo, J.-Hude, Gordon Ajonina, Didier Dibong, and Minette Tomedi. "Distribution, Devastating Effect, and Drivers of the Exotic Mangrove Nypa Fruticans Van Wurmb (Arecaceae) on the Mangroves of West and Central Africa." In *Biotechnological Utilization of Mangrove Resources*, pp. 49–78. Academic Press, 2020.

39. Fatoyinbo and Simard, 2013.

40. "Guidelines on Mangrove Ecosystem Restoration for the Western Indian Ocean Region Western Indian Ocean Ecosystem Guidelines and Toolkits." 2020. www.wiomn.org.

41. Juma Hamza, Amina, Luciana S. Esteves, and Marin Cvitanović. "Changes in Mangrove Cover and Exposure to Coastal Hazards in Kenya." *Land,* 11(10), 2022: 1714.

42. Taylor, Ravilious, and Green, 2003.

43. Adams, J. B., B. M., Colloty, and G. C. Bate. "The Distribution and State of Mangroves along the Coast of Transkei, Eastern Cape Province, South Africa." *Wetlands Ecology and Management,* 12(5), 2004: 531–541.

44. Hamza, Esteves, and Cvitanović, 2022.

45. J.-Hude, Ajonina, Dibong, and Tomedi, 2022.

46. Numbere, Aroloye O. "The Impact of Oil and Gas Exploration: Invasive Nypa Palm Species and Urbanization on Mangroves in the Niger River Delta, Nigeria." In *Threats to Mangrove Forests*, pp. 247–266. Cham: Springer, 2018.

47. James, Godstime K. et al. "Satellite-based Assessment of the Extent and Changes in the Mangrove Ecosystem of the Niger Delta." *Marine Geodesy,* 30(3), 2007: 249–267.

48. Adams., Colloty, and Bate, 2004.

49. Hamza, Esteves, and Cvitanović, 2022.

50. United State Agency for International Development. *A Policy Brief: Mangroves in West Africa*. Forest Carbon, Markets and Communities, 2014.

51. Ufuoma Efe, Akpovwovwo and Adeniyi Gbadegesin. "Species Composition and Distribution Patterns of the Mangrove Forests of the Western Niger Delta, Nigeria." *African Geographical Review,* 41(4), 2022: 468–482.

52. Intergovernmental Panel on Climate Change 2013: The Physical Science Basis. Contribution of Working Group I to the Fifth Assessment Report of the Intergovernmental Panel on Climate Change, 2013..

53. Clough, B. F.. "Primary Productivity and Growth of Mangrove Forests." *Tropical Mangrove Ecosystems,* 41, 1992: 225–249.

54. Ward, Raymond D., Daniel A. Friess, Richard H. Day, and Richard A. Mackenzie. "Impacts of Climate Change on Mangrove Ecosystems: A Region-by-region Overview." *Ecosystem Health and Sustainability,* 2(4), 2016: e01211.

55. Armah, Ayaa Kojo et al. "The Role and Work of African Mangrove Network (AMN) in Conservation and Sustainable Use of Mangroves in Africa." *Nature & Faune,* 24(1), 2009: 30–41.

56. Ong, J. E.. "Mangroves — A Carbon Source and Sink." *Chemosphere*, 27, 1993, 1097–1107.

57. Amadi, L. and C. U. M. Ogonor. "Climate Change, Environmental Security, and Displacement: Experience from the Niger Delta Flood Disaster." *African Journal of Environmental Science and Technology*, 9(1), 2012: 53–74.

58. Amadi and Ogonor, 2012.

59. Bateman, Ian J., Georgina M. Mace, Carlo Fezzi, Giles Atkinson, and R. Kerry Turner. "Economic Analysis for Ecosystem Service Assessments." In *Valuing Ecosystem Services*, pp. 23–77. Edward Elgar Publishing, 2014.

60. Johansen, Pernille Greve et al. "Occurrence and Importance of Yeasts in Indigenous Fermented Food and Beverages Produced in Sub-Saharan Africa." *Frontiers in Microbiology,* 10, 2019: 1789.

61. Armah et al., 2009.

62. Mmom, Prince C., and Samuel B. Arokoyu. "Mangrove Forest Depletion, Biodiversity Loss and Traditional Resources Management Practices in the Niger Delta, Nigeria." *Research Journal of Applied Sciences, Engineering and Technology,* 2(1), 2010: 28–34.

63. Armah et al., 2009.

64. Barbier, E. B. et al. "Coastal Ecosystem-based Management with Non-linear Ecological Functions and Values." *Science*, 319, 2008: 321–323.

65. Food and Agricultural Organization, Status and trends in mangrove area extent worldwide. Wilkie, M. L., and S. Fortuna (Eds.), Forest Resources Assessment Working Paper No. 63, Forest Resources Division. FAO, Rome, 2003. http://www.fao.org/docrep/007/j1533e/j1533e00.HTM. Accessed August 1, 2009.

66. Yessoufou, K. and G. H. Stoffberg. "Biogeography, Threats and Phylogenetic Structure of Mangrove Forest Globally and in South Africa: A Review." *South African Journal of Botany,* 107, 2016: 114–120.

67. Costanza, R. et al. The Value of the World's Ecosystem Services and Natural Capital." *Nature,* 387, 1997: 253–260.

68. Adams, Colloty, and Bate, 2004.

69. Numbere, Aroloye O. "Impact of Urbanization and Crude Oil Exploration in Niger Delta Mangrove Ecosystem and its Livelihood Opportunities: A Footprint Perspective." In *Agroecological Footprints Management for Sustainable Food System,* pp. 309–344. Singapore: Springer, 2021.

70. Wang et al., 2016.

71. Institute for Biodiversity and Protected Areas, 2021.

72. Dahdouh-Guebas, Farid et al. "Human-impacted Mangroves in Gazi (Kenya): Predicting Future Vegetation Based on Retrospective Remote Sensing, Social Surveys, and tree Distribution." *Marine Ecology Progress Series,* 272 2004: 77–92.

73. Arifanti, Virni Budi, et al. "Challenges and Strategies for Sustainable Mangrove Management in Indonesia: A Review." *Forests,* 13(5), 2022: 695.

74. Witsen, Deidre B. "Mangrove Restoration and Management in Djibouti." PhD dissertation, master's thesis. Virginia Polytechnic Institute and State University, 2012.

REFERENCES

Arifanti, Virni Budi, Frida Sidik, Budi Mulyanto, Arida Susilowati, Tien Wahyuni, Naning Yuniarti, Aam Aminah, Eliya suita, Endang Karlina, Sri suharti, Pratiwi, Maman Turjaman, Asep Hidayat, Henti Hendalastuti Rachmat, Rinaldi Immanudin, Irma Yeny, Wida Darwiati, Nilam Sari, Safinah Suryah Hakim, Whitea Yasmin Slamet and Nisa Novita. "Challenges and strategies for sustainable mangrove management in Indonesia: A review." Forests 13, no. 5 (2022): 695.

Adams, J. B., B. M. Colloty, and G. C. Bate. "The distribution and state of mangroves along the coast of Transkei, Eastern Cape Province, South Africa." *Wetlands Ecology and Management* 12, no. 5 (2004): 531–541.

Ajonina, Gordon, Abdoulaye Diamé, and James Kairo. "Current status and conservation of mangroves in Africa: An overview." *World Rainforest Movement Bulletin* 133 (2008): 1–6.

Akpovwovwo, Ufuoma Efe, and Adeniyi Gbadegesin. "Species composition and distribution patterns of the Mangrove forests of the Western Niger Delta, Nigeria." *African Geographical Review* 41, no. 4 (2022): 468–482

Amadi, L., and C.U.M. Ogonor. "African journal of environmental science and technology." *Climate Change, Environmental Security and Displacement: Experience from the Niger Delta Flood Disaster* 9, no. 1 (2012): 53–64.

Armah, Ayaa Kojo, Abdoulaye Diame, Gordon Ajonina, and James Kairo. "The role and work of African Mangrove Network (AMN) in conservation and sustainable use of mangroves in Africa." *Nature & Faune* 24, no. 1 (2009): 30–41.

Barbier, Edward B., Evamaria W. Koch, Brian R. Silliman, Sally D. Hacker, Eric Wolanski, Jurgenne Primavera, Elise F. Granek et al. "Coastal ecosystem-based management with nonlinear ecological functions and values." *science* 319, no. 5861 (2008): 321–323.

Bateman, Ian J.,Georgina M. Mace, Carlo Fezzi, Giles Atkinson, and R. Kerry Turner. "Economic analysis for ecosystem service assessments." In *Valuing Ecosystem Services*, K. N. Ninan (Ed.), 23–77. Cheltenham:Edward Elgar Publishing, 2014. https://doi.org/10.4337/9781781955161.00013

Bosire, J. O., Mwita M. Mangora, S. O. Bandeira, A. Rajkaran, R. Ratsimbazafy, C. Appadoo, and J. G. Kairo. *Mangroves of the Western Indian Ocean: Status and management*. 2015.

Clough, B. F. "Primary productivity and growth of mangrove forests." *Tropical Mangrove Ecosystems* 41, (1992): 225–249.

Costanza, Robert, Ralph d'Arge, Rudolf De Groot, Stephen Farber, Monica Grasso, Bruce Hannon, Karin Limburg et al. "The value of the world's ecosystem services and natural capital." *Nature* 387, no. 6630 (1997): 253–260.

Dahdouh-Guebas, Farid et al. "Human-impacted mangroves in Gazi (Kenya): Predicting future vegetation based on retrospective remote sensing, social surveys, and tree distribution." *Marine Ecology Progress Series* 272, (2004): 77–92.

Donato, Daniel C., J. Boone Kauffman, Daniel Murdiyarso, Sofyan Kurnianto, Melanie Stidham, and Markku Kanninen. "Mangroves among the most carbon-rich forests in the tropics." *Nature Geoscience* 4, no. 5 (2011): 293–297.

Duke, Norman C. "Mangrove floristics and biogeography." *Tropical Mangrove Ecosystems* 41, (1992): 63–100.

Ellison, J. C., and I. Zouh. "Vulnerability to climate change of mangroves: Assessment from Cameroon, Central Africa." *Biology* 1, (2012): 617–638.

Fatoyinbo, Temilola E., and Marc Simard. "Height and biomass of mangroves in Africa from ICESat/GLAS and SRTM." *International Journal of Remote Sensing* 34, no. 2 (2013): 668–681. https://doi.org/10.1080/01431161.2012.712224.

Feka, Zebedee N., and Isaac Morrison. "Managing mangroves for coastal ecosystems change: A decade and beyond of conservation experiences and lessons for and from west-central Africa." *Journal of Ecology and The Natural Environment* 9, no. 6 (2017): 99–123.

Food and Agricultural Organization. "Status and trends in mangrove area extent worldwide." In M. L. Wilkie and S. Fortuna (Eds.), *Forest resources assessment working paper no. 63*, Forest Resources Division, Rome: FAO, 2003..

Institute for Biodiversity and Protected Areas. *A la découverte des Mangroves en Afrique de l'Ouest – Manuel d'Education à l'Environnement.* Guinea-Bissau: The Restoration Initiative (TRI) – IBAP/UICN/GEF, 2021.

Giri, Chandra, Edward Ochieng, Larry L. Tieszen, Zq Zhu, Ashbindu Singh, Tomas Loveland, Jeff Masek, and Norman Duke. "Status and distribution of mangrove forests of the world using earth observation satellite data." *Global Ecology and Biogeography* 20, no. 1 (2011): 154–159.

Food and Agriculture Organization. *Main report: Global forest resources assessment* 2020. Rome: FAO, 2020. https://www.fao.org/3/ca9825en/CA9825EN.pdf.

Kairo, James G., Mwita M. Mangora, Western Indian Ocean Mangrove Network, and Western Indian Ocean Marine Science Association. "Guidelines on Mangrove Ecosystem Restoration for the Western Indian Ocean Region-Western Indian Ocean Ecosystem Guidelines and Toolkits." (2020).

Hamza, Amina Juma, Luciana S. Esteves, and Marin Cvitanović. "Changes in mangrove cover and exposure to coastal hazards in Kenya." *Land* 11, no. 10 (2022): 1714.

IPCC (2013), Climate Change 2013: The physical science basis Contribution of working group I to the fifth assessment Report of the intergovernmental panel on climate change. (Stocker, T. F., D. Qin, G. K, Plattner, M. Tignor, S. K, Allen, J. Boschung, A. Nauels, Y. Xia, V, Bex, and P.M, Midgley, (eds)). 1535 pp. Cambrigde University Press, Cambridge, United Kingdom and New York, NY, USA, 2013

Itzkin, Adela et al. "Approaches to enhance integration and monitoring for social-ecological systems." *Land* 11 (2022): 10. MDPI. https://doi.org/10.3390/land11101848.

James, Godstime K., Jimmy O. Adegoke, Ekechukwu Saba, Peter Nwilo, and Joseph Akinyede. "Satellite-based assessment of the extent and changes in the mangrove ecosystem of the Niger Delta." *Marine Geodesy* 30, no. 3 (2007): 249–267.

Johansen, Pernille Greve, James Owusu-Kwarteng, Charles Parkouda, S. Wilfrid Padonou, and Lene Jespersen. "Occurrence and importance of yeasts in indigenous fermented food and beverages produced in sub-Saharan Africa." *Frontiers in Microbiology* 10, (2019): 1789.

Kuenzer, C., A. Bluemel, S. Gebhardt, and T. Vo Quoc. S. Dech. *Remote Sensing of Mangrove Ecosystems: A Review." Remote Sensing* 3: 878–928.

Machava-António, Vilma, Alberto Fernando, Mariana Cravo, Mágda Massingue, Hamilton Lima, Célia Macamo, Salomão Bandeira, and José Paula. "A Comparison of Mangrove Forest Structure and Ecosystem Services in Maputo Bay (Eastern Africa) and Príncipe Island (Western Africa)." *Forests* 13, no. 9 (2022): 1466.

McIvor, A. L., Iris Möller, Tom Spencer, and Mark Spalding. "Reduction of wind and swell waves by mangroves." *Natural Coastal Protection Series: Report 1. Cambridge Coastal Research Unit Working Paper 40. ISSN 2050-7941.* (2012).

Mmom, Prince C., and Samuel B. Arokoyu. “Mangrove forest depletion, biodiversity loss and traditional resources management practices in the Niger Delta, Nigeria.” *Research Journal of Applied Sciences, Engineering and Technology* 2, no. 1 (2010): 28–34.

Moudingo, J-Hude, Gordon Ajonina, Didier Dibong, and Minette Tomedi. “Distribution, devastating effect, and drivers of the exotic mangrove Nypa fruticans Van Wurmb (Arecaceae) on the mangroves of West and Central Africa.” In *Biotechnological Utilization of Mangrove Resources*, pp. 49–78. Academic Press, 2020.

Numbere, Aroloye O. “The impact of oil and gas exploration: invasive nypa palm species and urbanization on mangroves in the Niger River Delta, Nigeria.” In *Threats to mangrove forests*, pp. 247–266. Springer, Cham, 2018.

Numbere, Aroloye O. “Impact of urbanization and crude oil exploration in niger delta mangrove ecosystem and its livelihood opportunities: A footprint perspective.” In *Agroecological Footprints Management for Sustainable Food System*, Springer, Singapore (2021) pp. 309–344.

Ong, J.E. “Mangroves — a carbon source and sink” Chemosphere, 27 (1993), pp. 1097–1107.

Otero, Viviana, Katrien Quisthoudt, Nico Koedam, and Farid Dahdouh-Guebas. “Mangroves at their limits: Detection and area estimation of mangroves along the Sahara Desert Coast.” *Remote Sensing* 8, no. 6 (2016): 512.

Riungu, Purity M., Justine M. Nyaga, Michael N. Githaiga, and James G. Kairo. “Value chain and sustainability of mangrove wood harvesting in Lamu, Kenya.” *Trees, Forests and People* 9 (2022): 100322. Saenger, Peter G. “Mangrove Ecology, Silviculture and Conservation.” *Springer Netherlands* (2003).

Shapiro, Aurélie C., Carl C. Trettin, Helga Küchly, Sadroddin Alavinapanah, and Salomão Bandeira. “The mangroves of the Zambezi Delta: Increase in extent observed via satellite from 1994 to 2013.” *Remote Sensing* 7, no. 12 (2015): 16504–16518.

Spalding Mark, Francois Blasco, and Colin Field. *World Mangrove Atlas*. Okinawa: International Society for Mangrove Ecosystems, 1997.

Taylor, Michelle, Corinna Ravilious, and Edmund P. Green. “Mangroves.” *UNEP WCMC (UNEP World Conservation Monitoring Centre), Cambridge* (2003).

United Nations Environmental Programme. *Mangroves of Western and Central Africa*. United Kingdom: United Nations Environment Programme – Regional Seas Programme, 2007.

United State Agency for International Development. A Policy Brief: Mangroves in West Africa. Forest Carbon, Markets and Communities, 2014.

Ward, Raymond D., Daniel A. Friess, Richard H. Day, and Richard A. Mackenzie. “Impacts of climate change on mangrove ecosystems: a region-by-region overview.” *Ecosystem Health and sustainability* 2, no. 4 (2016): e01211.

Witsen, Deidre B. “Mangrove Restoration and Management in Djibouti.” PhD dissertation, master’s thesis. Virginia Polytechnic Institute and State University, (2012).

Yessoufou, K., and G. H. Stoffberg. “Biogeography, threats and phylogenetic structure of mangrove forest globally and in South Africa: A review.” *South African Journal of Botany* 107 (2016): 114–120.

Chapter 16

Population Growth and Resource Exploitation

Implications for Environmental Security in Nigeria

Faith Iwejingi Sajini

The association between human activities such as resource exploration and exploitation and environmental security has long been established in literature. Enger and Smith, whose concept of environment is the sum total of all the conditions that affect an organism all through its lifetime, likewise stated that all organisms, chiefly human beings, impinge on many components in their environment.[1] The components in question are those he (human beings) explores and exploits to provide his critical needs, and they include natural environmental resources like soils, rocks, vegetation, forests, wildlife, and water bodies.

Also identified in the literature are a myriad of influences that humans exert on the environment to the detriment of the ecosystem. Perceptual increase in human population has not only placed heavy demand on environmental resources but has also degraded it, thereby creating the need for management and planning.[2] It is a verifiable fact that man has inadvertently influenced global climate through his numerous socioeconomic activities, thereby making climate change a major threat to environmental security in the twenty-first century. Climate change is an alteration that has taken place in the climate of an area for a long period of time of at least 100–150 years with a clear and permanent impact on the ecosystem.[3] The role of Homo sapiens in significantly raising the concentration of GHGs in the atmosphere has also been recognized. They pollute the stratosphere through the introduction of waste materials, which alter natural systems all over the world. They also plant crops, find cities, and build industries, all of which act in one way or another to affect the natural environment.[4] To cap it all, Botkin and Keller posited that the explosive increase in human population practically underlies virtually all environmental issues around the world; therefore, solving the problem of the environment with a view to achieving the much-desired environmental security must be preceded by first tackling the population factor.

From the foregoing, one can conveniently present environmental security as being concerned with the effects of human activities on the ecosystem, focusing on the endangerment that environmental events and trends pose to an identified group of people, societies, and nations. Manifestations of environmentally induced insecurity especially as is visible in low-resource countries as in Africa include but are not limited to: climate change, environmental degradation, loss of biodiversity, overuse or overexploitation of non-renewable environmental resources, resource scarcity, lack of or limited access to resources, deforestation, and many other factors, which have either directly or indirectly caused, sustained, or prolonged conflicts in different parts of the world and Africa in particular. Given that the environment is that segment of the globe that supports and provides necessary life-sustaining resources for human development, concerns about both natural and human impacts on the environment and their resultant effects on social and political order at all levels have to be addressed for the sake of sustainable well-being and development in territories.

Generally, the global population has been growing rapidly. However, for all of human demographic history, population was growing very slowly due to high birth rate and high death rate. This was much more so during the Paleolithic era when men were predominantly hunters and fruit gatherers. However, with the advent of agriculture and sedentary lifestyle, the population began to grow gradually. The first billionth was recorded in the year 1804, the second billionth in 1927, the third billionth in 1960, the fourth billionth in 1974, the fifth billionth in 1987, and the sixth billionth in 1999. On October 1, 2011, the birth of the seven billionth baby was announced, meaning that the world's population has already attained seven billion.[5] The eight billionth was achieved in the last quarter of 2022 and it is visualized to grow to its peak sometime in 2080. The breakdown of the world's population by continents shows that Africa's population stands at 1.4 billion, Asia 4.7 billion, Europe 750 million, North America 602 million, South America 439 million, and Oceania 44 million. It is envisaged that the next milestone, which is 9 billion, may be attained sometime in 2030.[6]

THEORETICAL FRAMEWORK

The Malthusian theory of population growth as propounded by Thomas Robert Malthus (1798) is adopted for this chapter. Malthus in his essay titled "the

Principle of Population" asserted that population tends to grow faster than the rate at which the Earth produces sustenance. This implies that the population grows at a geometrical progression, while the food supply increases at an arithmetic rate. Malthus then predicted that a time would come when the population will absolutely outstrip the food supply and that will lead to both physical, social, environmental, and health hazards. Malthus has been criticized because he failed to envisage that the population can grow and that living standards could improve concurrently over an extensive period. The criticism notwithstanding, the theory provides an understanding of the population dynamics as it affects resource utilization. This chapter rests on this theory because an increase in population engenders over-exploitation of the resources of the environment as the basic needs of the teeming population have to be supplied, which will eventually undermine the quality of the environment.[7] Scramble for the limited and diminishing resources in a grossly degraded environment by a large population will induce notable social and political conflicts, thereby exacerbating environmental insecurity.

POPULATION GROWTH IN NIGERIA

The Neolithic revolution characterized by the advent and rise of agriculture which necessitated a sedentary culture or lifestyle passes as the factor that engendered the initial explosive growth. At the global and national scales, recent improvements in medical health care, which led to the discovery of causes of diseases and other breakthroughs, such as the administration of vaccines to combat killer diseases especially among children, coupled with improvement in sanitation and transportation which aided food circulation from areas of surplus to areas of deficit, laid the foundation for population increase. This growth in population resulted from a decrease in the death rate without a commensurate fall in the birth rate.

Nigeria is the seventh among the top ten countries with the largest population in the world. Having a current population of about 216 million people, an annual growth rate of about 2.6%, and a total fertility rate (TFR) of about five children per woman, it is apparent that the population is not only large but also growing at an increasing rate capable of doubling itself in twenty-seven years.[8] Several factors are considered responsible for the current Nigeria's demographic situation.

Negative Attitude Toward Small Family Size

A major fraction of Nigerian populace favor large families along with early marriage and a polygamous family structure. This is particularly common among the rural populace who unfortunately constitute the larger proportion of the Nigerian population when considering the rural/urban distribution of the country's population.

Religion

Religion is a vital component of the cultural basics of communities with a considerable influence on decision-making. Religious beliefs and ideology have a way of shaping demographic processes. Approaches to fertility matters and uptake of family planning methods vary among practitioners of various religions. For instance, although birth control is permitted in marriage and not forbidden by the Quran, Muslims consider abortion detestable.[9]

Lack of Relevant Education

The education in question is not just that of formal western education, although it is part of it, but the dearth of education that relates to population education. Population education allows individuals to be knowledgeable about the causes of population growth and its imprint on the quality of their lives and the environment. More importantly, population education enables people to understand the existing demographic condition of their individual countries and that of the world. It also enhances their participation in bringing the demographic situation in their family, society, and country to a desirable one.[10] However the inclination to adopt population education is dependent upon the level of western/formal education acquired. Research have shown that a woman's educational attainment is inversely associated with her level of fertility.[11] Due to Nigeria's large population, poverty rates are being fueled by illiteracy, which is also limiting access to formal education. Poor enrollment and gender disparity in school enrollment are manifest in Nigerian educational system. Unfortunately, this invariably affects the quantity and quality of education received. A largely illiterate or semi-literate society will hardly adopt population education. According to Chandna, population geographers regard literacy as a consistent index of economic and sociocultural development. He further noted that literacy is essential for eradicating poverty and mental seclusion, for cultivating peaceful and responsive international relations and free play of demographic processes.[12]

Sex Preference

Sex preference, especially in favor of the male child, is a factor perpetuating high fertility in some settings in Nigeria. Male offspring are more valued than female offspring in many Nigerian cultures for a variety of reasons, including carrying on the family name and having stronger upper bodies for physical labor. For the families that are yet to have a male offspring, the tendency is there for the practice of continuous childbirth so they can have male children. This preferential customary practice perpetuates high fertility in societies, which eventually culminates into excessively large population with its various attendant problems.

Absence of Social Security for the Elderly

In a country like Nigeria where social security for the elderly citizens is lacking, having many children becomes necessary as they serve as insurance against old age. In virtually all Nigerian societies, children are the only source of care for the elderly.

Economic Reasons

In a largely agrarian economy, as in most rural parts of Nigeria, having so many children is an advantage as they are seen as assets and as a source of manpower for supplementing farm labor. High poverty rate also exacerbates economic reasons for more children as children in poor families become economically active at a very young age, generating income to support their families.[13]

High Infant Mortality

According to the UN report, Nigeria is one of the countries with the highest infant mortality rate (IMR) in the world. The current IMR is 56 deaths per 1,000 live births.[14] Although efforts are in place to combat this menace as it has featured as one major target of the sustainable development goals, the fact is that the rate is still high. This single factor will make people want to have plenty of children to ensure that some will survive till old age.

POPULATION GROWTH AND RESOURCE EXPLOITATION IN NIGERIA

A resource portends anything that is useful to man, or anything that can be converted into a useful product or that can be used to create a useful thing. Natural resources are so termed because they are obtained from nature, and they form the center of all life on planet Earth.[15] There exists a great disparity between the less developed and the developed countries in both the distribution of population and resources. The less developed countries are those that currently are unable to provide a minimum acceptable standard of living for the huge majority of their population. The developed countries are endowed with about 85 percent of the global resources, but they have only about 19 percent of the world's population to support, whereas the developing realm to which Nigeria belongs hosts about 75 percent of the world's population with just about 15 percent of the world's resources to support that quantum population.[16]

Having established the fact that Nigeria's population is large and is still growing, it suffices to say that this growth places a heavy demand on the various resources needed for sustenance with its attendant consequences on the environment. In the last four decades, activities by humans, particularly that of resource extraction pertaining to oil and gas sector development projects, have accelerated deforestation and loss of topsoil, loss of habitat, species, and biodiversity and ruin of wetlands. The effects of population increase on resources are presently visible in human congestion and energy consumption among others.[17]

As per human congestion, virtually all environmental challenges have population as a significant contributing component. The environment is under stress due to overpopulation. Man's misuse of the environment's resources is the starting point of environmental issues including degradation, erosion, and desertification among others. Nigerians are said to have traditionally resisted efforts to dramatically lower their large population figure because they see it as a sign of prominence, supremacy, and status. Currently, human congestion and excessive population pressure is pushing traditional cultures to give up long-standing production methods and resource-management strategies such as shifting cultivation and bush fallowing system, which permitted them to feed their own populations with little or no destructive environmental impact.[18] As reported by the Global Forest Watch, 567,371 out of 10,048,732 hectares of wooded area had given way to deforestation as of 2006[19]. Also, Nigeria is said to have lost 14 percent of its principal forest between 2002 and 2020.[20] This can be attributed to the fact that forest resources serve as sources of raw material for industries such as furniture, building, and paper industries. Timbers are exported and woods are collected for fuel in virtually all Nigerian rural communities. Forests are generally cleared to make way for agricultural activities and the expansion of human settlement to provide for the basic needs of the ever-increasing human population.

In terms of Nigerian energy consumption, it is obvious that as population increases and lifestyle becomes more complex and energy consumption per capita rises concurrently. Energy consumption per capita can be seen as a measure of the yearly consumption of saleable crucial energy such as coal, lignite, petroleum, natural gas, and electricity emanating from geothermal and nuclear power plants expressed in kilograms of oil equivalents per person.[21] Energy production and consumption and environmental problems are somewhat connected.

The production, transportation, and consumption of energy can rarely occur without leaving a significant and undesirable mark on the environment. The derivatives of these activities as they relate to energy production and consumption include water pollution, thermal pollution, air pollution, indiscriminate solid waste disposal, and climate change. The combustion of fossil fuel is a key source of GHG emissions, and it also accounts greatly for the perceptible urban air pollution. Numerous land and water pollution are also linked with energy handling. For instance, oil spills that result from oil utilization processes pollute land and water in the exploitation area where the spill occurred.

Also, solid wastes generated in the process of some forms of energy consumption contribute to land and water pollution depending on where they are dumped, particularly when they are toxic.[22] No doubt, a growing population increases energy consumption. In 2020, research conducted on population growth and disaggregated energy generation sources in Nigeria revealed that population growth has a positively significant impact on energy generation from natural gas, oil, and coal sources while exhibiting a negative impact on hydropower sources.[23]

ENVIRONMENTAL IMPLICATION OF OVEREXPLOITATION OF RESOURCES

Explosive population growth remains a major factor driving environmental insecurity via exploitation of resources in excess of what the environment can offer without stress.[24] Population growth, in particular, places increasing pressures on the planet's resources—water, forests, land, and the Earth's atmosphere, thereby making it a contributive factor of climate change and a challenge to environmental sustainability. The consequences of overpopulation and the environmental implication of the attendant resource exploitation include environmental degradation, wildlife losses, and pollution alongside other socioeconomic consequences like unemployment and underemployment, lower living standards, poor state of education and health care, social unrest, and political conflicts. Population growth inevitably results in a shrinking land base, arising from the extra demand for land required to grow more crops to feed the growing population—the end result of overexploitation of resources occasioned by a rapidly growing population. In addition to the attendant environmental challenges, there is poverty succession, which is a forerunner of social and political disorder.[25]

Nigeria is one of the sub-Saharan nations highly susceptible to the impacts of environmental problems. All through the year, the impact of climate change has become more visible. This ranges from desertification and drought in the northern fringe to excessive rainfall in the extreme south, which often leads to flooding, damaging cultivated plants and displacing people. The incessant cases of herders–farmers conflicts in the Middle Belt and southern region of Nigeria persist because of resource exploitation. These two occupational groups struggle for scarce resources occasioned by increase in population and environmental insecurity particularly desertification and drought in the northern region, which made sustenance unavailable for livestock. This induces southward migration of herders, which increase competition over use of land and water resources in the central and southern region of Nigeria, where population increase has already created pressure on land due to loss of farmland to construction of human settlements. This creates the "survival of the fittest" scenario and the resultant effects are conflicts, which lead to loss of lives and properties.[26]

CONCLUSION

Population growth and natural resources are elaborately connected, and they determine the state of the environment. Man's impact on the environment is made mainly when he explores and exploits it for its benefits. Therefore, the more the number, the greater the impact made on the environment. Having therefore established the fact that Nigerian population is growing at an increasing rate, given the current annual growth rate and the fertility level, it is safe to conclude that Nigerian population rate of progression and the attendant exploitation of the environmental resources needed for the socio-economic well-being of the populace pose a threat to environmental security in Nigeria. Therefore, mutually reinforcing efforts between the government and the populace to slow down population growth and to improve environmental protection and resource conservation via the reduction of unsustainable consumption of resources are ways forward in the pursuit of achieving environmental security.

RECOMMENDATIONS

Since Nigeria's urban and rural landscape has gotten worse due to environmental degradation occasioned by an increasingly large population, strategies that will halt and reverse the unconducive scenario are needed. The following are itemized for consideration and implementation, if we must alter the dynamics between population, natural resources, and environmental sustainability and security:

- Enactment and implementation of population policy; although Nigeria has an existing population policy, strict enforcement is required to make uptake of family planning methods compulsory for adult males and females in the country once the stipulated number of child births has been achieved. This may also involve fixing the age at the first marriage and stipulating the number of children per couple.
- Making birth control measures available and accessible.
- Provision of social security to elderly citizens; this will encourage people to have fewer children as it will reduce economic reasons for extra children.
- Enacting and strengthening healthcare policies aimed at reducing infant and child mortality rates. This will ensure a higher survival rate of children, thereby reducing the desire for more children among couples.
- Population education and public enlightenment: this will change the orientation of the populace as it will help them gain the awareness, skills, and change attitudes and values that will enable them to make knowledgeable decisions about population and issues, which will affect their quality of life and that of their society, both presently and prospectively. It will also enhance their understanding of the close interaction of population growth and the developmental process, especially for the purpose of raising the standard of living.

For the protection and preservation of the environmental resources to achieve environmental security, environmental education and resource conservation are required.

- Environmental education: It enables persons to explore and identify issues of environmental concern and how to participate in solving them.
- Resource conservation: This portends a very valuable way to guarantee the availability of useful materials for future sustenance. To conserve resources, the three R's model should be adopted viz: reduce, reuse, and recycle.
- Reduce implies to trim down the quantity consumed and use less of such resources. Reuse means to repeat the usage of the same resource by passing it on to someone while recycle involves transforming it so that it can be used in another form.
- Adoption of the principle of environmental impact assessment (EIA) will help assess the extent of physical and socioeconomic damage the environment has sustained through resource exploitation, thereby making way for the appropriate intervention.

NOTES

1. E. D. Enger and B. F. Smith, *Environmental science: a study of interrelationships* (New York: McGraw Hill Higher Education, 2004), 5.

2. Ade Akinbode, *Environmental resource management* (Ibadan: Daybis limited, 2002), 58–60.

3. J. O. Ayoade, *Climate change : a synopsis of its nature, causes, effects and management* (Ibadan: Vantage Publishers Ibadan, 2003), 1–2.

4. D. Botkin and E. Keller, *Environmental science, earth as a living planet* (Canada: John Wiley and Sons, 1998), 74.

5. John R. Weeks, *Introduction to Population* (Canada: Eleventh edition WADSWORTH Cengage Learning, 2012), 5–7.

6. Visualized, "The world's population at 8 billion As at 2022." Accessed September 27, 2022. http://www.visualcapitalist.com.

7. Thomas. R. Malthus, *An essay on the principle of population*, ed. J. Johnson (London: digitalizzazione Project Gutenberg, 1798).

8. Akanni I. Akinyemi and Jacob W. Mobolaji, "Nigeria's large, youthful population could be an asset or a burden." Accessed November 17, 2022. https://theconversation.com/nigerias-large-youthful-population-could-be-an-asset-or-a-burden-186574#.

9. Sunday A. Adedini, Stella Babalola, Charity Ibeawuchi, Olakunle Omotoso et al. "Role of religious leaders in promoting contraceptive use in Nigeria: evidence from the Nigerian Urban reproductive health initiative." *Global Health Science and Practice,* 6(3), 2018: 504.

10. O. E Osahon and M. E. Osahon, "Population education, purpose, content and problems." Retrieved on November 5, 2022. https://www.globalacademicgroup.com/journals/nard/Osahon.pdf.

11. J. Kim, "Female education and its impact on fertility." Accessed November 11, 2016. https://wol.iza.org/uploads/articles/228/pdfs/female-education-and-its-impact-on-fertility.pdf.

12. R. C. Chandna, *Geography of population, concepts, determinants and pattern* (India: Kalyani Publishers, 2016), 431.

13. Faith I. Sajini and Patrick Ijeh P. N., "Socio demographic effects of child labour and its communication in Southern Nigeria: a case study of port-Harcourt city, rivers State, Nigeria." *Review of International Geographic Education (RIGEO),* 2021, 1966; Business Day, "Nigeria's disturbing high fertility rate, causes impact and solution." Accessed November 24, 2022, archive.businessday.ng/.

14. Nigeria IMR 1950–2022, Accessed November 5, 2022, https://www.macrotrends.net/countries/NGA/nigeria/infant-mortality-rate#.

15. D. K. Ashtana and Meera Ashtana, *A textbook of Environmental Studies* (Delhi: S. Chand and Company Ltd. 139, 2006) 40–42.

16. R. C. Chandna, *Geography of Population, Concepts, Determinants and Pattern*, 573–574.

17. Theodore Ikenna Oramah, "The effects of population growth in Nigeria," *Journal of Applied Sciences,* 6, 2006: 1335.

18. Megistu M. Maja and Samuel F. Ayano, "The impact of population growth on natural resources and farmers' capacity to adapt to climate change in olw-Income countries," *Earth Systems and Environment,* 5, 2021: 271.

19. Yusuf. A. Ahmed and Razaq A. Olaitan, "The challenges of deforestation and management in Nigeria: suggestions for improvement," *Ghana Journal of*

Geography, 16(1), 2024: 199. See also; Mayowa J. Fasona, Akinlabi O. Akintuyi, Peter A. Adeonipekun, Tamarabrakemi M. Akoso, Samuel K. Udofia, Oludare O. Agboola, Gbenga E. Ogunsanwo et al. "Recent trends in land-use and cover change and deforestation in south–west Nigeria." *GeoJournal,* 87(3), 2022: 1411–1437.

20. P. O. Oyetunji, O. S. Ibitoye, G. O. Akinyemi, O. A. Fadele, and O. T. Oyediji. "The effects of population growth on deforestation in Nigeria: 1991–2016," *Journal of Applied Sciences and Environmental Management,* 24(8), 2020: 1329–1334. See also; Sajini Faith I., "The effects of population growth on forest resources in auchi and environs," *State and Society,* 3(3), 2017: 289.

21. Theodore Ikenna Oramah, "The effects of population growth in Nigeria," 1335.

22. European Environment Agency, "Environmental impact of energy," Accessed November 20, 2022. https://www.eea.europa.eu/help/glossary/eea-glossary/environmental-impact-of-energy#.

23. Umar Bala, Ayatu Yusuf, and Rabiu Maijama'a, "The impact of population growth on disaggregate energy generation source from (Hydro Power, Natural Gas, Oil and Coal Source) in Nigeria," *Asian Bulletin of Energy Economics and Technology,* 5(1), 2020: 7.

24. Megistu M. Maja and Samuel F. Ayano, "The impact of population growth on natural resources and farmers' capacity to adapt to climate change in low-Income countries," 271.

25. Faith I. Sajini, "Demography and unemployment in Nigeria: a geographical perspective," *Journal of Positive School Psychology,* 6(6), (2022: 6726. C. K. H. Allen and M. S. Robert, "Population and income change: recent evidence," World Bank Discussion Paper, 2014, no. 249.

26. Herders against Farmers: Nigeria's Expanding Deadly Crises. Accessed February 26, 2023, https:/www.crisesgroup.org.

REFERENCES

Adedini, Sunday A., Stella Babalola, Charity Ibeawuchi, Olukunle Omotoso, Akinsewa Akiode, and Mojisola Odeku. "Role of religious leaders in promoting contraceptive use in Nigeria: evidence from the Nigerian urban reproductive health initiative." *Global Health: Science and Practice* 6, no. 3 (2018): 500–514.

Akinbode, Ade. *Environmental resource management,* Ibadan: Daybis limited, 2002.

Akinyemi, Akanni I., and Jacob W. Mobolaji. "Nigeria's Large, Youthful Population could be an Asset or a Burden." Accessed on November 17, 2022. https://theconversation.com/nigerias-large-youthful-population-could-be-an-asset-or-a-burden-186574#.

Allen, C.K.H., and M.S. Robert. *Population and Income Change: Recent Evidence,* World Bank Discussion Paper, no. 249, Washington, DC: World Bank, 2014.

Ashtana, D.K., and Meera Ashtana. *A textbook of Environmental Studies,* Delhi: S. Chand and Company Ltd, 2006.

Ayoade, J.O. *Climate Change : A Synopsis of its Nature, Causes, Effects and Management*, Ibadan: Vantage Publishers Ibadan, 2003.

Bala Umar, Ayatu Yusuf, and Rabiu Maijama'a. "The Impact of Population Growth on Disaggregate Energy Generation Source from (Hydro power, Natural Gas, Oil and Coal Source) in Nigeria." *Asian Bulletin of energy Economics and Technology* 5, no. 1 (2020): 1–8.

Botkin, D., and Keller, E. *Environmental Science, Earth as a Living Planet*, Canada: John Wiley and Sons, 1998.

BussinessDay. "Nigeria's Disturbing High Fertility Rate, Causes Impact and Solution." Accessed on November 24, 2022. achive.bussinessday.ng/.

Chandna R.C. *Geography of Population, Concepts, Determinants and Pattern,* India: KALYANI PUBLISHERS, 2016.

Enger, E.D., and Smith, B.F. *Environmental Science, A Study of Interrelationships*, New York: McGraw Hill Higher Education, 2004.

Fasona, Mayowa J., Akinlabi O. Akintuyi, Peter A. Adeonipekun, Tamarabrakemi M. Akoso, Samuel K. Udofia, Oludare O. Agboola, Gbenga E. Ogunsanwo et al. "Recent Trends in Land-Use and Cover Change and Deforestation in South–West Nigeria." *GeoJournal* 87, no. 3 (2022): 1411–1437.

Herders against Farmers: Nigeria's Expanding Deadly Crises. Accessed on February 26, 2023. https:/www.crisesgroup.org.

Kim, J. "Female Education and its impact on Fertility." Accessed on November 11, 2016. https://wol.iza.org/uploads/articles/228/pdfs/female-education-and-its-impact-on-fertility.pdf.

Maja, Megistu M., and Samuel F. Ayano. "The Impact of Population Growth on Natural Resources and Farmers' Capacity to Adapt to Climate Change in Low-Income Countries." *Earth Systems and Environment* 5, Issue (2021): 271–283.

Nigeria IMR 1950–2022. Accessed on November 5, 2022 https://www.macrotrends.net/countries/NGA/nigeria/infant-mortality-rate#

Oramah Ikenna Theodore. "The Effects of Population Growth in Nigeria." *Journal of Applied Sciences* 6, (2006): 1332–1337.

Osahon O.E., and Osahon M.E. "Population Education, Purpose, Content and Problems." Retrieved on November 5, 2022. https://www.globalacademicgroup.com/journals/nard/Osahon.pdf.

Oyetunji, P.O., O.S. Ibitoye, G.O. Akinyemi, O.A. Fadele, and O.T. Oyediji. "The Effects of Population Growth on Deforestation in Nigeria: 1991–2016." *Journal of Applied Sciences and Environmental Management* 24, no. 8 (2020): 1329–1334.

Sajini, Faith I., and Patrick Ijeh P.N. "Socio Demographic Effects of Child Labor and its Communication in Southern Nigeria: A Case Study of Port-Harcourt City, Rivers State, Nigeria." *Review of International Geographic Education (RIGEO)* 11, no. 10 (2021):1964–1972.

Sajini, Faith I. "Demography and Unemployment in Nigeria: A Geographical Perspective." *Journal of Positive School Psychology* 6, no. 6 (2022): 6723–6729.

Sajini Faith I. "The Effects of Population Growth on Forest Resources in Auchi and Environs." *State and Society* 3, no. 3 (2017): 289.

Malthus, Thomas. R. *An Essay on the Principle of Population*, ed. J. JOHNSON, (London: digitalizzazione Project Gutenberg), 1798.

Visualized. "The World's Population at 8 Billion at 2022." Accessed on September 27, 2022. http://www.visualcapitalist.com.

Weeks John.R. Introduction *to Population.* (Canada: Eleventh edition WADSWORTH Cengage Learning, 2012): 5–7.

Yusuf. A. Ahmed, and Razaq, A. Olaitan. "The Challenges of Deforestation and Management in Nigeria: Suggestions for Improvement." *Ghana Journal of Geography* 16, no.1 (2024): 199.

Chapter 17

"Land Grabs" and the "Local Community"

The Search for Human Security between Participation and Patronage in South Sudan

Sara de Simone

South Sudan's independence on July 9, 2011, fostered extremely high expectations of development and stability for what had been one of the most troubled regions of the world for decades. The civil war between Southern and Northern Sudan raged on for half a century since the country's independence from Britain in 1955 with only a short break in the 1970s. Since 1983, the Southern rebel movement, Sudan People's Liberation Movement/Army (SPLM/A), had been fighting against the political, social, and economic marginalization of the Southern Sudanese people, a colonial legacy perpetrated by successive regimes in Khartoum.

The civil war exacted an extremely heavy toll on Southern Sudanese lives, livelihood means, and infrastructures. The line between the warring parties was often blurred: while the SPLM/A remained the major Southern protagonist fighting against the national Sudanese army, Southern Sudan pullulated with ethnic militias that allied alternatively and opportunistically with the rebel movement or with the government, increasing the level of violence and destruction.[1] Civilians were deliberately attacked as a result of scorched earth military tactics, particularly to clear areas of strategic economic interest such as the oil fields.[2] Four million people were forced to flee from their houses, leaving their land and livelihoods behind.[3]

When the Comprehensive Peace Agreement (CPA) was signed in 2005, security remained volatile in most of the country.[4] This was partly linked to the widespread availability of small arms in the region—with several failed disarmament attempts[5]—but also to continued local tensions over the access to land and natural resources that were vital for the survival of the vastly rural population. In a word, what made hard security even more volatile was the complete lack of human security.

Human security emerged as a concept in the 1990s with the express aim of replacing states as the referent object of security, focusing on the individual instead.[6] The focus on individuals and their lives meant that the scope of security needed to be redefined in much broader terms, to include economic security, food security, health security, environmental security, personal security, community security, and political security—anything that could represent a threat to human well-being.[7] While human security has been the object of an extremely rich and variegated debate both about its theoretical features and its practical applicability,[8] this chapter uses it to describe the multifaceted security threats faced by Southern Sudanese people after the end of the war, as well as to normatively point at the goal to be achieved through the postconflict land reform that attributed customary land rights to the local communities. Given the predominantly rural nature of African societies, land represents the major source of natural capital for the lives of millions of people. Regulating people's rights to own and access land and its related resources—water, trees, wildlife, minerals—and unequal land tenure systems have been linked to human insecurity as they have significant repercussions on social relations, economic opportunities, and environmental sustainability.[9]

This chapter thus focuses on the postconflict land tenure reform implemented by the newly established Government of Southern Sudan since 2005. It highlights that, despite the emphasis on the recognition of communal ownership of the land on a customary basis that should have guaranteed a more equal access to land to the rural communities, the reform has failed to increase people's overall human security, resulting in increasing social fragmentation. Drawing on field research conducted in South Sudan between 2010 and 2016, the chapter particularly investigates how participative processes set up to protect the rights to land of local communities ended up overshadowing land alienation from the most vulnerable segments of these communities, ultimately strengthening internal inequalities and contributing to the reproduction of human insecurity and environmental injustice.

LAND MANAGEMENT IN SOUTHERN SUDAN AFTER THE CIVIL WAR

Control of and access to land and natural resources represent a key element in the explanation of Southern Sudan's conflicts. Clashes for the control of the oil-producing areas and guerrilla attacks perpetrated by the SPLM/A to block or sabotage extractive or other infrastructures—such as those against Chevron's oil facilities and the Jonglei Canal[10]—were frequent. Land laws promulgated by successive governments in Khartoum since the colonial times had always attributed to the government the possession of unregistered land, something that made local communities extremely vulnerable to arbitrary and uncompensated expropriation.[11] The growing intolerance for such centralizing policies became one of the drivers of the SPLM/A for the mobilization of support among the local population and a key aspect of the rebel movement's posture during the peace negotiations in the early 2000s.

"The land belongs to the people"[12] was a popular motto through which the SPLM/A claimed the Southern people's right to administer their own land through a decentralized land management system that included the recognition of customary and communal ownership of the land. Indeed, not only was the control of land and natural resources at the center of the conflict between Northern and Southern Sudan, it was also among the most important causes of intra-southern intercommunal conflicts. Increasingly militarized ethnic communities often clashed among themselves for access to grazing land and water, enabling the Sudanese government to play into these fault lines to weaken the SPLM/A and its political objectives.[13] Mass displacement caused by the fighting increased tensions over access to natural resources: at the end of the war, the management of internally displaced people and returnees, and their (often land-based) relations with host and home communities was one of the major problems that the newly established Government of Southern Sudan had to face in terms of local security management.[14]

While the CPA established a Southern Sudan Land Commission in charge of dealing with land issues at the national level, local land management was referred to the Interim Constitution of Southern Sudan and to the promulgation of several laws. Between 2005 and 2011, these laws progressively formalized the recognition of communal rights to land on a customary basis for all the land that was not registered as government land or private land. These forms of property, under a freehold or leasehold status, predominated in gazetted areas from colonial times (national parks and forest reserves) and in urban areas, whereas the rural areas were almost entirely recognized as belonging to the various local communities inhabiting them.[15]

The recognition of communal rights on a customary basis was in line with a more general trend of formalizing local rights to land in Africa under the administration of variously defined local chiefs.[16] This trend was encouraged by international organizations such as the World Bank and several UN agencies with the idea that customary land tenure would provide a more equal and effective system for land allocation at individual and household levels, improving food security and other forms of human security. Also, the formalization of customary rights to land would increase tenure security, thus creating a more conducive environment for foreign investments. The combination of these benefits would provide a powerful tool for poverty reduction and economic development.[17]

The benefits of such a system of land administration for Southern Sudan seemed to be crystal clear: at the time of independence, the country had a population of 9.8 million[18] and extremely low development indicators. With 51 percent of its population living below the absolute poverty line,[19] 3.3 million people suffering from severe food insecurity,[20] and 98 percent of its budget depending on fluctuating oil revenues,[21] the country was in dire need of reducing poverty and improving food security, while at the same time deactivating local conflict drivers. Communal land ownership promised to make access to land more secure for individuals and families: it shifted decision-making over land from state bureaucracies and markets to arguably more legitimate—and context-knowledgeable—local authorities; it increased the likelihood that benefits deriving from productive investments on the land or from the commodification of ecosystem services would be shared with local communities;[22] it increased the latter's stakes in safeguarding their own environment;[23] and, last but not least, it provided broader room for maneuver to local authorities for implementing local, flexible negotiation and mediation practices in resolving land disputes.[24] Against this promising picture, a detail was overlooked: despite the emphasis placed on the local community as a right-bearing subject, no clear definition of how the local community could be identified was included in the legal documents.

While a number of international experts have engaged with this issue, proposing strategies to make local communities "legible" and compliant with human rights and the rights of vulnerable social groups such as women and youth,[25] no clear-cut solution could be identified.[26] This was accentuated by the fact that private capital quickly stepped into the picture, with very limited interest in sorting out these kinds of issues, but rather looking forward to making the system work for its own profit. The newly appointed Government of Southern Sudan needed to promote economic development while reducing the country's oil dependence. Given the widespread perception of Southern Sudanese land being severely underused—with FAO estimating that in 2011 only 4.5 percent of the total land was put to productive use[27]the government turned to foreign private investors to foster the development of the agricultural and forestry sector. The legal framework recognizing customary rights to land, developed with the support of international donors such as USAID and the European Commission, also contained rules and procedures for the negotiation of land investments, including the involvement of local communities in the negotiation of land concessions, time limits, and obligations to conduct environmental and social impact assessments and to compensate evicted communities.[28] And indeed, foreign investments quickly flocked into the region, attracted by the abundance of land and natural resources, the inevitable gaps in the legal framework, the inexperience of the ruling elite—and its proneness for under-the-counter deals. Notwithstanding the continued volatility of the region in terms of security even after the signing of the CPA, in the years 2007–2010 over 2.64 million hectares of land were estimated to have been leased to foreign companies for agriculture, biofuel, timber production, carbon credits, and tourism projects.

LAND GRABS AND PARTICIPATORY *MAQUILLAGE*

Thanks to NGOs' advocacy campaigns, participation of local affected communities in the negotiation of land deals has become a relatively standardized practice, just like the employment of participatory mechanisms in the implementation of development projects. Local communities are usually involved, through representation by some sort of "community leaders," in mechanisms that include consultation workshops, during which villages are mobilized to meet state and government officials together with representatives of the private company involved in the deal; steering committees, made of "members of the community," in charge of supervising social development projects or funding promised by the company; and hand-over ceremonies, which usually include the slaughtering of a bull or other acts of friendship upon the payment of a lump sum to the local chiefs for the organization of the event.[29] These processes typically result in the signing of side protocols by "community leaders"—usually the traditional chiefs—to testify of the participation of the local community in the negotiation of land deals involving communal land. Even though these mechanisms have been described as a "formality to be undertaken only after the government and the investor have already come to an understanding on the important elements of the investment"[30], the evaluation of their effectiveness in influencing the decision-making process is beyond the scope of this chapter. It is, however, worth acknowledging that these mechanisms have become routinized enough as to be claimed by local actors every time they feel sidelined, indeed formally achieving at least one of the objectives of the recognition of customary rights to land: that of a better involvement of local communities in processes concerning productive investments on their land.

Yet, in 2011, the NGO Norwegian People's Aid (NPA) published a report in which most of the land investments in Southern Sudan since 2007 were described as "land grabs."[31] According to the report, most of the land deals concluded in the years 2007–2011 involved communal land, which is administered by customary authorities on behalf of the community, which should have guaranteed the community's involvement. In fact, in many cases documented by the two reports, local communities had been excluded. These cases were framed as conflicts between a company misbehaving—acquiring land through illegal deals either with the wrong level of government or with big men in the political or military elite with virtually no involvement of the local community[32]—and a local "affected" community, portrayed through the victimizing terms of loss and destruction of livelihood assets and opportunities. As a solution to the land grabbing problem highlighted by the report, NPA advocated for the completion of the legal framework and for equipping the local population with the appropriate knowledge for resisting abuses and welcome development—something the NGO did by supporting local Civil Society Land Alliances.[33] One "success story" about a local community that managed to use the existing legal framework to resist land grabbing was that of Lainya County (Central Equatoria State), where "the people" obtained the cancellation of a lease agreement signed by the government with a foreign company with no community consultation by virtue of communal land ownership.[34]

Both the participatory rhetoric supporting local communities' involvement in the negotiation of land deals and the land grabbing narrative, however, fail to recognize the internal diversity and hierarchical organization of local communities. In both cases, a homogenizing tendency emerges: the local community is taken for granted as a right-bearing subject that can be either involved in consultations or protected from land grabbing. As Gilfoy shows with regard to advocacy campaigns against land grabs in Liberia, this homogenizing rhetoric often overshadows local communities' dis-homogeneity and internal conflicts.[35] Matched with the unclear definition of what a local community actually is, this contributes to turning participatory mechanisms into simple maquillage: despite the good intentions in moving the formulation of the legal framework and the land tenure reform, it neither contributes to increasing human security nor to improving social relations within a highly fragmented society. More specifically, the two case studies presented below provide examples of different patterns of negotiation on the allocation and use of communal land involving third parties. In both cases, discourses of "community engagement" and participation were used to conceal divisions within the very community and the actual impossibility of separating the rural "grassroots" from its urban patrons.

"THE COMMUNITY MEMBER IS SOMEONE WHO UNDERSTANDS": PARTICIPATION DYNAMICS IN TINDILO PAYAM

In 2012, when fieldwork was conducted, Tindilo Payam was a very remote rural area comprising of five bomas (villages), mostly inhabited by Mudari pastoralist communities and situated in Terekeka County, in the northern part of Central Equatoria State. In 2007, a Norwegian company active in timber trade and carbon credit schemes started negotiating a lease agreement of 179,000 hectares of communal land with the Government of Central Equatoria State.[36] The company's major interlocutor was the Ministry of Agriculture and Forestry (MAF) of Central Equatoria State. The state minister was delighted that the company had followed the "right procedures" for the negotiation of the lease agreement and interpreted it as a recognition of his authority over the land in Central Equatoria:

> They are the only corporation that followed the right procedures: they came to us, we went to the community to have consultation, they engaged the community with Corporate Social Responsibility, they are doing really good.[37]

To support the company in the process, the state minister appointed a ministerial focal point from his staff to work with the company. The appointed officer was a native of Tindilo Payam. He encouraged the company to consult the Central Equatoria State Investment Authority, chaired by another native from Tindilo. After an assessment, unsurprisingly, the company ended up investing in Tindilo Payam and employed the focal point from the Ministry of Agriculture as a Plantation Manager.[38] At the same time, however, the Ministry of Agriculture and Forestry at national level was dissatisfied with the company's behavior. A senior official complained that the company had overlooked the national ministry in the lease negotiation process and accused it of having kicked him out of the board of directors of the subsidiary branch that the company had registered in Southern Sudan because he opposed the ninety-nine-year length of the lease agreement.[39]

With the aim of building closer ties with the government and the community, the board of directors of the subsidiary initially included one representative of the national MAF, one of the Ministry of Agriculture of Central Equatoria State, one member from the Norwegian company, two members from the local subsidiary, one lawyer, and one "member of the community." However, after alleged criticism from international media for the involvement of corruptible government officials on the board, the latter was dissolved, and a new one was installed formed only of members of the company. The member of the community was also confirmed in his position. "The community member is someone who understands this kind of issues," explained the Plantation Manager. "He was chosen by the community [. . .]. He represents the community."[40] Putting great emphasis on the connectedness of this person and on his vicinity to circles of power, the Plantation Manager revealed that the member of the community was, in fact, the chairman of the Central Equatoria State Investment Authority, who was also appointed as the chair of a local Steering Committee formed in Tindilo Payam following the first round of consultations with the local community. The Steering Committee was in charge of negotiating for the inclusion of development projects in the land agreement. It comprises five members: the chair, a Member of the Central Equatoria State Parliament who hailed from Tindilo, an employee of the Investment Authority, a traditional chief based in Juba, and the local Paramount Chief, who was, in fact, the only committee member based in Tindilo. The steering committee obtained the signing of a MoU with the company that provided for several development projects for the community, including the construction of a school, a health center, and a football field.[41] The MoU was incorporated into the agreement as a Community Support Programme signed by the "Chairperson of Tindilo Payam Community."[42] However, since 2010, after having started a trial plantation and hired about a hundred workers, the company had "scaled down" its activities for unclear reasons.[43] The Paramount Chief and the Plantation Managers mentioned "interferences" by the national government, with the Ministry of Justice failing to validate the company's title deed, as a possible form of retaliation for the way in which the whole leasing process had been conducted.

Irrespective of its tense relations with the national MAF, the Norwegian company was often cited as one of the best examples of community involvement, in contrast with narratives around land grabbing.[44]

Did participatory mechanisms really work, then? At first glance, yes. Participatory mechanisms and bodies had been put in place, community needs were taken into account, and local residents were generally happy about the conduct of the company—aside from its apparent withdrawal and unfulfilled promises.[45] The Paramount Chief, an illiterate old man who, at the time of my visit, was very ill also expressed a positive opinion about how the negotiation of the agreement had been conducted, emphasizing that he was even taken to Juba to sign the agreement as a member of the Steering Committee.[46] Even though the participation of the local community was extremely indirect and completely mediated by the members of the Juba-based elite, grievances against the land deal seemed to be very limited.

LAINYA COUNTY: REPRESENTATION OR PATRONAGE?

Lainya County is located between Juba and Yei Counties in Central Equatoria State and inhabited by people predominantly belonging to the Bari-speaking agriculturalist group of Pojulu. During colonialism, the government initiated teak plantation in Lainya and Yei areas, which were then kept as forest reserves under postindependence government control. Memories of evictions among the local population endured to the present day: "In the 1920s [. . .] people were evicted and displaced at the borders [of the plantation]. Some people participated in planting the forest but saw no benefits and they also lost their land."[47] While there was no questioning of government ownership of the forest reserve, the local community is very jealous of the customary ownership of all the land that falls outside the "gazetted" areas.

In 2007, a private company negotiated a lease agreement with the government to exploit the already existing teak plantation and to expand it into a vast area of customary land. The company started negotiations with the national MAF, as the forest reserve formally fell under its jurisdiction. Neither the local community nor state and local government institutions were involved. In fact, the lease agreement signed in 2007 between the national MAF and the company was mostly about customary land. Of the total 52,000 hectares involved, 50,000 were natural forested land owned by the local community. The signing of this lease agreement created great discontent at the state level[48] and among local residents of the three payams in Lainya and Yei counties involved in the agreement. Backed by the Lainya County Commissioner and the payam administrator, local residents from Kenyi Payam occupied the forest to prevent the company from starting its operations. The company was thus compelled to turn to participatory mechanisms organizing stakeholders' workshops to inform the population of its project and to make promises about the construction of social facilities to benefit the community.[49] These workshops were held in the three payam headquarters with the participation of company members, national government officials, county and payam representatives, and traditional authorities. Each payam received USD 2,000 for the organization of a handover ceremony for the forest plantation in their respective areas. Three local Steering Committees—plus one with representatives from the three payams—were created to supervise the spending of the USD 200,000 Social Fund, which the company had pledged for "community development projects," plus USD 5 per cubic meter of "sawn board exported."[50] The members of the steering committee were chosen from among the most authoritative and respected personalities in the communities, with the participation of the local chiefs, women, and youth representatives.[51] In Kenyi Payam, the Steering Committee was chaired by a self-declared "son of Loka community"—the village where both the teak plantation and the natural forest included in the concession to the company were located. He was also a government officer, working in Lainya County Education Department, and considered himself a "local public authority."[52] He was satisfied with the degree of community involvement brought about by protests, although he was upset because of the delays in the project implementation. When asked about the involvement of the community actually living in the forest, he argued that it had been residing in the forest only temporarily and the people were ready to move away to allow the investors to bring development and job opportunities.[53] A short visit to one of the household conglomerates characterizing the forest landscape in Loka Boma, however, gave a rather different picture. One local headman claimed:

> We settled here in 1972. [. . .] During the war, we were forced to move to the hill, but we came back in 2001 and we have no intention of leaving this place. [. . .] We know this area has been given to [the company], but we already told them that it is not acceptable for them to come and operate where people live. If they really want to expand into this area, they will have to come and sit with the community, so we can identify together an area for the plantation.[54]

Technically, however, the company had already sat with the community or, more exactly, with the most visible part of the community. The poorer and more vulnerable households living far away from the payam headquarters had instead been excluded from the negotiation process. Besides the somewhat obvious remark that "whatever the collective rhetoric [. . .] it is rare for the whole community to take part equally" (White, 1996, p. 13) in "participatory" or consultative processes, this situation unveils once again the tricky nature of the idea of "local community" as a right-bearing subject.

CONCLUSION

Land grabbing can be considered as a form of environmental injustice, but it does not always occur in a clear and visible way. The two case studies presented above refer to situations that could not be technically referred to as land grabs: in both situations, some sort of involvement of the local community had been put in place, and the grievances against the land deals cut with the companies were relatively limited. Yet, even though environmental justice scholars typically focus on protests and conflicts to identify environmental injustices,[55] we must be aware that in many instances the most vulnerable have limited resources to organize themselves into such forms of active and visible resistance. Situations of extreme vulnerability are much more likely to result in adaptation to environmental injustice,[56] or in the adoption of passive forms of resistance—the "weapons of the weak."[57]

In the two case studies presented, the identification of the "local community" as a right-bearing subject is problematic. In the case of the forest community in Loka Boma, the cleavage within the community is clearly visible: a part of the community is simply excluded from the interaction with the investor and government authorities, and its interests are not taken into account. The case of Tindilo is less straightforward: various kinds of members of the community—including random people sitting at the market inquired about the project—seemed not to have had major problems even though they knew relatively little about it.[58] Nevertheless, we are left wondering if the complete control of the participation process by the Juba-based elite members—and, particularly, of government officials in charge of facilitating private investments—acting as representatives of the local community really contributes to strengthening the rural people's rights to land.

The importance of acknowledging the non-homogeneous character of communities when dealing with land tenure systems recognizing communal land rights that are supposed to strengthen rural peoples' human security and environmental rights needs to be taken seriously. If it is true that land deals are likely to limit rural communities' access to water and livelihoods, have costs for the environment,[59] and pose a threat to food security,[60] the analysis of rural communities' internal inequalities both in access to wealth and in terms of participation in decision-making processes over the land needs to be given a more prominent role.

NOTES

1. Johnson, *The Root Causes of Sudan's Civil Wars.*
2. Human Rights Watch, *Sudan, oil, and human rights.*
3. Reliefweb, "1.9 million Dead from Sudan's Civil War; More Than 70,000 Deaths in 1998, Report Estimates." December 10, 1998. https://reliefweb.int/report/sudan/19-million-dead-sudans-civil-war-more-70000-deaths-1998-report-estimates (last accessed on February 19, 2023).
4. International Crisis Group, *Jonglei's Tribal Conflicts: Countering Insecurity in South Sudan.*
5. O'Brien, *Shots in the Dark*; Young, *The South Sudan Defence Forces in the Wake of the Juba Declaration.*
6. UNDP, *Human Development Report 1994.*
7. Ibid.
8. de Simone, "Beyond normativity and benchmarking: applying human security to the production of context-specific knowledge in refugee hosting areas in Africa."
9. Clover e Eriksen, "The effects of land tenure change on sustainability."
10. Johnson, *The Root Causes of Sudan's Civil Wars*; Shanmugaratnam, *Between War & Peace in Sudan & Sri Lanka.*
11. Komey, *Land, Governance, Conflict, and the Nuba of Sudan.*
12. For a detailed discussion of how the original claim that "the land belongs to the communities" was turned into the more general "the land belongs to the people," see de Simone, *State-Building South Sudan.*
13. Johnson, *The Root Causes of Sudan's Civil Wars*; Hutchinson, *Nuer Dilemmas.*

14. Pantuliano et al., "The long road home. Opportunities and obstacles to the reintegration of IDPs and refugees returning to Southern Sudan and the Three Areas"; Badiey, *The State of Post-Conflict Reconstruction*.

15. Government of Southern Sudan, The Land Act.

16. In 2003, the WB acknowledged that 90 percent of African land was administered under some form of customary regime. See: World Bank, *Land Policies for Growth and Poverty Reduction*. In 2013, the number of African countries that had formally recognized customary ownership had doubled since 2002. See the WB website: http://www.worldbank.org/en/news/pressrelease/2013/07/22/how-africa-can-transform-land-tenure-revolutionize-agricultureend-poverty (last accessed on 14/10/2022).

17. De Wit, Tanner, and Norfolk, "Land policy development in an African context. Lessons learned from selected experiences."

18. WB data: https://data.worldbank.org/indicator/SP.POP.TOTL?locations=SS (last accessed on October 5, 2022).

19. South Sudan National Bureau of Statistics, *South Sudan Statistical Yearbook*.

20. Data from the Integrated Food Security Phase Classification: https://www.ipcinfo.org/ipc-country-analysis/details-map/fi/c/459567/?iso3=SSD (last accessed on November 25, 2022).

21. Robinson, "Special Report: In South Sudan, a state of dependency."

22. Byamugisha, *Securing Africa's Land for Shared Prosperity*.

23. Ganz, "Indigenous Peoples and Land Tenure."

24. These practices were successfully employed also during the civil war to negotiate seasonal agreements between different communities with competing land use needs. For a thorough analysis of those experiences, see Bradbury et al., *Local Peace Processes in Sudan*.

25. De Wit, Tanner, and Norfolk, "Land policy development in an African context. Lessons learned from selected experiences"; Marzatico, "Challenges of land governance in the making of a new nation: experiences from South Sudan."

26. de Simone, *State-Building South Sudan*.

27. FAO website: http://www.fao.org/news/story/it/item/81693/icode/ (last accessed on October 05, 2022).

28. Government of Southern Sudan, The Land Act; Government of Southern Sudan, Local Government Act; Government of Southern Sudan, Investment Promotion Act.

29. de Simone, *State-Building South Sudan*.

30. Deng, "'Land belongs to the community'. Demistifying the 'global land grab' in Southern Sudan."

31. Deng, *The New Frontier. A Baseline Survey of Large-scale Land-based Investment in Southern Sudan*. Land grabbing as a global phenomenon gained attention in the 2000s. It can be defined as the acquisition of large tracts of land and other natural resources by global capital through mechanisms that lead to intensive exploitation of local resources to the detriment of local populations. Not only are the latter excluded from sharing the benefits of development, but also, they are often evicted from their land, losing their livelihood means, and forced to bear the environmental and social

consequences of intensive resource exploitation. Borras et al., "Land Grabbing and Global Capitalist Accumulation." Land grabbing represents one form on environmental injustice: not only does it have a heavy environmental and social impact, it also takes place where communities are more vulnerable. Busscher, Parra, and Vanclay, "Environmental Justice Implications of Land Grabbing for Industrial Agriculture and Forestry in Argentina."

32. Interview with South Sudanese university lecturer, 2013.

33. de Simone, "State-building South Sudan. Discourses, practices and actors of a negotiated project (1999–2013)."

34. Radio Miraya. "Lanya County committee push for cancellation of land lease contract." August 12, 2011. http://www.mirayafm.org/index.php/southsudan/6606-mukaya-payam-lanya-county-committee-push-for-annulment-of-land-lease-contract; NPA. "Striving to secure rights for 89.000 land users". June 6, 2012. http://www.npaid.org/en/news/?module=Articles&action=Article.publicShow&ID=19266 (last accessed on 01/01/2015).

35. Gilfoy, "Land grabbing and NGO advocacy in Liberia."

36. At that time, the Land Act had not yet been enacted, but the recognition of customary ownership of land on a communal basis had been stated in the Interim Constitution of Southern Sudan (2005).

37. Interview with senior state government officer, Juba, 2012.

38. Interview with senior national staff of private company, 2012; interview with senior state government officer, Juba, 2012. In fact, the Investment Authority denied having influenced the choice and suggested that the company would have been brought to Tindilo by NPA, which had previously run a development project in the area. A staff of the NGO denied any link with the company and rather stated that they had pulled out of the area after the arrival of the company to avoid "being confused" with a private profit-seeking firm. Interviews, Juba, 2012.

39. Interview with senior national government officer, Juba, 2012.

40. Ibid.

41. Ibid.

42. "Community Support Program between Tindilo Payam and TreeFarms Sudan Ltd" (07/05/2011). Appendix to the Land Lease Agreement.

43. Several people, including the Investment Authority Chairperson, suggested that the company had "run out of funds" due to problems with "their donors." This reveals the great confusion between business and aid work, which emerged time and again in virtually all my interviews on this topic, mostly expressed through the borrowing of development aid terminology to talk about private companies and investments.

44. Interview with senior state government officer. Juba, 2012.

45. Collective interview with residents, Tindilo Payam headquarters, 2012.

46. Interview with the traditional authority, Tindilo, 2012.

47. Interview with senior government officer, Juba, 2012.

48. Interview with senior national government officer, Juba, 2012.

49. Interview with traditional authority, 2012.

50. Interviews with the local government officer, Yei, 2012; with a senior national government officer, Juba, 2012; and with a senior local government officer, Yei, 2012.

51. Interview with local government officer, Loka, 2012; Government of Southern Sudan (Ministry of Agriculture and Forestry), and Central Equatoria Teak Company Limited (23/11/2007). *Management and Development of Teak Plantation Agreement*, 2007.

52. Interview with local government officer, Loka, 2012.

53. Ibid.

54. Interview with community member, Loka, 2012.

55. Schlosberg and Carruthers, "Indigenous struggles, environmental justice, and community capabilities."

56. Busscher, Parra, and Vanclay, "Environmental justice implications of land grabbing for industrial agriculture and forestry in Argentina."

57. Scott, *Weapons of the Weak. Everyday Forms of Peasant Resistance.*

58. Personal communications with people at Tindilo market, 03/04/2012.

59. Anseeuw et al., *Land Rights and the Rush for Land.*

60. De Schutter, "How not to think of land-grabbing."

REFERENCES

Anseeuw, Ward, Liz Alden Wily, Lorenzo Cotula, and Michael Taylor. *Land Rights and the Rush for Land: Findings of the Global Commercial Pressures on Land Research Project*. ILC, 2012.

Badiey, Naseem. *The State of Post-Conflict Reconstruction: Land, Urban Development and State-Building in Juba, Southern Sudan*. Martlesham: Boydell & Brewer Ltd, 2014.

Borras, Saturnino M., Cristóbal Kay, Sergio Gómez, and John Wilkinson. "Land Grabbing and Global Capitalist Accumulation: Key Features in Latin America." *Canadian Journal of Development Studies/Revue Canadienne d'études Du Développement* 33, no. 4 (2012): 402–416.

Bradbury, Mark, John Ryle, Michael Medley, and Kwesi Sansculotte-Greenidge. *Local Peace Processes in Sudan: A Baseline Study*. Nairobi: Rift Valley Institute, 2006.

Busscher, Nienke, Constanza Parra, and Frank Vanclay. "Environmental Justice Implications of Land Grabbing for Industrial Agriculture and Forestry in Argentina." *Journal of Environmental Planning and Management* 63, no. 3 (2020): 500–522.

Byamugisha, Frank. *Securing Africa's Land for Shared Prosperity: A Program to Scale up Reforms and Investments*. Washington, D.C: World Bank, 2013.

Clover, Jenny, and Siri Eriksen. "The Effects of Land Tenure Change on Sustainability: Human Security and Environmental Change in Southern African Savannas." *Environmental Science & Policy* 12, no. 1 (2009): 53–70.

De Schutter, Olivier. "How not to Think of Land-Grabbing: Three Critiques of Large-scale Investments in Farmland." *The Journal of Peasant Studies* 3, no. 2 (2011): 249–279.

de Simone, Sara. "Beyond Normativity and Benchmarking: Applying Human Security to the Production of Context-Specific Knowledge in Refugee Hosting Areas in Africa." *Third World Quarterly* 41, no. 1 (2020): 168–183.

de Simone, Sara. "State-building South Sudan. Discourses, Practices and Actors of a Negotiated Project (1999–2013)." Doctoral Thesis, Università degli Studi di Napoli L'Orientale—Université de Paris 1 Panthéon-Sorbonne, 2016.

de Simone, Sara. *State-Building South Sudan: International Intervention and the Formation of a Fragmented State*. Leiden: Brill, 2022.

De Wit, Paul, Christopher Tanner, and Simon Norfolk. "Land Policy Development in an African Context. Lessons Learned from Selected Experiences." Land Tenure Working Paper. FAO, 2009.

Deng, David K. "'Land Belongs to the Community'. Demistifying the 'Global Land Grab' in Southern Sudan." Paper presented at the International Conference on Global Land Grabbing, Apr. 6, 2011.

Deng, David K. *The New Frontier. A Baseline Survey of Large-Scale Land-Based Investment in Southern Sudan*. Oslo: Norwegian People's Aid, 2011.

Ganz, Beth. "Indigenous Peoples and Land Tenure: An Issue of Human Rights and Environmental Protection Note." *Georgetown International Environmental Law Review* 9, no. 1 (1997–1996): 173–194.

Gilfoy, K. "Land Grabbing and NGO Advocacy in Liberia: A Deconstruction of the 'Homogeneous Community'." *African Affairs* 114, no. 455 (2015): 185–205.

Government of Southern Sudan. Investment Promotion Act (2009).

Government of Southern Sudan. Local Government Act (2009).

Government of Southern Sudan. The Land Act (2009).

Human Rights Watch. *Sudan, Oil, and Human Rights*. New York: Human Rights Watch, 2003.

Hutchinson, Sharon Elaine. *Nuer Dilemmas: Coping with Money, War, and the State*. Berkeley: University of California Press, 1996.

International Crisis Group. *Jonglei's Tribal Conflicts: Countering Insecurity in South Sudan*. Africa Report 154, 2009.

Johnson, Douglas H. *The Root Causes of Sudan's Civil Wars*. Bloomington; Kampala: Indiana University Press; Fountain Publishers, 2003.

Komey, Guma Kunda. *Land, Governance, Conflict, and the Nuba of Sudan*. Woodbridge: James Currey, 2010.

Marzatico, Francesca. «Challenges of Land Governance in the Making of a New Nation: Experiences from South Sudan». Presentation at World Bank Conference on Land and Poverty, Washington DC, Mar. 24–27, 2014.

O'Brien, Adam. *Shots in the Dark: The 2008 South Sudan Civilian Disarmament Campaign*. Geneva: Small Arms Survey, 2009.

Pantuliano, Sara, Margie Buchanan-Smith, Paul Murphy, and Irina Mosel. "The Long Road Home. Opportunities and Obstacles to the Reintegration of IDPs and Refugees returning to Southern Sudan and the Three Areas." Synthesis Paper. Humanitarian Policy Group—Overseas Development Institute, 2008.

Robinson, Simon. "Special Report: In South Sudan, A State of Dependency." *Reuters*, July 10, 2012. https://www.reuters.com/article/us-south-sudan-aid-idUSBRE86909V20120710.

Schlosberg, David, and David Carruthers. "Indigenous Struggles, Environmental Justice, and Community Capabilities." *Global Environmental Politics* 10, no. 4 (2010): 12–35.

Scott, James C. *Weapons of the Weak. Everyday Forms of Peasant Resistance*. New Haven: Yale University Press, 1985.
Shanmugaratnam, N. *Between War & Peace in Sudan & Sri Lanka: Deprivation & Livelihood Revival*. Oxford: James Currey, 2008.
South Sudan National Bureau of Statistics. *South Sudan Statistical Yearbook*, 2011.
UNDP. *Human Development Report 1994*. New York: Oxford University Press, 1994.
White, S. "Depoliticising Development: The Uses and Abuses of Participation." *Development in Practice* 6, no. 1 (1996): 6–15.
World Bank. *Land Policies for Growth and Poverty Reduction*. World Bank Policy Research Report. The World Bank, Oxford University Press, 2003.
Young, John. *The South Sudan Defence Forces in the Wake of the Juba Declaration*. Geneva: Small Arms Survey, 2006.

Conclusion

Environmental Security in Africa: The Way Forward

Elisha Jasper Dung, Leonard Sitji Bombom, and Augustine Avwunudiogba

Environmental security is a prerequisite for economic development and social well-being of the population of any region, country, or continent. Conversely, environmental insecurity contributes to economic stagnation or slow economic development and reduction in the overall societal well-being. This edited volume examines most of the elements and drivers of environmental insecurity in Africa, such as global climate change, deforestation, population explosion, and pollution and degradation of the natural environment. These elements and drivers of environmental insecurity lead to food insecurity and resource scarcity, and by extension, ethnic clashes and communal conflicts in Africa.[1] Although these conditions are not endemic to Africa, they have been particularly exacerbated on the African continent because of political inactions/instability and low institutional capacity to adapt to the challenges brought about by environmental insecurity.

ENVIRONMENTAL SECURITY AND RESOURCE UTILIZATION

The concept of environmental security is at once both basic and complex. In its basic form, the environment is conceptualized as the component of the earth that sustains human life and living. In its complex form, the environment encompasses the physical, social, cultural, and political spheres of human existence. In either conception, environmental security speaks to the health and well-being of the environment and the concomitant effects of changes that may be occasioned by unhealthy interactions within the environmental system. Although natural forces and processes affect and cause changes, ample evidence shows that humans are the most potent agents of change in environmental systems.[2] Indeed, the scientific community generally agrees that we now live in the era of the Anthropocene, an era marked by human domination of the earth systems at local, regional, and global scales.[3] There is hardly any part of our earth system that does not have a human imprint in one way or another. Thus, human use of environmental resources has literally amplified the rate of environmental change, instability, and insecurity.

One of the ramifications of this environmental change is global climate change. Although climate change is a feature of our Earth's geologic history, the scientific consensus is that current global climate change is mainly due to the profound effects of anthropogenic forces and processes on our earth systems, with potential consequential global effects if no serious and urgent action is undertaken to reverse the trend. Global temperature is expected to rise by between 2.0 and4.9 degrees Celsius, based on current observations and predictions from global climate change models.[4] The model developed by Raftery et al. in 2017 predicted a medium global temperature increase of 3.2 degrees Celsius and concluded that there was only a 1 percent chance that the earth will warm at less than 1.5 degrees Celsius, which will require a rapid decline in carbon intensity compared to the recent past for this lower global temperature increase to occur. This in turn will lead to the melting of continental glaciers and snowcaps, rise in sea levels, and submerged shores, islands, and other low-lying coastal areas around seas and oceans. Additional impacts include an increase in the frequency and severity of high-magnitude droughts and floods.

As case studies from this edited volume show, all these consequences are already posing threats to human habitation in many regions of Africa. Other effects of climate change on African environmental systems include deforestation, desertification, and loss of biodiversity among many others.[5] This resembles a list of catastrophic events predicted in the days of Armageddon which in the African continent have triggered conflicts over scarcity or access to dwindling environmental resources as the economies of many countries in Africa continue to witness stagnation as reflected in the low and sometimes decline in GDP. For example, available World Bank data confirm that in the last decade, most African countries south of the Sahara have not experienced significant improvements in GDP in comparison to other regions of the world with concomitant disposable income in many indices or measures of economic well-being,[6] a situation that can be linked to the deteriorating environmental resource base of the continent.

Unfortunately, many scholars would argue, not without justification, that environmental Armageddon either already exists in Africa or is just around the corner. Africa has struggled and continues to struggle with many of the issues on the laundry list of global environmental concerns. Deforestation and desertification, droughts and floods, loss of biodiversity, and environmental conflicts over access to resources are old problems that have bedeviled the continent and seemingly defied solutions. For centuries, Africans have grappled with these challenges in various forms year in and year out.

The Sahara, Namib, Kalahari, and other desert frontiers account for more than 20 percent of the African continent's land area. They represent some of the most challenging environmental conditions in the world. Frighteningly, they have been expanding in scope as human negligence and inefficiency in environmental management and the effects of global climate change have led to the conversion of the marginal land areas around them to deserts. The East African and the Sahelian subregions have undergone traditional cycles of drought and floods for decades due to their unique geographic location, but many other regions of the continent with predictable climatic patterns are beginning to undergo similar experiences.[7]

New environmental challenges have been added to the old and familiar problems. The effects of global climate change are uncharitably disproportionately skewed against the African continent. Africa bears the brunt of the impacts of the phenomenon of global climate change, although it contributes the least to the emission of global GHGs[8] the major anthropogenic culprit and driver of the various facets of global climate change, including global warming, current and projected sea level rise, desertification, and drought,[9] to mention a few. With weak economies and political institutions, Africa has been judged to have the highest vulnerability and the least adaptive capacity and resilience to manage the impacts of global climate change. Droughts and floods have become more intense, the rate of environmental changes is more rapid, migration has become impelled and more fluid—leading to higher numbers of violent conflicts, and food insecurity is now a more real and urgent threat to many communities and countries. The spillover effects of nations attempting to deal with these problems have manifested in international tension. All these environmental challenges have stymied development on the continent. Therefore, environmental security is a major concern and existential threat in Africa.

ENVIRONMENTAL SECURITY AND DEVELOPMENT

Development is commonly accepted to be inextricably linked to the environment's health. The development of a country or continent does not occur in a vacuum; rather, it is the product of a conscious and efficient management of resources. These resources, both physical and human, come from the environment and are products of the environment. Therefore, Africa's quest for development rests on the security of its environment. However, for the desired development to materialize, Africa must resolve, or at least better manage, many of its existing and emerging environmental security concerns. This requires more creative and innovative solutions. Based on the breadth and depth of information presented in the case studies in this edited volume, we make the following suggestions for formulating action-driven policies geared toward environmental security on the African continent.

(1) African nations must be more proactive in engaging environmental problems. There is a seeming negligence of environmental concerns until they become intractable. Like all global systems, many of the environmental problems evolve over time. Initiatives should be taken and duly pursued at the incubation stage to nip it in the bud. It is hereby granted that some of the problems appear to be spontaneous or provide little warning time, such as flash floods. Nevertheless, they are predictable, and much can be done to safeguard against them and mitigate their impacts. For example, improvements in weather forecasting and citizen warning systems can go a long way in mitigating the debilitating effects of flash floods in many African countries. Other environmental security challenges such as the effects of global climate change have long incubation periods that provide opportunities for the development of long-term plans and strategies to mitigate their negative impacts on the economy and livelihoods of African populations. Generally, the failure to plan against environmental hazards produces disasters of seemingly monumental proportions, which take a huge toll on the people, communities, and often entire countries. African countries must formulate executable plans that are adequate, technologically feasible, economically viable, and socially acceptable to the populace. To effectively do this, existing, and if need be, new local, national, and transnational African institutions should be developed and strengthened through capacity building. This will help to improve the adaptive and resilience capacities of people and communities in Africa.

(2) As shown in this edited volume, some of the major environmental security concerns, especially those associated with global climate change, such as drought and desertification, are international in nature. Therefore, regional cooperation and collaboration among countries to find lasting solutions to common environmental problems are expedient. Rather than adopt individualistic and often nationalistic positions that antagonize each other and render or neutralize the efforts and interests of others to resolve similar problems in their realms, regional blocs may create pacts based on common environmental concerns that require two or more countries to work together for resolution. African countries should identify common environmental problems that demand collective attention and invest time, effort, and resources to

produce sustainable solutions. Indeed, such environmental security collaborations can be implemented by developing specific environmental security programs that can be executed within the organs of existing regional blocs such as the Economic Community of West African States (ECOWAS), the South African Development Community (SADC), Community of Sahel–Saharan States (CEN-SAD), Economic Community of Central African States (ECCAS), the EAC, and the Intergovernmental Authority on Development (IGAD).

(3) Some environmental security concerns affecting African countries are global in nature and, therefore, require transcontinental partnerships between the African continent and the rest of the world. The AU is Africa's continental body that oversees the collective political, economic, and social interests of member states on the continent. The African Charter on Human and Peoples Rights contains provisions for ensuring environmental security.[10] However, it requires legally binding contractual agreements among all countries to make for more effective monitoring and implementation of environmental laws and agreements between and among African countries. An African continental court should be constituted with eminent jurists to oversee the implementation of the African Charter, especially its environmental law provisions. This may be more effective in encouraging cooperation between African countries on resolving common environmental issues and adjudicating areas of disagreement rather than relying on international bodies such as the International Court of Justice in the Hague that are far away from the continent.

(4) Environmental security concerns in Africa are multidimensional, multidisciplinary, and multisectoral. The formulation and implementation of appropriate strategies to combat environmental security challenges on the African continent require the knowledge base and mobilization of African expertise in various academic and professional fields. The AU and other regional bodies in Africa should facilitate the mobilization of African expertise residing on the continent and in the diaspora. A comprehensive database of African experts in various fields should be created, maintained, and used to develop a volunteer environmental security corp. Such environmental corps can be deployed as required to various parts of the continent to help develop community-based programs and projects that are geared toward addressing environmental security concerns. In addition to resolving the environmental security issues in Africa, this strategy has the potential to save African countries millions of USD in contracts awarded to international agencies and corporations to address environmental problems on the continent.

It is imperative that both old and new environmental challenges in the continent be engaged in more constructive ways. The problems are not insurmountable but there is a need for new approaches to be developed to address them. One way to do this is to beam the searchlight on the problems in order that they may be viewed and appreciated with new lenses. This is one of the aims of this book on environmental security in Africa. The book addresses the lack of political will by most African states to deal with the potential for conflict stemming from Africa's increasing resource scarcity and natural disasters, particularly because of prolonged drought, desertification, and deforestation. Contributors to this edited volume examine several case studies of environmental conflicts, as well as the threats to security and resource scarcity stemming from various parts of Africa. Several authors examine the role that environmental changes and scarcity of and inhibited access to environmental resources play in the many conflicts and theaters of war in Africa. The kernel of their argument is that environmental resources are at the core or have aided, abetted, and employed to support many of the conflicts on the continent. Other authors draw attention to the insensitive and negligent ways in which the African environment is treated by "entrepreneurs" who exploit the weak political and governance institutions in Africa to their advantage and detriment of the environment. They call for greater vigilance and strengthening of institutions to conserve the environment to ensure sustainable development. The book therefore highlights major issues of environmental security concern in Africa and how they have challenged the development process in the continent.

With a burgeoning population and dwindling resources aided by inefficient management, the future of Africa appears to tether on the brink of an abyss of uncertainty.

NOTES

1. Ajala, Olayinka. "New drivers of conflict in Nigeria: an analysis of the clashes between farmers and pastoralists." *Third World Quarterly* 41, no. 12 (2020): 2048–2066.

2. An, Li, Volker Grimm, Abigail Sullivan, B. L. Turner II, Nicolas Malleson, Alison Heppenstall, Christian Vincenot et al. "Challenges, tasks, and opportunities in modeling agent-based complex systems." *Ecological Modeling* 457 (2021): 109685.

3. Biermann, Frank. "Politics for a new earth: governing in the 'Anthropocene'." *New Earth Politics* (2016): 405–420.

4. Raftery, Adrian E., Alec Zimmer, Dargan MW Frierson, Richard Startz, and Peiran Liu. "Less than 2 C warming by 2100 is unlikely." *Nature Climate Change* 7, no. 9 (2017): 637–641.

5. Hailu, Fekadu. “Climate change as a trigger for desertification and possible alternatives to reduce biodiversity loss.” *Journal of the Selva Andina Biosphere* 11, no. 1 (2023): 94–111.

6. Simola, Antti, Ole Boysen, Emanuele Ferrari, Victor Nechifor, and Pierre Boulanger. “Economic integration and food security–The case of the AfCFTA.” *Global Food Security* 35 (2022): 100651. See also; Jayne, Thomas S., and Pedro A. Sanchez. “Agricultural productivity must improve in sub-Saharan Africa.” *Science* 372, no. 6546 (2021): 1045–1047.

7. Ayugi, Brian, Emmanuel Olaoluwa Eresanya, Augustine Omondi Onyango, Faustin Katchele Ogou, Eucharia Chidinma Okoro, Charles Obinwanne Okoye, Chukwuma Moses Anoruo et al. “Review of meteorological drought in Africa: historical trends, impacts, mitigation measures, and prospects.” *Pure and Applied Geophysics* 179, no. 4 (2022): 1365–1386.

8. Leal Filho, Walter, Ulisses Miranda Azeiteiro, Abdul-Lateef Balogun, Andréia Faraoni Freitas Setti, Serafino AR Mucova, Desalegn Ayal, Edmond Totin, Adeleke Mosunmola Lydia, Felix Kanungwe Kalaba, and Nicholas Otienoh Oguge. “The influence of ecosystems services depletion to climate change adaptation efforts in Africa.” *Science of The Total Environment* 779 (2021): 146414.

9. Hermans, Kathleen, and Robert McLeman. “Climate change, drought, land degradation and migration: exploring the linkages.” *Current Opinion in Environmental Sustainability* 50 (2021): 236–244.

10. Kasimbazi, Emmanuel, and Jean-Claude N. Ashukem. “The emergence of a human right to a healthy environment and the pursuit of sustainability in Africa.” In *Human Rights and the Environment in Africa*, pp. 39–58. Routledge, 2023.

REFERENCES

Ajala, Olayinka. “New drivers of conflict in Nigeria: An analysis of the clashes between farmers and pastoralists.” *Third World Quarterly* 41, no. 12 (2020): 2048–2066.

An, Li, Volker Grimm, Abigail Sullivan, B. L. Turner II, Nicolas Malleson, Alison Heppenstall, Christian Vincenot et al. “Challenges, tasks, and opportunities in modeling agent-based complex systems.” *Ecological Modelling* 457, (2021): 109685.

Ayugi, Brian, Emmanuel Olaoluwa Eresanya, Augustine Omondi Onyango, Faustin Katchele Ogou, Eucharia Chidinma Okoro, Charles Obinwanne Okoye, Chukwuma Moses Anoruo et al. “Review of meteorological drought in Africa: Historical trends, impacts, mitigation measures, and prospects.” *Pure and Applied Geophysics* 179, no. 4 (2022): 1365–1386.

Biermann, Frank. “Politics for a new earth: Governing in the ‘Anthropocene’.” In *New Earth Politics: Essays from Anthropocene*, Simon Nicholson and Sikina Jinnah (Eds.), 40–420, 2016.

Hailu, Fekadu. “Climate change as a trigger for desertification and possible alternatives to reduce biodiversity loss.” *Journal of the Selva Andina Biosphere* 11, no. 1 (2023): 94–111.

Hermans, Kathleen, and Robert McLeman. "Climate change, drought, land degradation and migration: exploring the linkages." *Current opinion in environmental sustainability* 50, (2021): 236–244.

Jayne, Thomas S., and Pedro A. Sanchez. "Agricultural productivity must improve in sub-Saharan Africa." *Science* 372, no. 6546 (2021): 1045–1047.

Kasimbazi, Emmanuel, and Jean-Claude N. Ashukem. "The emergence of a human right to a healthy environment and the pursuit of sustainability in Africa." In Jean-Cloude N. Ashukem and Semie M. Sama (Eds.), *Human Rights and the Environment in Africa*, 39–58. Routledge, 2023.

Leal Filho, Walter, Ulisses Miranda Azeiteiro, Abdul-Lateef Balogun, Andréia Faraoni Freitas Setti, Serafino AR Mucova, Desalegn Ayal, Edmond Totin, Adeleke Mosunmola Lydia, Felix Kanungwe Kalaba, and Nicholas Otienoh Oguge. "The influence of ecosystems services depletion to climate change adaptation efforts in Africa." *Science of The Total Environment* 779, (2021): 146414.

Raftery, Adrian E., Alec Zimmer, Dargan M. W Frierson, Richard Startz, and Peiran Liu. "Less than 2 C warming by 2100 is unlikely." *Nature Climate Change* 7, no. 9 (2017): 637–641.

Simola, Antti, Ole Boysen, Emanuele Ferrari, Victor Nechifor, and Pierre Boulanger. "Economic integration and food security–The case of the AfCFTA." *Global Food Security* 35 (2022): 100651.

Index

About the Editors

Elisha Jasper Dung is Professor and Coordinator of Geography in the Department of Advancement Studies at Alabama State University, Montgomery, Alabama. He holds a BS in Geography and an MS in Environmental Resources Planning from the University of Jos, Nigeria. Dr. Dung holds a PhD in Geography from Oklahoma State University, Stillwater, Oklahoma, and an MA in Geography from the University of Northern Iowa, United States. Dr. Dung teaches courses in World Regional Geography, Cultural Geography, Regional Geography of North America, Computer Applications in the Social Sciences, and the Geography of Africa. He has published on African Diaspora, bilateral and multilateral cooperation between Africa, Latin America and the Caribbean, African migration, Caribbean out-migration, the effects of gas flaring on crops in the Niger Delta, Nigeria, and the Implications of Land cover and Landscape change Dynamics in the Central High Plains (CHP) of the United States. His coedited book *Human Trafficking: Global History and Perspectives* was published in 2021.

Leonard Sitji Bombom has a BSc degree in Geography and Planning and an MSc in Environmental and Resources Planning, both from the University of Jos, Nigeria. He also has an MA in Geography from the University of Northern Iowa (UNI), Cedar Falls, United States, a Certificate in Geographic Information System (GIS), and PhD in Geography from the Oklahoma State University, Stillwater, Oklahoma, United States. He is currently a Senior Lecturer in the Department of Geography and Planning, University of Jos; an Associate Senior Lecturer in the Department of Urban and Regional Planning, University of Jos; and Head of Programmes at the Anthony Nyong Climate Centre of Excellence (ANCCE), University of Jos. He was also a Transportation Scholar at the University of Northern Iowa, United States. He has a wide range of teaching and research interests including Environmental security, Environmental Planning and Management, Conflict Analysis, Transportation, GIS and remote sensing, and Tourism Studies. He has published extensively in reputable international journals and contributed several book chapters. He has also presented papers at many international conferences.

Augustine Avwunudiogba is Professor of Geography and Environmental Resources in the Department of Anthropology, Geography and Environmental Resources, California State University Stanislaus. He holds a BSc (Hons) and MSc in geography from the University of Benin, Benin City, Nigeria, and an MA in geographical studies from Southern Illinois University, Edwardsville, Illinois. Dr. Avwunudiogba holds a PhD in Geography from the University of Texas at Austin, Texas. He teaches courses in Geographic Problems of the Developing World, World Regional Geography, Regional Geography of Africa, and Regional Geography of Mexico and Central America. His current research focuses on the development issues and challenges in Africa and Latin America. His most recent publications include a coedited book on Human Trafficking: Global History and Perspectives (2021), coauthored chapter/chapters in edited volumes: *China in Africa: Between Imperialism and Partnership in Humanitarian Development* (2021), *African Migrants and the Refugee Crisis* (2021), *The Challenges of Refugees and Internally Displaced Persons in Africa* (2021), *Fidel Castro and Africa's Liberation Struggle* (2020), and *Africa, Latin America, and the Caribbean: The Case for Bilateral and Multilateral Cooperation* (2018). Dr. Avwunudiogba is a member of the African Specialty, and Latin America Specialty groups of the American Association of Geographers.

About the Contributors

Temple Probyne Abali is a lecturer and researcher in the Department of Geography and Environmental Management, Rivers State University, Port Harcourt, Nigeria. He holds BSc (Hons), MSc, and PhD degrees in Geography and Environmental Management, with focus on Fluvial Geomorphology, Soil studies and Environmental Management from the University of Port Harcourt, Rivers State, Nigeria. Dr. Abali teaches courses in Geomorphology, Soil Studies, Hydrology and Environmental Resource Management/ Conservation. He is a consultant to the World Bank-Assisted SEEFOR/Third National Fadama Development Project (NFDP III), having facilitated the funding of the project as a Community Facilitator in Rivers State, Nigeria, between 2008 and 2013, through the preparation of EIAs, Business Plans (BPs), Work Plans (WPs), and so forth as contained in the Local Development Plans (LDPs) of the World Bank. He is an associate member of the Nigeria Association of Hydrological Sciences (NAHS), an editorial board member of *CARI Journals Limited*, Washington DC, United States; *Global Scientific Journals* (GSJ), India; and International Multidisciplinary Academic Research (IMAR), New Delhi. He has published several scholarly articles in both local and international journals and contributed to several book chapters as well as participated in local and international conferences. He was a recipient of a meritorious postgraduate (PhD) scholarship award by the Rivers State Government of Nigeria in 2009 and an award for the most sociable male student by the Nigerian Geography Students' Association (NUGSA) millennium award (2000), University of Port Harcourt chapter, Rivers State, Nigeria.

Ahmad Hamza Abdullahi is a lecturer in the Department of Geography and Environmental Management, Ahmadu Bello University, Zaria, Nigeria. He teaches courses in Man's Physical Environment, Systematic Geography of Nigeria, Tropical Climatology, and Introduction to Remote Sensing and GIS in the Department. He holds a Diploma in data processing and information technology from DOT COM computer training school; BSc Geography and Postgraduate Diploma (PGD) in Banking and Finance from Bayero University, Kano; PGD in Education from University of Maiduguri; and MSc in Remote sensing and GIS from Ahmadu Bello University, Zaria. Abdullahi also holds a PhD in Geography from Ahmadu Bello University, Zaria. Dr. Abdullahi has published many peer-reviewed papers in national and internationally reputable journals. Some of his publications include Role of Rainfall on the Spatio-Temporal Density of Vegetation in Sudano-Sahelian Ecological Zone of Nigeria and Perception on the Impacts of Climate Variability on Water Supply and Management in Kano State. He also wrote a book chapter on GIS-based analysis of the relationship between urban vegetation and land surface temperature of Kano metropolis, Nigeria (Ecosystem Dynamics and Disaster Risk Management, 2020).

Ufuoma Efe Akpovwovwo is a lecturer in the Department of Geography and Regional Planning, Delta State University, Abraka, Nigeria. She holds a PhD in Geography from the University of Ibadan, Nigeria. Her PhD research centered on Mangrove Forest Regeneration and Management in the Niger Delta of Nigeria. In the search for more formidable and current tools for research, she obtained a professional masters in GIS from the University of Ibadan, Nigeria. Her research focus spans the following themes: Biogeography, Soil Geography, Environmental Resource Management, Forest Management, Mangrove Forest Management, Agroforestry, GIS, and remote sensing. Her stint while working with an Environmental-based NGO, Nigerian Conservation Foundation, provided the requisite experience and knowledge for research in Environmental Management. This further served as a basis for more research, while also serving as an impetus for knowledge transference within the academic setting. Her research cuts across Nigeria, with a major focus on the Niger Delta of Nigeria.

Racheal Yemisi Amupitan is a Human Rights Lawyer and Advocate. She is also an Associate Professor and Deputy Director of the Centre for Conflict Management and Peace Studies, University of Jos, Nigeria. She holds a PhD in Human Rights Law from the University of Jos. Dr. Amupitan has published papers in several journals including the International Journal of Comparative Law. One of Dr. Amupitan's recent publications, "The impact of restorative justice on post-conflict peace building process in Plateau state," was published in the *International Journal of Comparative Law and Legal Philosophy*, 3 (1), 2021. Dr. Amupitan has presented at conferences and seminars both locally and internationally.

Augustine Avwunudiogba is a Professor of Geography and Environmental Resources in the Department of Anthropology, Geography and Environmental Resources, California State University Stanislaus, California. He holds a BSc (Hons) and MSc in Geography from the University of Benin, Benin City, Nigeria, and an MA in Geographical Studies from Southern Illinois University, Edwardsville, Illinois. Dr. Avwunudiogba holds a PhD in Geography from the University of Texas at Austin, Texas. He teaches courses in Geographic Problems of the Developing World, World Regional Geography, Africa South of Africa, and Geography of Mexico and Central America. His current research focuses on the development issues and challenges in Africa and Latin America. His most recent publications include chapters in the edited volume *Comparative Assessment of Social Issues in Africa, Latin America, and the Caribbean* (2023), coedited the book *Human Trafficking: Global History and Perspectives* (2021), and coauthored chapter/chapters in edited volumes: *China in Africa: Between Imperialism and Partnership in Humanitarian Development* (2021), *African Migrants and the Refugee Crisis* (2021), *The Challenges of Refugees and Internally Displaced Persons in Africa* (2021), *Fidel Castro and Africa's Liberation Struggle* (2020), and *Africa, Latin America, and the Caribbean: The Case for Bilateral and Multilateral Cooperation* (2018). Dr. Avwunudiogba is a member of the African Specialty and Latin America Specialty groups of the American Association of Geographers.

Fagdéba Bakoyéma is an ABD in Educational Leadership, Policy, and Law at Alabama State University (ASU). Fagdéba a professional librarian (over twenty years) serving as the information literacy and learning assistance coordinator at the ASU library, and the liaison to faculty in all colleges. Additionally, Fagdéba serves as the reference librarian for the College of Business Administration. Fagdéba holds several master's degrees: MA in English (West Virginia University, Morgantown); MA in French (Indiana University, Bloomington); MLS in Library Science (Indiana University, Bloomington), and MIS in Information Science (Indiana University) with a concentration in usability research. His research interests include the netnographic exploration of local and global educational leadership on policies on marginalized groups, critical pedagogy, critical race theory, critical information literacy, faculty research impact evaluation, research analytics, and the importance of online teaching in times of global socioeconomic and political disruptions.

Leonard Sitji Bombom has a BSc degree in Geography and Planning and MSc in Environmental and Resources Planning, both from the University of Jos, Nigeria. He also has an MA in Geography from the University of Northern Iowa (UNI), Cedar Falls, United States, and Certificate in GIS and PhD in Geography from the Oklahoma State University, Stillwater, Oklahoma, United States. He is currently a Senior Lecturer in the Department of Geography and Planning, University of Jos; an Associate Senior Lecturer in the Department of Urban and Regional Planning, University of Jos; and Head of Programmes at the Anthony Nyong Climate Centre of Excellence (ANCCE), University of Jos. He was also a Transportation Scholar at the University of Northern Iowa, United States. He has a wide range of teaching and research interests including Environmental Security, Environmental Planning and Management, Conflict Analysis, Transportation, GIS and Remote Sensing, and Tourism Studies. He has published extensively in reputable international journals and contributed several book chapters. He has also presented papers at many international conferences.

Muhammad Kabiru Dahiru, PhD, is currently a Senior Lecturer in the Department of Geography, Federal University of Lafia, Nigeria. He had also been a Lecturer in the Department of Geography, Nasarawa State University, Keffi. He had earlier worked in various capacities with both Federal Civil Service of Nigeria and World Bank-assisted development projects and involved in the monitoring and evaluation of projects, ensuring environmental and social compliance in line with World Bank's safeguards policies and community needs. Dr. Dahiru is a passionate scholar with many scholarly publications to his credit.

Michael Uzodianya Didia is a Lecturer and Researcher in the Department of Geography and Environmental Management, Rivers State University, Port Harcourt, Nigeria. He holds BSc (Hons) and MSc in Geography and Environmental Management from the University of Port Harcourt, Nigeria. Dr. Didia holds a PhD in Environmental Management and Control from the Rivers State University, Nigeria. Dr. Didia teaches courses in Environmental Resource Management/Conservation, Environmental health, and Geography. He is an Associate Member of the NAHS, where he holds the position of Deputy Secretary. He is also an Associate Member of the Nigerian Environmental Society. He has published several scholarly articles in both local and international journals and participated in several local conferences.

Elisha Jasper Dung is a Professor and Coordinator of Geography in the Department of Advancement Studies at ASU, Montgomery, Alabama. He holds a BS in Geography and an MS in Environmental Resources Planning from the University of Jos, Nigeria. Dr. Dung holds a PhD in Geography from Oklahoma State University, Stillwater, Oklahoma, and an MA in Geography from the University of Northern Iowa, United States. Dr. Dung teaches courses in World Regional Geography, Cultural Geography, Regional

Geography of North America, Computer Applications in the Social Sciences, and the Geography of Africa. He has published on African Diaspora, bilateral and multilateral cooperation between Africa, Latin America and the Caribbean, African migration, Caribbean out-migration, the effects of gas flaring on crops in the Niger Delta, Nigeria, and the Implications of Land cover and Landscape change Dynamics in the Central High Plains (CHP) of the United States. His coedited book *Human Trafficking: Global History and Perspectives* was published in 2021.

Marcus Ogwu Edino earned a BSc degree in Geography from the University of Jos, Nigeria, and an MSc in Geography with specialization in Environmental Management from the Federal University of Technology, Minna, Nigeria. He also has MSc in Energy and Environmental Management and PhD in Sustainable Development both from the School of Engineering and the Built Environment, Glasgow Caledonian University, United Kingdom. He had his postdoctoral fellowship in environmental security at Colgate University, New York, United States. He teaches courses on environmental policy and management, and human dimensions of environmental change at the Department of Earth and Atmospheric Sciences, University of Alberta, Edmonton, Canada. He has worked extensively on environmental issues in the Niger Delta of Nigeria, as well as in public and private sector organizations on indigenous communities, disaster risk management, environmental public health science, and natural resource development in Canada and published many peer-reviewed research articles in reputable scholarly journals and made many conference presentations. He consults on environment and social development and environmental security.

Onianwa Oluchukwu Ignatus is a Lecturer and Coordinator of the Department of History, Strategic and International Studies, Dennis Osadebay University, Asaba, Delta State, Nigeria. Dr. Ignatus is a Nigerian Historian specializing in the Transnational Diplomatic and Military History of the Nigerian Civil War and Biafran Studies. He obtained his PhD from the University of Ibadan in 2020. He is the author of the book: Britain's Injurious Peace Games in the Nigerian Civil War Washington DC and London: Academica Press, 2018, and Once Upon a Time in Biafra: Memories, Foreign Visitations, and Life Experiences in the Nigerian Civil War, 1967–1970, Washington DC and London: Academica Press. He had worked with the Swiss Military Academy ETH Zurich, Switzerland, on a book project titled "Captivity in War"; "Visions of African Unity" with the International Studies Department University of Utrecht, Netherland and International Studies Group at the University of the Free State, Bloemfontein South Africa; "A Tight-Embrace: Euro-Africa Relations" with the Department of Political and Social Science, the University of Bologna, Italy; and "Human Rights Breakthrough of 1970" with the Department of International Studies, University of Trento, Italy. Onianwa is currently working with the Department of History and Political Science ASU on "Africa: A Globalizing World, and the Challenges of Peace, Security, and Development." He is a research team member with Department of History, Philosophy and Religious Studies Eastern Kentucky University, Richmond, Kentucky, United States on the book "African Experience." Onianwa is a Research Team Member on the Palgrave Handbook on Nigeria and Its History and Routledge Encyclopedia of African Studies (REAS) and Palgrave Handbook on Social Problems and a Contributor to the Strife Journal of the Department of War Studies, Kings College, London.

Balla M. Keita is an Expert in International Relations and Comparative Politics. His diverse experience reaches the political, corporate, and diplomatic domains. Dr. Keita currently teaches political science at the ASU with a focus on comparative politics and international development. Dr. Keita obtained a PhD in Political Science with a concentration on International Relations and Comparative Politics from Howard University. A master's degree in history from Howard University and a bachelor's degree in history from the University of Houston, Texas. His publications include *Balla M. Keita, Politics of Ethnicity, Region and Religion in Post Houphouët-Boigny Côte d'Ivoire: Political History 1936–2011* (2013); Balla M. Keita, "The 2010 Presidential Election Crisis in Côte d'Ivoire," in Olayiwola Abegunrin, ed., Africa in the New World Order: Peace and Security Challenges in the Twenty First Century (2014); and Balla M. Keita, Professionalization of Political Parties: A Precondition to Peace and Stability in Africa. Brookings Institution. January 2014.

Reginald Chikere, Keke, PhD, is a Historian in African history, gender studies, diplomacy, and international studies. He presently teaches History, Diplomacy, and International Studies at Admiralty University of Nigeria, Delta State, Nigeria. He is also the Director of Research and Planning at Onyoma Research Publications in Port Harcourt, Rivers State, Nigeria.

Domagoj Krpan is a Croatian historian and philosopher. He received his MA in History and Philosophy at the Faculty of Humanities and Social Science, University in Rijeka, Croatia. He was a Graduate Student Assistant in different capacities within the Department of History and was involved as a translator for the Student International Film Festival in Rijeka. He is currently the Secretary of the Centre for Interreligious Dialogue, contributing to the national and international interreligious and intercultural projects. He has a broad area of interest in history, philosophy, and cultural studies. He has presented papers at various international scientific conferences discussing racial, ethnic, religious, and cultural relations in a multicultural society. Currently, he is enrolled as a PhD researcher in the field of political philosophy at the University of Rijeka.

Abdulkadir Maharazu is one of the pioneer academic staff, from 2019 to date, at the Department of Geography, Federal University Gusau, Nigeria. Abdulkadir holds BSc Geography (1997) and master's in development studies (2005) from Bayero University Kano (BUK), PGD in Education from Federal College of Education Kano (2015), and MSc in GIS and Remote Sensing (2017) from Ahmadu Bello University Zaria. Abdulkadir is a PhD student in Natural Resource Management and Climate Change (NRM and CC) at BUK. Abdulkadir is a registered member of Association of Nigerian Geographers (ANG), Science Teachers Association of Nigeria (STAN), and Teachers Registration Council of Nigeria (TRCN). He has nine published works (journal articles and book chapters) and has eight conference papers. Prior to his coming to the University as a Lecturer, Abdulkadir spent 14 years working with Kano State Secondary Schools Management Board as a Classroom Teacher. His hobbies include Reading, Traveling, and Research activities.

Zainab Nuhu is a lecturer at the Department of Environmental Management, Bayero University, Kano. Dr. Nuhu obtained a BSc in Geography from Bayero University Kano (2008), an MSc in Land Resources Development (2012), and a PhD in Natural Resources Management and Climate Change (2021). Dr. Nuhu has worked in both public and private sectors in Nigeria, providing consultancy services with specialty in research (data capture, entry, and analysis) as well as monitoring and evaluation. Worked on projects funded by local and international organizations few of which included National Primary healthcare, Presidential Committee on Small Arms and Light Weapons, FAO research team member on stock routes identification for farmers'/herders' conflict resolution and natural resources conflicts research works, DFID, DELPHI, and British Council. She is a member of ANG and the current welfare person of Environmental Management Association of Nigeria (EMAN).

Emmanuel Ezi Obuah is currently an Assistant Professor in the Department of History and Political Science at Tuskegee University, Alabama. He has served as a Professor of International Relations and Diplomacy in the Department of History and Diplomatic Studies, University of Port Harcourt, Nigeria, where he taught History, Diplomacy, and International Relations. He was a former Associate Professor in the Political Science and History

Department at Alabama A&M University, United States (2004—2012) and ASU, Montgomery, United States (2012—2014). He holds a PhD in International Relations. He was a former Sir Adam Thomson fellow, 1987—1988, and Commonwealth fellow 1992—1996 University of Sussex, UK. He has published books and a number of articles in trail-blazing peer-reviewed international journals, including *Understanding the dynamics of Sino—Africa Relations: Building communities of practice* (2013); *Glossary of key terms in International Relations* (2016); Combating Global trafficking in Persons: The Role of the United States Post-September 2001, *International Politics*, 42, 2006; and Outsourcing prosecution of the Somali pirates to Kenya, *Africa Security Review volume 21 (13)*, 2012, and *the role of the United States Post-September, In. Dung & Avwunudiogba, eds. Human Trafficking: Global History & Perspectives, 2021.*

Ogutu Moses Onyango is a PhD student in Political Science at the Maxwell School of Citizenship and Public Affairs, Syracuse University. He previously served as an International Business and Trade faculty at the African Leadership University (ALU), Rwanda. He holds an MA in International Relations and an MPhil in Inclusive Innovation from the University of Cape Town, South Africa, as well as a BSc (Hons) in Political Science from the University of Mauritius. His current research explores African Agency and Africa's engagement with the international system, focusing on trade, climate change, sustainable development, and science and technology collaboration. In addition to his academic pursuits, Moses works as a consultant in trade and development policy and as a researcher on social entrepreneurship, inclusive innovation, and leadership development in Africa.

Jonathan Ali Ogwuche is a Professor of Environmental Management, and former Dean, Faculty of Environmental Sciences, Benue State University, Makurdi, Nigeria. He holds a master's degree and Doctor of Philosophy degree in Environmental Management from Enugu State University of Science and Technology, Enugu. He is a Registered Surveyor, and a member of professional bodies, including Nigerian Institution of Surveyors (NIS), Nigerian Environmental Society (NES), EMAN, Institute of Development Administration of Nigeria (IDAN), and International Society for Development and Sustainability (ISDS) (Japan). He has published widely in local and international journals. His research interests include geospatial information technology and management, environmental management, gender, and development.

Brian Ochieng Okoth is a research and policy analyst with over eight years of experience in environment and climate change. He holds a bachelor's degree in environmental studies (Environmental Resource Conservation) from Kenyatta University. His research areas of interest include climate change governance, mitigation, and adaptation, energy transition, and the role of subnational climate finance in addressing the impacts of climate change in the rural setup.

Clifford Collins Omondi Okwany is a senior researcher at the Institute of Public Policy and International Affairs (IPPIA). His research work is based in the Horn of Africa. He is also an expert in international security and foreign policy, focusing on ontological security, conflict, and violence theories such as the frontier and spatial theory (territoriality). More specifically, researching counterterrorism, counter-violent extremism, radicalization, deradicalization, community-based armed groups, and non-state-armed groups. He has been part of the EU Horizon 2020 research project on Community Policing (ICT4COP) as part of the team for the African package group.

Sebastian Okechukwu Onah is a Catholic Priest of the Catholic Diocese of Nsukka, Nigeria. He holds bachelor's degrees in Philosophy and Theology from Urban University Rome and a master's degree in Pastoral Theology from the Catholic Institute of West Africa/University of Calabar, Nigeria. He has a Licentiate in Pastoral Theology and Catechetics from Lumen Vitae/Universitee Catholique de Louvain La Neuve, Belgium, a master's degree, an Advanced Certificate of Specialization, and a Doctorate Degree in Social and Cultural Anthropology from the Katholieke Universiteit, Leuven, Belgium. He is an academic staff in the Department of Sociology and Anthropology, University of Nigeria, Nsukka. He is currently a Senior Lecturer, with Sociocultural Anthropology, Eco-Environmental Anthropology, and Environmental Sociology as focus areas of academic interest. He has published extensively in these areas of studies, and more. Ping-pong, football, athletics, drama, and the observation of culture and nature are his highly cherished hobbies.

Evelyne Atieno Owino is a Research Fellow at Bonn International Center for Conflict Studies (BICC). Her work cuts across thematic areas of development, peace, and conflict within Kenya. Her recent publications include the implications of large-scale infrastructure projects on pastoral communities, future visions, and present conflicts: the ethnicized politics of anticipation surrounding an infrastructure corridor in northern Kenya, Water Governance, Institutions and Conflicts in the Maasai Rangelands, and The Nature of Community Armed Groups in Northern Kenya: a framework of Territoriality and Ethnicity. She is now pursuing a PhD at the University of Bonn in Germany.

Tasi'u Yalwa Rilwanu is a Senior Lecturer at the Department of Geography, Bayero University, Kano. He obtained a BSc in Geography at Bayero University, Kano, (1997); He subsequently proceeded to complete a Master's degree program in Land Resources Development in the same university in 2008. He obtained a PGD in Education (2003) at Federal College of Education (F.C.E), Kano, and earned his PhD in Geography at Ahmadu Bello University, Zaria in 2015. Dr. Tasi'u worked as a Classroom Teacher in Girls Science and Technical College, Kano; Lecturer in Sa'adatu Rimi College of Education, Kano, from 2000 to 2005 and 2005 to 2015, respectively. He served as a Government-appointed Caretaker Councilor for community development in his local government from April 2003 to September 2004. Dr. Tasi'u has over forty local and international publications in fluvial geomorphology, groundwater potential, surface and groundwater quality, geomorphology, hydrology, surface and groundwater management, sediments and discharge studies, flood studies, and agroclimatology, among others. He attended many conferences and authored a book "Geomorphology: An Abridge Text." He is a registered member with the Nigerian TRC, ANG, and NAHS. Dr. Tasi'u participated in many projects including NEWMAP, Climate Change, Water Conservation, and Erosion Control.

Faith Iwejingi Sajini is Associate Professor and the Acting Head of Department of Geography and Regional planning at Delta State University, Abraka, Delta State, Nigeria. Dr. Sajini holds a PhD in Geography, specializing in Population Studies. Dr. Sajini has a strong passion for teaching, research, and community development with a good research measurable output. Her research interest centers around socioeconomic and environmental implications of population change and she has several publications in peer-reviewed national and international journals.

Sara de Simone is a postdoctoral research fellow at the School of International Study of the University of Trento (Italy). She holds a PhD in African Studies and Political Sciences from the University of Naples L'Orientale (Italy) and the University of Paris 1 Panthéon-Sorbonne (France). She has worked as Adjunct Professor on African International Relations at the University of Naples L'Orientale and as a Consultant for several development organizations. Her research interests include state-building and state formation processes in fragile contexts, international aid, refugee management, and processes of knowledge production. She published on international peer-reviewed journals and authored *State-building South Sudan. International Intervention and the Formation of a Fragmented State* (2022).

www.ingramcontent.com/pod-product-compliance
Ingram Content Group UK Ltd.
Pitfield, Milton Keynes, MK11 3LW, UK
UKHW041831031026
15587UKWH00002B/687

9 781666 936353

